CELL BIOLOGY ASSAYS
PROTEINS

CELL BIOLOGY
ASSAYS
PROTEINS

Edited by

GERI KREITZER
Weill Medical College of Cornell University

FANNY JAULIN
Weill Medical College of Cornell University

CEDRIC ESPENEL
Weill Medical College of Cornell University

AMSTERDAM • BOSTON • HEIDELBERG • LONDON • NEW YORK • OXFORD
PARIS • SAN DIEGO • SAN FRANCISCO • SINGAPORE • SYDNEY • TOKYO
Academic Press is an imprint of Elsevier

Academic Press is an imprint of Elsevier
32 Jamestown Road, London NW1 7BY, UK
30 Corporate Drive, Suite 400, Burlington, MA 01803, USA
525 B Street, Suite 1900, San Diego, CA 92101-4495, USA

First edition 2010

Material in this work originally appeared in *Cell Biology, Third Edition,* edited by Julio E. Celis,
(Elsevier, Inc. 2006).

British Library Cataloguing-in-Publication Data
A catalogue record for this book is available from the British Library

Library of Congress Cataloging-in-Publication Data
A catalog record for this book is available from the Library of Congress

ISBN : 978-0-12-375692-3

For information on all Academic Press publications
visit our website at elsevierdirect.com

Typeset by Macmillan Publishing Solutions
www.macmillansolutions.com

Printed and bound in United States of America

10 11 12 13 14 15   10 9 8 7 6 5 4 3 2 1

Contents

IV

PROTEIN–DNA INTERACTIONS

V

MASS SPECTROMETRY METHODS FOR DETERMINATION OF PROTEIN IDENTITY AND PROTEIN MODIFICATIONS

Preface

Over the past several decades the range and diversity of experimental approaches used by biologists has expanded dramatically. Today, cell-based analyses of biological processes are often linked intimately with the use of molecular biology, biochemistry, proteomics and genomics. The research goals of modern cell biology have been facilitated tremendously by advances in laboratory techniques and the imaginative application of a broad spectrum of experimental approaches. The relative ease with which these approaches can now be used has led to an evolution of scientists across varied disciplines from being specialists to jacks-of-all-trades. While extremely exciting, this has also raised the bar with respect to cross-disciplinary approaches that can be used, and are often expected, when addressing the ever-growing and complex biological questions posed by researchers around the globe. Thus proficiency in, or at least a thorough understanding of, such diverse approaches has become an attribute necessary to the modern biologist. The chapters included in this volume, taken from *Cell Biology: A Laboratory Handbook, 3rd edition*, describe a variety of methods for identification of proteins of interest, detection and analysis of these proteins, determination of protein-protein interactions and DNA–protein interactions. In combination with cell-based experiments, each may factor critically in studies aimed at elucidating detailed molecular mechanisms as to the inner-workings of cells.

The methods described in these chapters can be broadly separated into five groups. In the first, incorporation of radioactive amino acids, ions and phosphates or chemical modification with fluorophores to label known or unknown proteins are outlined. Sensitive methods for detecting proteins with silver and fluorescent stains are also described. The second group of methods lays out procedures for the separation and detection of proteins by isoelectric focusing and mobility in 2D gels for proteomic analysis. The third section includes chapters describing the use of immunoprecipitation, with and without chemical cross-linking, affinity binding of soluble proteins to others immobilized on solid supports, yeast-2-hybrid, and co-transport assays in cells to identify and characterize protein-protein interactions. In the fourth group of chapters, methods to identify interactions of proteins with DNA using chromatin immunoprecipitation, electrophoretic mobility shift assays and oligonucleotide trapping are described. And finally, an extensive set of mass spectrometry approaches are described in detail for the identification of proteins and protein post-translational modifications.

Using experimental examples, the chapters in this volume provide detailed methods for a wide range of techniques that can be applied in most laboratories. Some are relatively straightforward, require little special equipment and can be adapted to accommodate commonly available tools and supplies. The chapters describing mass spectrometry methods also enunciate important considerations regarding the usefulness and applicability of each technique for assessing specific questions related to protein identification and characterization. This is particularly helpful for researchers that aim to use mass spectrometry on occasion, rather than routinely. Together, the step-by-step instructions with detailed discussion of each technique make this laboratory handbook an essential resource. In this era of cell biology, technical breadth serves an enabling function, allowing researchers to address, at a molecular level, the many questions associated with complex biological events.

List of Contributors

Ruedi Aebersold The Institute for Systems Biology, 1441 North 34th Street, Seattle, WA 98103-8904

Natalie G. Ahn Department of Chemistry & Biochemistry, University of Colorado, 215 UCB, Boulder, CO 80309

Jens S. Andersen Protein Interaction Laboratory, University of Southern Denmark—Odense, Campusvej 55, Odense M, DK-5230, DENMARK

Ron D. Appel Swiss Institute of Bioinformatics, CMU, Rue Michel Servet 1, Geneva 4, CH-1211, SWITZERLAND

Christopher M. Armstrong Dana Faber Cancer Institute, Harvard University, 44 Binney Street, Boston, MA 02115

Jiri Bartek Department of Cell Cycle and Cancer, Danish Cancer Society, Strandboulevarden 49, Copenhagen, DK-2100, DENMARK

Thomas M. Behr Department of Nuclear Medicine, Philipp's-University of Marburg, Baldingerstraße, Marburg, D-35043, GERMANY

Martin Béhé Department of Nuclear Medicine, Philipp's-University of Marburg, Baldingerstraße, Marburg/Lahn, D-35043, GERMANY

Blagoy Blagoev Protein Interaction Laboratory, University of Southern Denmark—Odense, Campusvej 55, Odense M, DK-5230, DENMARK

Gérard Bouchet Swiss Institute of Bioinformatics Proteome Informatics group, CMU-1, rue Michel Servet, Geneva 4, CH-1211, SWITZERLAND

Rosemary Boyle The Institute for Systems Biology, 1441 North 34th St., Seattle, WA 98109

Julio E. Celis Danish Cancer Society, Institute of Cancer Biology and Danish Centre for Translational Breast Cancer Research, Strandboulevarden 49, Copenhagen O, DK-2100, DENMARK

Aaron Ciechanover Center for Tumor and Vascular Biology, The Rappaport Faculty of Medicine and Research Institute, Technion-Israel Institute of Technology, POB 9649, Efron Street, Bat Galim, Haifa, 31096, ISRAEL

Anne Dell Department of Biological Sciences, Biochemistry Building, Imperial College of Science, Technology & Medicine, Biochemistry Building, London, SW7 2AY, UNITED KINGDOM

Bart Devreese Department of Biochemistry, Physiology and Microbiology, University of Ghent, K.L. Ledeganckstraat 35, Ghent, B-9000, BELGIUM

Lynda J. Donald Department of Chemistry, University of Manitoba, Room 531 Parker Building, Winnipeg, MB, R3T 2N2, CANADA

Harry W. Duckworth Department of Chemistry, University of Manitoba, Room 531 Parker Building, Winnipeg, MB, R3T 2N2, CANADA

Christoph Eckerskorn Protein Analytics, Max Planck Institute for Biochemistry, Klopferspitz 18, Martinsried, D-82152, GERMANY

Leonard J. Foster Protein Interaction Laboratory, University of Southern Denmark, Odense, Campusvej 55, Odense M, DK-5230, DENMARK

Masanori Fujimoto University of Tokyo, Department of Veterinary Pharmacology, 7-3-1 Hongo, Bunkyo-ku, Tokyo 113-0033, JAPAN.

Kris Gevaert Dept. Medical Protein Research, Flanders Interuniversity Institute for Biotechnology, Faculty of Medicine and Health Sciences, Ghent University, Instituut Rommelaere—Blok D, Albert Baertsoenkaai 3, Gent, B-9000, BELGIUM

Martin Gotthardt Department of Nuclear Medicine, Philipp's-University of Marburg, Baldingerstraße, Marburg/Lahn, D-35043, GERMANY

Angelika Görg Fachgebiet Proteomik, Technische Universität München, Am Forum 2, Freising Weihenstephan, D-85350, GERMANY

Pavel Gromov Institute of Cancer Biology and Danish Centre for Translational Breast Cancer Research, Danish Cancer Society, Strandboulevarden 49, Copenhagen, DK-2100, DENMARK

Irina Gromova Department of Medical Biochemistry and Danish Centre for Translational Breast Cancer Research, Danish Cancer Society, Strandboulevarden 49, Copenhagen, DK-2100, DENMARK

Martin Guttenberger Zentrum für Molekulariologie der Pflanzen, Universitat Tübingen, Entwicklungsgenetik, Auf der Morgenstelle 3, Tübingen, D-72076, GERMANY

Klaus Hansen

William Hayes McDonald Department of Cell Biology, The Scripps Research Institute, 10550 North Torrey Pines Rd, La Jolla, CA 92370

Kai Hell Adolf-Butenandt-Institut fur Physiologische Chemie, Lehrstuhl: Physiologische Chemie, Universitat Munchen, Butenandtstr. 5, Gebäude B, Munchen, D-81377, GERMANY

Andrew N. Hoofnagle School of Medicine, University of Colorado Health Sciences Center, Denver, CO 80262

Harry W. Jarrett Department of Biochemistry, University of Tennessee Health Sciences Center, Memphis, TN 38163

Antonius Koller Department of Cell Biology, Torrey Mesa Research Institute, 3115 Merryfield Row, San Diego, CA 92121

Yasuhiro Kuramitsu Department of Biochemistry and Biomolecular Recognition, Yamaguchi University School of Medicine, 1-1-1 Minami-kogushi, Ube, Yamaguchi, 755-8505, JAPAN

Irina Kratchmarova Protein Interaction Laboratory, University of Southern Denmark—Odense, Campusvej 55, Odense M, DK-5230, DENMARK

Martin R. Larsen Department of Biochemistry and Molecular Biology, University of Southern Denmark, Campusvej 55, Odense M, DK-5230, DENMARK

Sabrina Laugesen Department of Biochemistry and Molecular Biology, University of Southern Denmark, Campusvej 55, Odense M, DK-5230, DENMARK

Thomas Lee Dept of Chemistry and Biochemistry, Univ of Colorado, 215 UCB, Boulder, CO 80309-0215

Siming Li Dana Faber Cancer Institute, Harvard University, 44 Binney Street, Boston, MA 02115

Adam J. Liska Max Planck Institute of Molecular Cell Biology and Genetics, Pfotenhauerst 108, Dresden, D-01307, GERMANY

Jiri Lukas Department of Cell Cycle and Cancer, Danish Cancer Society, Strandboulevarden 49, Copenhagen, DK-2100, DENMARK

Matthias Mann Protein Interaction Laboratory, University of Southern Denmark, Odense, Campusvej 55, Odense M, DK-5230, DENMARK

Jill Meisenhelder Molecular and Cell Biology Laboratory, The Salk Institute, 10010 North Torrey Pines Road, La Jolla, CA 92037

Suchareeta Mitra Department of Biochemistry, University of Tennessee Health Sciences Center, Memphis, TN 38163

Dejana Mokranjac Adolf-Butenandt-Institut fur Physiologische Chemie, Lehrstuhl: Physiologische Chemie, Universitat Munchen, Butenandtstr. 5, Gebäude B, Munchen, D-81377, GERMANY

Peter L. Molloy CSIRO Molecular Science, PO Box 184, North Ryde, NSW, 1670, AUSTRALIA

Robert A. Moxley Department of Biochemistry, University of Tennessee Health Sciences Center, Memphis, TN 38163

Kazuyuki Nakamura Department of Biochemistry and Biomolecular Recognition, Yamaguchi University School of Medicine, 1-1-1 Minami-kogushi, Ube, Yamaguchi, 755-8505, JAPAN

Dobrin Nedelkov Intrinsic Bioprobes, Inc., 625 S. Smith Road, Suite 22, Tempe, AZ 85281

Randall W. Nelson Intrinsic Bioprobes Inc., 625 S. Smith Road, Suite 22, Tempe, AZ 85281

Walter Neupert Adolf-Butenandt-Institut fur Physiologische Chemie, Lehrstuhl: Physiologische Chemie, Universitat Munchen, Butenandtstr. 5, Gebäude B, Munchen, D-81377, GERMANY

Trine Nilsen Department of Biochemistry, Institute for Cancer Research, The Norwegian Radium Hospital, Montebello, Oslo, N-0310, NORWAY

Ole Nørregaard Jensen Protein Research Group, Department of Biochemistry and Molecular Biology, University of Southern Denmark, Campusvej 55, Odense M, DK-5230, DENMARK

Sjur Olsnes Department of Biochemistry, The Norwegian Radium Hospital, Montebello, Oslo, 0310, NORWAY

Shao-En Ong Protein Interaction Laboratory, University of Southern Denmark—Odense, Campusvej 55, Odense M, DK-5230, DENMARK

Wayne F. Patton Perkin-Elmer LAS, Building 100-1, 549 Albany Street, Boston, MA 02118

Patricia M. Palagi Swiss Institute of Bioinformatics, CMU, 1 Michel Servet, Geneva 4, CH-1211, SWITZERLAND

Eric Quéméneur Institut de Biologie Environnementale et Biotechnologie, service de Biochimie et toxicologie Nucléaire, BP17171 – Centre de Marcoule 30207 Bagnols-sur-Céze

Andreas S. Reichert Department of Physiological Chemistry, University of Munich, Butenandtstr. 5, München, D-81377, GERMANY

Katheryn A. Resing Dept of Chemistry and Biochemistry, University of Colorado, 215 UCB, Boulder, CO 80309-0215

Peter Roepstorff Department of Biochemistry and Molecular Biology, University of Southern Denmark, Campusvej 55, Odense M, DK-5230, DENMARK

Rhys C. Roberts

David Schieltz Department of Cell Biology, Torrey Mesa Research Institute, 3115 Merryfield Row, San Diego, CA 92121

Andrej Shevchenko Max Planck Institute for Molecular Cell Biology and Genetics, Pfotenhauerstrasse 108, Dresden, D-01307, GERMANY

Camilla Skiple Skjerpen Department of Biochemistry, Institute for Cancer Research, The Norwegian Radium Hospital, Montebello, Oslo, N-0310, NORWAY

Joël Smet Department of Pediatrics and Medical Genetics, University Hospital, De Pintelaan 185, Ghent, B-9000, BELGIUM

Kenneth G. Standing Department of Physics and Astronomy, University of Manitoba, 510 Allen Bldg, Winnipeg, MB, R3T 2N2, CANADA

Mark Sutton-Smith Burnham Institute, center for Cancer Research, San Diego, CA

Kazusuke Takeo Department of Biochemistry and Biomolecular Recognition, Yamaguchi University School of Medicine, 1-1-1, Minami-kogushi, Ube, Yamaguchi, 755-8505, JAPAN

Signe Trentemølle Institute of Cancer Biology and Danish Centre for Translational Breast Cancer Research, Danish Cancer Society, Strandboulevarden 49, Copenhagen, DK-2100, DENMARK

Joël Vandekerckhove University of Ghent, Albert Baertsoenkaai 3, 9000 Ghent

Jozef Van Beeumen Department of Biochemistry, Physiology and Microbiology, University of Ghent, K.L. Ledeganckstraat 35, Ghent, B-9000, BELGIUM

Rudy N. A. Van Coster Department of Pediatrics and Medical Genetics, University Hospital, University of Ghent, De Pintelaan 185, Ghent, B-9000, BELGIUM

Peter van der Geer Department of Chemistry and Biochemistry, University of California, San Diego, 9500 Gilman Dr., La Jolla, CA 92093-0601

John Venable Department of Cell Biology, Scripps Research Institute, 10550 North Torrey Pines Road, La Jolla, CA 92037

Marc Vidal Cancer Biology Department, Dana-Farber Cancer Institute, 44 Binney Street, Boston, MA 02115

Sonja Voordijk Geneva Bioinformatics SA, Avenue de Champel 25, Geneva, CH-1211, SWITZERLAND

Daniel Walther Swiss Institute of Bioinformatics (SIB), CMU, rue Michel-Servet 1, Genève 4, 1211, SWITZERLAND

Gerhard Weber Department of Anthropology University of Vienna, Althanstr. 14, A-1090 Vienna, AUSTRIA.

Peter J. A. Weber Proteomics Division, Tecan Munich GmbH, Feldkirchnerstr. 12a, Kirchheim, D-, 85551, GERMANY

Walter Weiss Fachgebiet Proteomik, Technische Universität Muenchen, Am Forum 2, Freising Weihenstephan, D-85350, GERMANY

John R. Yates III Department of Cell Biology, Scripps Research Institute, 10550 North Torrey Pines Road, La Jolla, CA 92037

Huilin Zhou Department of Cellular and Molecular Medicine, Ludwig Institute for Cancer Research, University of California, San Diego, 9500 Gilman Drive, CMM-East, Rm 3050, La Jolla, CA 92093-0660

LABELING AND DETECTING PROTEINS

CHAPTER

1

Protein Determination

Martin Guttenberger

I. INTRODUCTION

The protein content of tissues or samples can serve a number of purposes: It can be a research topic of its own (e.g., in nutritional studies; Hoffmann *et al.*, 2002), a loading control in gel electrophoresis (Ünlü *et al.*, 1997), or a reference quantity in biochemical (e.g., yields in protein purification) or physiological (e.g., specific activities of enzyme preparations; Guttenberger *et al.*, 1994) investigations. In addition, with the advent of proteomics, there is an increasing need for protein quantitation in complex sample buffers containing detergents and urea as potentially interfering compounds (Ünlü *et al.*, 1997). In any case, care should be taken to obtain correct results. This article focuses on three techniques and outlines the specific pros and cons.

II. MATERIALS AND INSTRUMENTATION

The following reagents are from the indicated suppliers. All other reagents are of analytical grade (Merck):

A. Lowry Assay

From Lowry *et al.* (1951): Folin–Ciocalteu phenol reagent (Merck, Cat. No. 1.09001). A detergent-compatible modification of the Lowry assay is available as a kit (Bio-Rad 500-0116).

B. Bradford Assay

From Bradford (1976): Coomassie brilliant blue G-250 (Serva Blue G, Serva, Cat. No. 35050). The reagent for this assay is available commercially from Bio-Rad (Cat. No. 500-0006).

C. Neuhoff Assay (Dot-Blot Assay)

From Guttenberger *et al.* (1991) and Neuhoff *et al.* (1979): Ammonium sulfate for biochemical purposes (Merck, Cat. No. 1.01211), benzoxanthene yellow (Hoechst 2495, Merck Biosciences, Cat. No. 382057, available upon request), cellulose acetate membranes (Sartorius, Cat. No. SM 11200), glycine, and SDS (Serva, Cat. Nos. 23390 and 20763, respectively). Commercially available ammonium sulfate frequently contains substantial amounts of undefined UV-absorbing and fluorescing substances. These lead to more or less yellowish solutions. Use only colourless solutions to avoid possible interference in fluorometry.

Solutions are prepared from bidistilled water. Bovine serum albumin (BSA, fraction V, Roche, Cat. No. 735086) is used as a standard protein. Ninety-six-well, flat-bottomed polystyrene microtiter plates (Greiner, Cat. No. 655101) are used for the photometric tests.

III. PROCEDURES

With respect to convenience and speed, microplate reader assays are described where appropriate. These assays can be read easily in conventional instruments by employing microcuvettes or by scaling up the volumes (fivefold).

The composition of the sample (extraction) buffer requires thought with respect to the avoidance of artifactual alterations of the protein and to the compatibility with the intended experimental procedures. The former requires strict control of adverse enzyme activities (especially proteases and phenol oxidases) and, in the case of plant tissues, of interactions with secondary metabolites. A convenient, semiquantitative assay for proteolytic activities allowing for the screening of suitable inhibitors was described by Gallagher *et al.* (1986). There is some uncertainty as to which assay gives the most reliable results in combination with extracts from plant tissues rich in phenolic substances. The influence of such substances can never be predicted. It is therefore imperative to minimize interaction of these substances with protein in the course of sample preparation. For a more detailed discussion of this problem, see Guttenberger *et al.* (1994).

A frequent source of ambiguity is the use of the term "soluble protein." Soluble as opposed to membrane-bound proteins stay in solution during centrifugation for 1 h at $105,000\,g$ (Hjelmeland and Chrambach, 1984).

All assays described in this article quantitate protein relative to a standard protein. The choice of the standard protein can markedly influence the result. This requires special attention for proteins with a high content of certain amino acids (e.g., aromatic, acidic, or basic amino acids). For most accurate results, choose a standard protein with similar amino acid composition or, if not available, compare different assays and standard proteins. Alternatively, employ a modified Lowry procedure that allows for absolute quantitation of protein (Raghupathi and Diwan, 1994).

The most efficient way to prepare an exact dilution series of the standard protein employs a handheld dispenser (e.g., Eppendorf multipette). Typically a six-point series is pipetted according to Table 1.1. In any case, avoid a concentration gradient of the sample buffer.

TABLE 1.1 Pipetting Scheme for Preparation of a Standard Dilution Series[a]

Concentration	Blank	0.2×	0.4×	0.6×	0.8×	1.0×
Water	5	4	3	2	1	0
Standard protein (2×)	0	1	2	3	4	5
Buffer (2×)	5	5	5	5	5	5

[a] To prepare 1 ml of each concentration, 1 volume corresponds to 0.1 ml.

Usually samples and standards may be kept at −20°C for a couple of weeks. For longer storage intervals, keep at −80°C.

A. Lowry Assay

See Lowry *et al.* (1951).

Solutions

Note: For samples low in protein (0.02 mg·ml^{-1} or less), prepare reagents A and B at double strength.

1. *Reagent A*: 2% (w/v) sodium carbonate (Na$_2$CO$_3$) in 0.10N NaOH. To make 1 litre of reagent A (5000 determinations), dissolve 20 g Na$_2$CO$_3$ in 1 litre 0.10M NaOH. Keep at room temperature in tightly closed screw-cap plastic bottles.
2. *Reagent B*: 0.5% CuSO$_4$·5H$_2$O in 1% sodium or potassium tartrate. To make 20 ml of reagent B, dissolve 0.1 g CuSO$_4$·5H$_2$O in 20 ml 1% tartrate (0.2 g sodium or potassium tartrate dissolved in 20 ml water). Keep at room temperature.
3. *Reagent C (alkaline copper solution)*: Mix 25 ml of reagent A and 0.5 ml of reagent B. Prepare fresh each day.
4. *Reagent D (Folin–Ciocalteu phenol reagent)*: Dilute with an equal volume of water just prior to use

Steps

1. Place 40 μl of sample (protein concentration 0.02–1 mg·ml^{-1}) or blank into cavities of a microplate or into appropriate test tubes.
2. Add 200 μl of reagent C and mix. Allow to stand for at least 10 min.
3. Add 20 μl of reagent D and mix *immediately*. Allow to stand for 30 min or longer.
4. Read the samples in a microplate reader or any other photometer at 750 nm.

Modifications

1. The sample volume may be raised to 140 μl when samples are low in protein (0.02 mg·ml^{-1} or less). In this case, employ double-strength reagent C.
2. If samples have been dissolved in 0.5M NaOH (recommended for resolubilization of acid precipitates), omit NaOH from reagent A.

B. Bradford Assay

See Bradford (1976).

Solutions

1. *Protein reagent stock solution*: 0.05% (w/v) Coomassie brilliant blue G-250, 23.8% (v/v) ethanol, 42.5% (w/v) phosphoric acid. To make 200 ml of stock solution (5000 determinations), dissolve 0.1 g Serva blue G in 50 ml 95% ethanol (denatured ethanol works as well), add 100 ml 85% phosphoric acid, and make up to 200 ml by adding water. The stock solution is available commercially (Bio-Rad). Keep at 4°C. The reagent contains phosphoric acid and ethanol or methanol. Handle with due care (especially when employing a dispenser)!
2. *Protein reagent*: Prepare from the stock solution by diluting in water (1:5). Filter immediately prior to use.

Steps

1. Place 4 μl of sample (protein concentration 0.1–1 mg·ml⁻¹) or blank into cavities of a microplate or into appropriate test tubes.
2. Add 200 μl of protein reagent and mix. Allow to stand for at least 5 min.
3. Read the samples within 1 h in a microplate reader or any other photometer at 595 nm.

Modifications

1. For improved linearity and sensitivity, compute the ratio of the absorbances, 590 nm over 450 nm (Zor and Selinger, 1996).
2. Microassay: For diluted samples (less than 0.1 mg·ml⁻¹), proceed as follows: Employ 200 μl of sample and add 50 μl of protein reagent stock.

C. Dot-Blot Assay

See Guttenberger *et al.* (1991). Do not change the chemistry of the membranes. Nitrocellulose will dissolve in the staining solution; PVDF membranes develop a strong background.

Solutions

1. *Benzoxanthene stock*: To prepare the stock solution add 1 ml of water to 0.5 g of the fluorescent dye (as supplied, weighing not necessary); keep at −20°C. The toxicity of benzoxanthene is not thoroughly studied, it might be mutagenic!
2. *Destaining solution*: Methanol/acetic acid (90/10, v/v). To make 1 litre, mix 100 ml acetic acid and 900 ml methanol.
3. *Staining solution*: To obtain 100 ml, dilute 80 μl benzoxanthene stock in 100 ml destaining solution. Be sure to pour the destaining solution onto the stock solution to prevent the latter from clotting. Keep staining and destaining solutions in tightly closed screw-cap bottles at 4°C in the dark. They are stable for months and can be used repeatedly. Take

due care in handling the highly volatile solutions containing methanol!

4. *SDS stock*: To make 30 ml of 10% (w/v) SDS stock solution, dissolve 3 g SDS in approximately 20 ml of water, stir, and make up to 30 ml (allow some time for settling of foam). Keep at room temperature; it is stable for at least 1 year.
5. *Elution buffer*: 0.25 M glycine–sulfuric acid buffer (pH 3.6) and 0.02% (w/v) SDS. To prepare 1 litre, dissolve 18.8 g glycine in approximately 900 ml water and add 15 ml of 0.5 M sulfuric acid. Slight deviations from pH 3.6 are tolerable. Add 2 ml SDS stock and make up to 1 litre. Keep at room temperature; it is stable for months.
   The following solutions are not needed for the standard protocol.
6. *Washing solution A*: Saturated ammonium sulfate, adjust to pH 7.0 with Tris. To make 1 litre, stir ammonium sulfate in warm water (do not heat excessively). Let the solution cool to room temperature overnight and titrate to pH 7.0 with a concentrated (approximately 2 M) solution of Tris (usually approximately 1 ml is required). Keep at room temperature. As ammonium sulfate tends to produce lumps in the storage bottle it might be easier to weigh the entire bottle, add some water, remove the resulting slurry, and weigh the empty bottle again. To produce a saturated solution (53.1%, w/v), dissolve 760 g ammonium sulfate in 1 litre water.
7. *Washing solution B*: Methanol/acetic acid/water (50/10/40, v/v). To make 1 litre, mix 100 ml acetic acid and 500 ml methanol; make up to 1 litre. Keep at 4°C.
8. *Drying solution*: 1-Butanol/methanol/acetic acid (60/30/10, v/v). To make 0.1 litre, mix 10 ml acetic acid, 30 ml methanol, and 60 ml butanol. Keep at 4°C; use up to six times.

Steps

The dot-blot assay is a versatile tool; its different modifications enable one to cope with almost every potentially interfering substance. In the following description the steps for all modifications are included.

1. Preparation of filter sheets (cellulose acetate membrane). *Handle the sheets with clean forceps and scissors, do not touch!* Cut one corner to aid in orientation during processing of the sheet. Mark the points of sample application (see later). Mount the membrane in such a way that the points of sample application are not supported (otherwise a loss of protein due to absorption through the membrane may be encountered). There are two different ways to achieve these requirements.

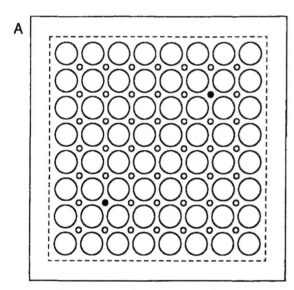

B

FIGURE 1.1 Dot-blot apparatus. (A) Top view. (B) Section along the diagonal. The apparatus has not been drawn to scale. Dashed lines indicate the position of the cellulose acetate membrane. Large circles correspond to the application points, small ones to the holes that are used for piercing the membrane (arrows in B), and solid small ones to the position of the pins that hold together the apparatus.

a. For routine assays it is recommended to mount the sheets in a special dot-blot apparatus (Fig. 1.1). Mark dot areas by piercing the sheets through small holes in the upper part of the device.

b. For occasional assays, mark the application points by impressing a grid (approximately 1-cm edge length) onto the filter surface (use a blunt blade and a clean support, preferably a glass plate covering a sheet of graph paper). Mount the sheets on a wire grating (preferably made from stainless steel, fixation by means of adhesive tape is recommended; cut off the taped areas prior to staining).

2. Apply samples ($0.01–10\,mg\cdot ml^{-1}$) to the membrane sheets in aliquots of $2\,\mu l$ (piston pipettes are highly recommended; well-rinsed capillary pipettes may be used instead). Leave to dry for a couple of minutes. Dilute samples may be assayed by applying samples repeatedly (let the sample dry prior to the next application).

3. Perform heat fixation. *Note: This step is imperative for samples containing SDS whereas it might prove deleterious to samples lacking SDS!* Bake the dot-blot membranes on a clean glass plate for 10 min at 120°C (oven or heating plate).

4. Remove interfering substances. *Note: This step is optional! Its use depends on the presence of potentially interfering substances (mainly carrier ampholytes, but also peptides and the buffer PIPES).* Remove interfering substances prior to protein staining by vigorous shaking in washing solution A ($3 \times 5\,min$), followed by gentle agitation in washing solution B ($3 \times 2\,min$).

5. Stain and destain. Perform staining (10 min) and destaining (5, 5, and 15 min) in closed trays (polyethylene food boxes work very well) on a laboratory shaker at ambient temperature. For the last destaining bath, employ fresh destaining solution; discard the first destaining bath. The incubation

times given here represent the minimal time intervals needed. As long as the vessels are closed tightly, each of these steps may be delayed according to convenience (in case of the last destaining bath, rinse in fresh destaining solution before proceeding).

6. Dry the stained membrane sheets. To facilitate cutting dot areas from the sheets, the following drying step is recommended. Shake the membranes in drying solution for exactly 2 min, mount them between two clamps[1] (Fig. 1.2), and leave them to dry in a fume hood. The dried sheets may be stored in the dark for later analysis.

7. Elute. Prior to elution, cut the dots from the membrane sheet. Perform elution (45 min in 2 ml of elution buffer) in glass scintillation vials on a laboratory shaker at ambient temperature (bright illumination should be avoided). Dried sheets have to be rewetted in destaining solution prior to immersion in elution buffer. It is recommended to dispense the destaining solution (25 μl) and the elution buffer with appropriate repetitive devices (e.g., Eppendorf multipette and Brand dispensette, respectively).

8. Take readings in a fluorometer (e.g., Luminescence Spectrometer LS 50B; Perkin-Elmer; Beaconsfield, UK) at 425 (excitation) and 475 (emission) nm.

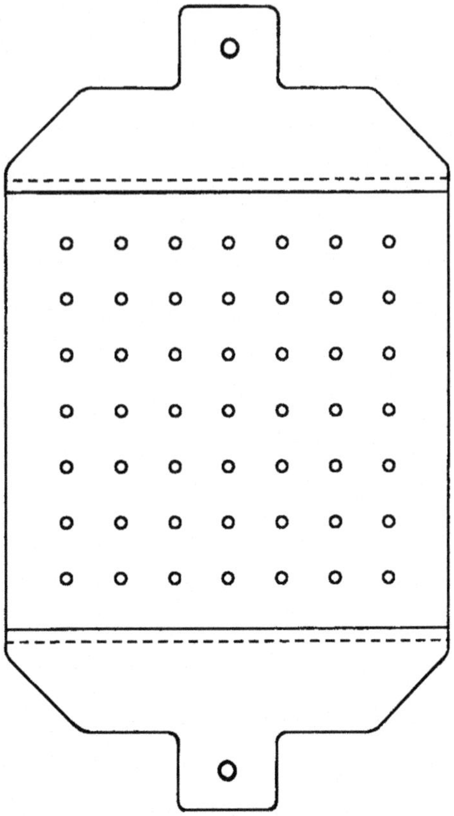

FIGURE 1.2 Membrane mounted for drying. Be sure to mount the drying membranes between two clamps of sufficient size to prevent distortion by uneven shrinkage. The weight of the lower clamp should keep the membrane spread evenly.

Modification

Skip elution and take readings directly from the wet membrane sheets (step 6) with a video documentation system (e.g., DIANA, Raytest GmbH, Straubenhardt, Germany; Hoffmann et al., 2002). Depending on the choice of filters, there might be considerable deviation from linearity.

IV. COMMENTS

With the exception of protein solutions, most stock solutions have a long shelf life. Discard any stock solution that changed its original appearance (e.g., got cloudy or discoloured).

Calculate standard curves according to the method of least squares. Appropriate algorithms are provided with scientific calculators

---

[1] Test for chemical resistance prior to first use: The edges of the clamp can be protected by a piece of silicon tubing cut open along one side.

and most spreadsheet programs for personal computers. It is better to compute standard curves employing single readings instead of means. Be aware of the basic assumptions made in regression analysis. For additional reading on the statistics of standard curves, compare Sokal and Rohlf (1995).

A. Lowry Assay

Pros: The Lowry assay exhibits the best accuracy with regard to absolute protein concentrations due to the chemical reaction with polypeptides. It is also useful for the quantitation of oligopeptides. This contrasts with the other two methods, which, as dye-binding assays, exhibit more variation depending on the different reactivity of the given proteins (standards as well as samples).

Cons: High sensitivity to potentially interfering substances; least shelf life of the reagents employed.

Recommendation: Employ where absolute protein contents are of interest.

B. Bradford Assay

Pros: The assay is widespread because of its ease of performance (only one stable reagent is needed, low sensitivity to potentially interfering substances, unsurpassed rapidity), its sensitivity, and its low cost.

Cons: High blank values, requires dual-wavelength readings for linearity, and possibly rather high deviations from absolute protein values (depending on the choice of standard protein).

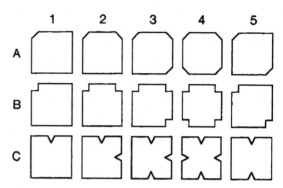

FIGURE 1.3 Useful incision patterns employed for marking membrane sheets prior to reprobing. Additional patterns may be generated by combination.

Recommendation: Employ where relative protein contents are sufficient (in most cases such as electrophoresis) and where the assay shows no interference by sample constituents (compare Bradford, 1976).

C. Dot-Blot Assay

Pros: The dot-blot assay combines high sensitivity, an extended range of linearity (20 ng to 20 μg), and high tolerance to potentially interfering substances. The sample is not used up during assay. Hence, it may be reprobed[2] (Fig. 1.3) for immunological tests or detection of glycoproteins (Neuhoff *et al.*, 1981).

Cons: More demanding and time-consuming than the other assays and rather expensive (chemicals and instrumentation).

Recommendation: Employ where (1) the other assays show interference, especially with complex sample buffers used in one-dimensional[3]

---

[2] Sheets containing single dot areas can be marked conveniently by cutting the edges (Fig. 1.3, Neuhoff *et al.*, 1979).

[3] Sample buffer according to Laemmli (1970): 62.5 mM Tris–HCl (pH 6.8), 2% (w/v) SDS, 10% (v/v) glycerol, 5% (v/v) 2-mercaptoethanol, and 0.001% (w/v) bromphenol blue. Range of the assay: 0.04 to 10 mg·ml^{-1}, i.e., 80 ng to 20 μg in the test.

and two-dimensional[4] electrophoresis; (2) the amount of sample is limited and/or reprobing of the dotted samples is desirable; or (3) the mere detection of protein in aliquots, e.g., from column chromatography, is needed (spot 0.2–2 µl onto membrane, process according to standard protocol, prevent evaporation by covering the destained membrane with a thin glass plate, view under UV light).

V. PITFALLS

1. Solutions containing protein exhibit an altered surface tension. Avoid foaming and pipette slowly and steadily.
2. Extraction or precipitation steps to eliminate interfering substances should be carefully controlled for complete recovery of protein (Lowry et al., 1951). The more demanding dot-blot assay frequently is a good alternative because of a considerable gain of convenience and accuracy with respect to a simplified sample preparation.
3. Omission of known interfering buffer components from just those samples that are intended for protein determination is strongly discouraged as the solubility of proteins might be influenced (carrier ampholytes, e.g., enhance solubilization of membrane proteins in two-dimensional electrophoresis sample buffer; for references, see Guttenberger et al., 1991).
4. In the case of photometers/fluorometers operating with filters (usually microplate readers), the correct wavelength may not be available. Instead, a similar wavelength may be employed [Lowry assay: 530–800 nm, Bradford assay: 540–620 nm, dot-blot.

assay: 366–450 nm (excitation), 450–520 nm (emission)]. In the case of fluorometry, allow for a sufficient wavelength interval between excitation and emission (consult the operating instructions of your instrument). Be aware that considerable deviations from the standard wavelengths will be at the expense of linearity and sensitivity.
5. In microplates it is important to achieve uniform menisci: Prick air bubbles with a thin wire and mix the plates on a gyratory shaker.
6. Analysis of dilute samples by application of larger sample volumes also increases the amount of potentially interfering substances. Include appropriate controls.

A. Lowry Assay

1. Many reagents used commonly in protein extraction interfere with this assay. The main groups of interfering substances are reductants (e.g., sulfhydryl compounds such as mercaptoethanol, reducing sugars such as glucose), chelating agents (e.g., EDTA), amine derivatives (many common buffering substances such as Tris), and detergents (e.g., Triton, SDS). A detailed list of interfering substances, along with remedies and tolerable limits, is provided by Petersen (1979).
2. Reagent D is not stable at a basic pH. Immediate mixing after the addition of reagent D is imperative. In microplates the use of a small plastic spatula is convenient for this purpose (change or rinse between samples).
3. The colour reaction takes about 80 min to come to completion. Prior to this, reading

[4]Sample (lysis) buffer according to O'Farrell (1975): 9.5 M urea, 2% (w/v) Nonidet P-40, 5% (v/v) 2-mercaptoethanol, and 2% (w/v) carrier ampholytes. Standards are prepared by a stepwise dilution of the BSA stock solution in a modified sample buffer lacking carrier ampholytes. These are added from a doubly concentrated stock solution (4%, w/v) in sample buffer. Range of the assay: 0.02 to 8 mg·ml^{-1}, i.e., 40 ng to 16 µg in the test.

of samples over an extended period of time will give rise to experimental error (more than 20%; Kirazov *et al.*, 1993). Keep the reading interval to a minimum. Alternatively, both incubation steps can be cut to 3 min by raising the incubation temperature to 37°C (Shakir *et al.*, 1994). As the time to reach thermal equilibration will depend on the experimental setup, a test run in comparison to the original method is recommended.

B. Bradford Assay

1. The commonly used standard protein BSA is highly reactive in this dye-binding assay. As a consequence the protein content of the samples is underestimated. This systematic error does not matter in comparative analyses but brings about wrong absolute values. Bovine γ-globulin is a preferable standard.
2. The standard curves are not strictly linear in the original version of the assay. If the necessary equipment for the recommended dual-wavelength ratio is not available, do not extend the range of standard concentrations beyond one order of magnitude or do not calculate standard curves by means of linear regression.
3. Samples containing detergents (1% will interfere) must be diluted (if possible) or precipitated (compare Section V.2) prior to analysis.
4. The protein–dye complex is insoluble and will precipitate with time (Marshall and Williams, 1992). For highest accuracy, take readings within an interval between 5 and 20 min after addition of the reagent. With crude extracts (e.g., from mycelia of certain fungi), this interval may be considerably shorter—too short to take meaningful readings. In this case, alter the way of sample preparation or use another assay.

5. Plastic and glassware (especially quartz glass) tend to bind dye. Remove the resulting blue colour by one of the following procedures: (1) Rinse with glassware detergent (avoid strongly alkaline detergents with cuvettes; rinse thoroughly to remove detergent again), (2) rinse with ethanol or methanol, or (3) soak in 0.1 M HCl (takes several hours).

C. Dot-Blot Assay

1. Generally, it is imperative to prevent the membrane sheets from drying during one of the transfer steps (residual acetic acid will destroy the filter matrix).
2. In case of highly variable results, inspection of the stained filters (last destaining bath or dried) under UV illumination may be helpful: Background staining resulting from improper handling of the membranes will be visible (do not use UV-irradiated membranes for quantitative analyses).
3. After the washing procedure, thorough rinsing in washing solution B is imperative. Ammonium sulfate accumulating in the staining solution will interfere with the assay.
4. Although the dot-blot assay is extremely insensitive to potentially interfering substances, it is advisable to include appropriate controls (at least blank buffer and buffer plus standard).
5. In the case of buffers containing detergent plus carrier ampholytes, the storage conditions and the number of freeze–thaw cycles may prove important. Use fresh solutions or run appropriate controls.
6. If membrane sheets turn transparent upon drying, they have not been equilibrated properly in the drying solution (keep in time: 2 min) or the drying solution has been diluted by accumulation of destaining solution (do not reuse the drying solution too often).

References

Bradford, M. M. (1976). A rapid and sensitive method for the quantitation of microgram quantities of protein utilizing the principle of protein-dye binding. *Anal. Biochem.* **72**, 248–254.

Gallagher, S. R., Carroll, E. J., Jr., and Leonard, R. T. (1986). A sensitive diffusion plate assay for screening inhibitors of protease activity in plant cell fractions. *Plant Physiol.* **81**, 869–874.

Guttenberger, M., Neuhoff, V., and Hampp, R. (1991). A dot-blot assay for quantitation of nanogram amounts of protein in the presence of carrier ampholytes and other possibly interfering substances. *Anal. Biochem.* **196**, 99–103.

Guttenberger, M., Schaeffer, C., and Hampp, R. (1994). Kinetic and electrophoretic characterization of NADP dependent dehydrogenases from root tissues of Norway spruce (*Picea abies* [L.] Karst.) employing a rapid one-step extraction procedure. *Trees* **8**, 191–197.

Hjelmeland, L. M., and Chrambach, A. (1984). Solubilization of functional membrane proteins. *Methods Enzymol.* **104**, 305–318.

Hoffmann, E. M., Muetzel, S., and Becker, K. (2002). A modified dot-blot method of protein determination applied in the tannin-protein precipitation assay to facilitate the evaluation of tannin activity in animal feeds. *Br. J. Nutr.* **87**, 421–426.

Kirazov, L. P., Venkov, L. G., and Kirazov, E. P. (1993). Comparison of the Lowry and the Bradford protein assays as applied for protein estimation of membrane-containing fractions. *Anal. Biochem.* **208**, 44–48.

Laemmli, U. K. (1970). Cleavage of structural proteins during the assembly of the head of bacteriophage T4. *Nature* **227**, 680–685.

Lowry, O. H., Rosebrough, N. J., Farr, A. L., and Randall, R. J. (1951). Protein measurement with the Folin phenol reagent. *J. Biol. Chem.* **193**, 265–275.

Marshall, T., and Williams, K. M. (1992). Coomassie blue protein dye-binding assays measure formation of an insoluble protein-dye complex. *Anal. Biochem.* **204**, 107–109.

Neuhoff, V., Ewers, E., and Huether, G. (1981). Spot analysis for glycoprotein determination in the nanogram range. *Hoppe-Seyler's Z. Physiol. Chem.* **362**, 1427–1434.

Neuhoff, V., Philipp, K., Zimmer, H.-G., and Mesecke, S. (1979). A simple, versatile, sensitive and volume-independent method for quantitative protein determination which is independent of other external influences. *Hoppe-Seyler's Z. Physiol. Chem.* **360**, 1657–1670.

O'Farrell, P. H. (1975). High resolution two-dimensional electrophoresis of proteins. *J. Biol. Chem.* **250**, 4007–4021.

Peterson, G. L. (1979). Review of the Folin phenol protein quantitation method of Lowry, Rosebrough, Farr and Randall. *Anal. Biochem.* **100**, 201–220.

Raghupathi, R. N., and Diwan, A. M. (1994). A protocol for protein estimation that gives a nearly constant color yield with simple proteins and nullifies the effects of four known interfering agents: Microestimation of peptide groups. *Anal. Biochem.* **219**, 356–359.

Shakir, F. K., Audilet, D., Drake, A. J., III, and Shakir, K. M. M. (1994). A rapid protein determination by modification of the Lowry procedure. *Anal. Biochem.* **216**, 232–233.

Sokal, R. R., and Rohlf, F. J. (1995). "Biometry." Freeman, New York.

Ünlü, M., Morgan, M. E., and Minden, J. S. (1997). Difference gel electrophoresis: A single gel method for detecting changes in protein extracts. *Electrophoresis* **18**, 2071–2077.

Zor, T., and Selinger, Z. (1996). Linearization of the Bradford protein assay increases its sensitivity: Theoretical and experimental studies. *Anal. Biochem.* **236**, 302–308.

2

Phosphopeptide Mapping: A Basic Protocol

Jill Meisenhelder and Peter van der Geer

I. INTRODUCTION

Peptide mapping is a technique in which a radioactively labeled protein is digested with a sequence specific protease. The resulting peptides are separated in two dimensions on a thin-layer cellulose (TLC) plate by electrophoresis and chromatography. The peptides are visualized by autoradiography, giving rise to a peptide map.

Peptide maps of ^{35}S-labeled proteins are used most often to find out whether two polypeptides are related. Peptide maps of ^{32}P-labeled proteins are used to obtain information about the phosphorylation of the protein under investigation. Proteins can be labeled *in vivo* by incubating cells in the presence of [^{32}P]orthophosphate or by incubating them *in vitro* with an appropriate protein kinase in the presence of [γ-^{32}P]ATP. Proteins are usually separated from other contaminating proteins by SDS–PAGE and then subjected to phosphopeptide mapping or phosphoamino acid analysis.

II. MATERIALS AND INSTRUMENTATION

HTLE 7000 electrophoresis system (CBS Scientific, Del Mar, CA) *pH 1.9 electrophoresis buffer*: 50 ml formic acid (88%, w/v), 156 ml glacial acetic acid, and 1794 ml deionized water

pH 3.5 electrophoresis buffer: 100 ml glacial acetic acid, 10 ml pyridine, and 1890 ml deionized water

pH 4.72 electrophoresis buffer: 100 ml *n*-butanol, 50 ml pyridine, 50 ml glacial acetic acid, and 1800 ml deionized water

pH 8.9 buffer: 20 g $(NH_4)_2CO_3$ and 2000 ml deionized water

Regular chromatography buffer: 785 ml *n*-butanol, 607 ml pyridine, 122 ml glacial acetic acid, and 486 ml deionized water

Phospho-chromatography buffer: 750 ml *n*-butanol, 500 ml pyridine, 150 ml glacial acetic acid, and 600 ml deionized water

Isobutyric acid buffer: 1250 ml isobutyric acid, 38 ml *n*-butanol, 96 ml pyridine, 58 ml glacial acetic acid, and 558 ml deionized water

Phosphoamino acid stocks: 1 mg/ml each in deionized water is stable for years at $-20°C$.

50 mM NH_4HCO_3 pH 7.3–7.6. Make up fresh; lower the pH by bubbling CO_2 through it if necessary. The pH of this buffer will drift overnight toward pH 8.0.

1.5-ml microfuge tubes with plastic pestles can be obtained from Kimble Kontes (Vineland, NJ). These pestles fit nicely into Sarstedt screw-cap microcentrifuge tubes.

RNase A is dissolved in deionized water at 1 mg/ml, boiled for 5 min, and stored at $-20°C$.

TPCK-treated trypsin (Worthington Lakewood, NJ) can be dissolved at 1 mg/ml in 1 mM HCl and is stable at $-70°C$ for years.

III. PROCEDURES

A. Phosphopeptide Mapping

1. Separate the ^{32}P-labeled protein of interest from other contaminants by resolving the sample by SDS–PAGE. Dry the gel onto Whatman 3 MM paper, mark the paper backing around the gel with radioactive or fluorescent ink, and expose the gel to X-ray film. Line the gel up with the film using the markings on the paper backing and autorad, localize the protein of interest, and cut the protein band out of the gel with a clean, single edge razor or a surgical blade. Remove the paper backing from the gel slices by scraping gently with a razor blade.

2. Extract the protein from the gel by grinding the gel into small fragments. Place the gel slice(s) in a 1.7-ml screw-cap tube and hydrate briefly in 500 µl 50 mM NH_4HCO_3, pH 7.3–7.6. Grind the gel to small pieces using a fitted plastic, disposable pestle. Add 500 µl more NH_4HCO_3, 10 µl 10% SDS, and 10 µl βME, vortex, boil for 5 min, and extract for at least 4 h on an agitator at room temperature.

3. Spin down the gel bits by centrifugation in a microfuge for 5 min at 2000 rpm, transfer the supernatant to a new microfuge tube, and store at 4°C. This supernatant represents volume × (µl). Add (1300−×) µl more NH_4HCO_3 to the gel bits, vortex, and extract again for at least 4 h on an agitator at room temperature. Spin down the gel bits and combine this supernatant with the first extract.

4. Clear the (combined) extract by centrifugation. Spin 15 min at 15,000 rpm in a microfuge at room temperature. Transfer the supernatant to a new tube, leaving the final 20 µl behind to avoid transfer of particulate material. Repeat this step one or two more times. It is important that the final extract is free of any particulate materials (gel and paper bits).

5. Concentrate the protein by TCA precipitation. Add 20 µl RNase A (1 mg/ml) to the protein extract, mix, and incubate 20 min on ice. Add 250 µl ice-cold 100% TCA, mix, and incubate 1 h on ice. Spin 15 min at 15,000 rpm in a microfuge at 4°C and remove the supernatant. Add 0.5 ml 100% cold

ethanol to the pellet, invert the tube, and spin 10 min at 15,000 rpm in a microfuge and remove the supernatant. Spin again briefly, remove residual ethanol, and briefly air dry the pellet.

6. To avoid the formation of oxidation-state isomers, oxidize the protein to completion by incubation in performic acid. Performic acid is formed by incubating 9 parts 98% formic acid with 1 part H_2O_2 for 30–60 min at room temperature and then cool on ice. Resuspend the TCA pellet in 50 μl cold performic acid, incubate for 1 h on ice, add 400 μl deionized water, freeze, and lyophilize.

7. In order to analyze the different phosphorylation sites, digest the protein with a sequence-specific protease. We routinely use trypsin because it works well on denatured protein and its specificity is well characterized. Resuspend the oxidized protein in 50 μl 50 mM NH$_4$HCO$_3$, pH 8.0. Add 10 μl 1 mg/ml TPCK-treated trypsin, vortex, and incubate for 4–16 h at 37°C. Add a second aliquot of trypsin, vortex, and incubate again for 4–16 h at 37°C.

8. Now subject the sample to several rounds of lyophilization to remove the ammonium bicarbonate. Add 400 μl deionized water to the sample, mix, and lyophilize. Repeat this procedure two to three times and then spin the final rinse for 5 min at 15,000 rpm in a microfuge and transfer the supernatant to a new tube and lyophilize.

The peptides are now ready for application onto a 20 × 20 TLC plate. Electrophoresis will be used for separation in the first dimension. Three different buffer systems are commonly used for electrophoresis (pH 1.9, pH 4.72, and pH 8.9). All three buffer systems should be tried to determine which will best separate the tryptic phosphopeptides of a particular protein.

9. Dissolve the final pellet in 5–10 μl of the electrophoresis buffer to be used; use deionized water instead of pH 8.9 buffer.

Spot the peptide mix using a gel-loading tip fitted to an adjustable micropipette. Keep the sample on as small an area on the plate as possible by spotting the samples 0.3–0.5 μl at a time, drying the sample between spottings. Spot the sample 3 cm from the bottom of the plate and 5 cm from the left side for electrophoresis at pH 1.9 or 4.72 or 10 cm from the left hand side for electrophoresis at pH 8.9 (see Fig. 2.1a). Mark origins on the plate with a blunt, soft pencil. We like to spot 1 μl of marker dye mixture 2 cm from the top of the plate above the sample origin (dye origin first dimension, Fig. 2.1a). The mobilities of marker dyes can be used as standards when comparing different maps.

In our laboratories we use the HTLE 7000 electrophoresis system. This system should be connected to a power supply, cooling water, and an air line with a pressure regulator and should be set up according to the manufacturer's directions (Fig. 2.2).

10. Wet the TLC plate with electrophoresis buffer immediately before placing it onto the electrophoresis apparatus as described in Fig. 2.1. The plate should be damp with no puddles present. Shut off the air on the HTLE 7000, remove the securing pins, the neoprene pad, the Teflon insulator, and the upper polyethylene protector sheet and fold the electrophoresis wicks back over the buffer tanks. Wipe excess buffer from the upper and lower polyethylene protector sheets and place the TLC plate onto the lower polyethylene sheet. Fold the electrophoresis wicks so that they overlap ~1 cm onto the TLC plate. Reassemble the apparatus, insert the securing pins, and adjust the air pressure to 10 lbs/in^2. Turn on the cooling water and switch on the high voltage power supply. We normally run maps for 20–30 min at 1000 V. When the run is finished, disassemble the apparatus and air dry the plate.

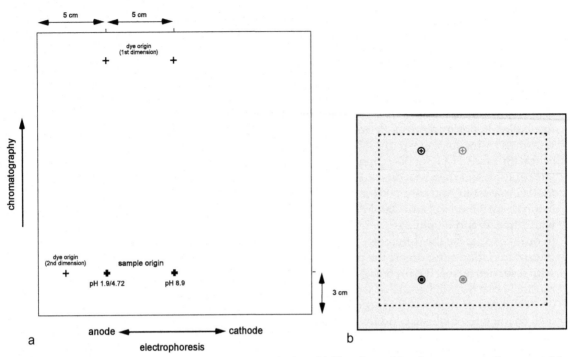

FIGURE 2.1   Applying the peptide mixture onto a TLC plate. (a) Phosphopeptide mixtures are usually separated by electrophoresis in the horizontal dimension and chromatography in the vertical dimension. The mixture is spotted 3 cm from the bottom of the plate and 5 cm from the left side for electrophoresis at pH 1.9 or pH 4.72 or in the center of the plate for electrophoresis at pH 8.9 (with the anode on the left and the cathode on the right after placing the plate on the HTLE 7000). We usually mark the sample and dye origin on the TLC plate with a soft blunt pencil. (b) Plates are wetted with electrophoresis buffer using a blotter composed of two layers of Whatman 3MM paper. Separate blotters are used for different electrophoresis buffers and each blotter can be used many times. A blotter contains two 1.5-cm-diameter holes that correspond to the sample and dye origin. The blotter is soaked in electrophoresis buffer and excess buffer is removed by blotting with a piece of 3MM paper. The blotter is then placed on the TLC plate so that the sample and dye origins are in the center of the two holes. The blotter is pushed onto the plate with the palm of the hand so that buffer is transferred from the blotter onto the plate. The edge of the holes is pushed onto the plate so that buffer moves from the blotter toward the center of the holes. This will concentrate the sample and dye on their origins. When the plate is completely wet, the blotter is removed and the plate is placed on the HTLE 7000. The electrophoresis apparatus is reassembled (Fig. 2.2) and electrophoresis is started.

Running times and origins can be adjusted for individual proteins. We recommend increasing the running time rather than the voltage to get better separation of the peptides on the map.

11. Separate peptides in the vertical dimension by ascending chromatography. A plastic tank (57 × 23 × 57 cm) is available from CBS Scientific (Del Mar, CA 92014). These tanks can hold up to eight TLC plates. Tanks need to be equilibrated with chromatography buffer for at least 24 h before the run. Three different types of chromatography buffer are commonly used in our laboratories (regular chromatography, phosphochromatography, and isobutyric acid buffers). We recommend trying the phospho-chromatography buffer first.

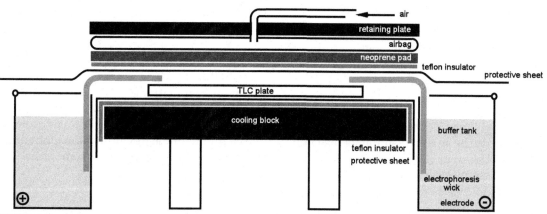

FIGURE 2.2 **Setting up the HTLE 7000.** Fill the buffer tanks with 600 ml freshly prepared electrophoresis buffer. Cover the cooling plate and fitted Teflon insulator with a polyethylene protector sheet that can be tucked between the cooling plate and the buffer tanks. Insert the wet electrophoresis wicks (14 × 20-cm double-layer Whatman 3 MM paper) into the buffer tanks and fold them over the cooling plate. Cover the cooling plate and electrophoresis wicks with the top polyethylene protector sheet, which should extend over the buffer tanks. Add the Teflon protector sheet, the neoprene pad, and close the cover. Insert the two pins to secure the cover before turning up the air pressure to 10 lbs/in^2. Immediately before starting electrophoresis, wet a plate as described in Fig. 2.1. Then turn off the air pressure, take out the pins, open the apparatus, and remove the neoprene pad, the Teflon protector, and the polyethylene sheet. Fold the electrophoresis wicks backward over the buffer tanks. Dry the bottom and top protector sheets with a Kimwipe, place the TLC plate on the apparatus, and fold the electrophoresis wicks over the cooling plate so that they overlap ~1 cm with the TLC plate. Place the polyethylene protector sheet on top of the TLC plate and reassemble the apparatus. Adjust the air pressure to 10 lbs/in^2., turn on the cooling water, and start electrophoresis.

We like to spot 1.0 µl of marker dye in the right or left margin of the plate at the same level as the sample (dye origin second dimension, Fig. 2.1a). Place all plates in the tank at the same time, leaning them at the same angle; replace the lid and do not open the tank again while chromatography is in progress. The front advances more slowly as it climbs higher on the plate. Let chromatography proceed until the buffer front reaches 1–2 cm from the top of the plate. This can take from 8 to 16 h depending on the lot of cellulose plates, the ambient temperature, and the age of the buffer. At this point remove all plates from the tank and let them air dry. Mark the plates with fluorescent or radioactive ink and expose to X-ray film in the presence of an intensifier screen or to a PhosphorImager screen.

B. Phosphoamino Acid Analysis

1. To analyze the phosphoamino acid content of a phosphoprotein, the protein can be isolated exactly as described for phosphopeptide mapping (Section III,A, steps 1–6). Transfer a fraction of the sample resuspended in performic acid (step 6) to a screw-cap microfuge tube and lyophilize. If the entire sample is to be used for phosphoamino acid analysis, use the ethanol-washed TCA precipitate (step 5).
2. Hydrolyze the protein to liberate the individual phosphoamino acids. Dissolve the sample in 50 µl 6M HCl and incubate for 1 h at 110°C.
3. Remove the hydrochloric acid by lyophilization.

4. Dissolve the amino acid mixture in 5–10 µl pH 1.9 buffer containing unlabeled phosphoamino acids as standards (70 µg/ml of each) by vortexing vigorously. Spin for 5 min at maximal speed in a microfuge to pellet particulate matter.

5. Spot the samples on a TLC plate as described for phosphopeptide mapping (step 9). Four samples can be analyzed simultaneously on a single TLC plate (Fig. 2.3a).

6. After all samples are spotted, wet the plate with pH 1.9 buffer using a blotter containing five holes that correspond to the four sample origins and a dye origin, as described for phosphopeptide mapping (step 10, Fig. 2.3b). Place the wetted plate immediately on the HTLE 7000 containing pH 1.9 buffer in the tanks and reassemble the apparatus. Resolve the samples in the first dimension by electrophoresis for 20 min at 1.5 kV (Fig. 2.3a, lower left) and air dry the plates.

7. Before running electrophoresis in the second dimension, wet the plate with pH 3.5 buffer containing ~0.1 mM EDTA using three rectangular blotters 10, 6, and 4 cm wide. Place blotters on the plate so that a ~1.5-cm strip of the plate containing the sample separated in the first dimension remains uncovered (Fig. 2.3c). By pressing the blotters onto the plate the buffer will migrate from the blotters and concentrate the samples on a line midway between the blotter edges (Fig. 2.3c). Turn the plate 90° counterclockwise to place it on the HTLE 7000. Reassemble the apparatus, now containing pH 3.5 buffer in the tanks, and carry out electrophoresis for 16 min at 1.3 kV (Fig. 2.3a, upper left). After separation in the second dimension, disassemble the apparatus and dry the plate.

8. To visualize the standard phosphoamino acids, spray the plate with 0.25% ninhydrin in acetone. Develop the ninhydrin stain by placing the plate for 5–15 min at 65°C.

9. Mark the plate around the edges with radioactive or fluorescent ink and expose it to X-ray film in the presence of an intensifier screen. After developing, align the X-ray film with the TLC plate using the radioactive or fluorescent markings. Phosphoamino acids are identified by comparing the mobility of the radioactive phosphoamino acids seen on the film with the mobility of the ninhydrin-stained phosphoamino acid standards.

IV. COMMENTS

Peptide maps provide information about the phosphorylation status of a protein. Only radioactively labeled peptides are visualized by autoradiography. In principle, every spot on the map represents a phosphorylation site; there are, however, exceptions (Boyle et al., 1991; Meisenhelder et al., 1999). Occasionally, two phosphorylation sites are close together and consequently are contained within a single peptide. If this is the case, a single spot may represent two phosphorylation sites. It is important to realize that single and doubly phosphorylated forms of a particular peptide will be resolved from each other because adding a phosphate group will add a negative charge to the peptide. It is also possible that as a consequence of partial digestion, multiple peptides/spots are generated that represent a single phosphorylation site.

We have described the basic protocol in which proteins are isolated from an SDS–PAGE gel before digestion with proteases. An alternative protocol exists in which proteins are transferred from the gel to a membrane. Membrane-bound proteins are then digested or hydrolyzed (Luo et al., 1990, 1991). Peptides and phosphoamino acids detach from the membrane and are analyzed as decribed here. A possible problem with this protocol is that not all peptides detach from the membrane, resulting in loss of information.

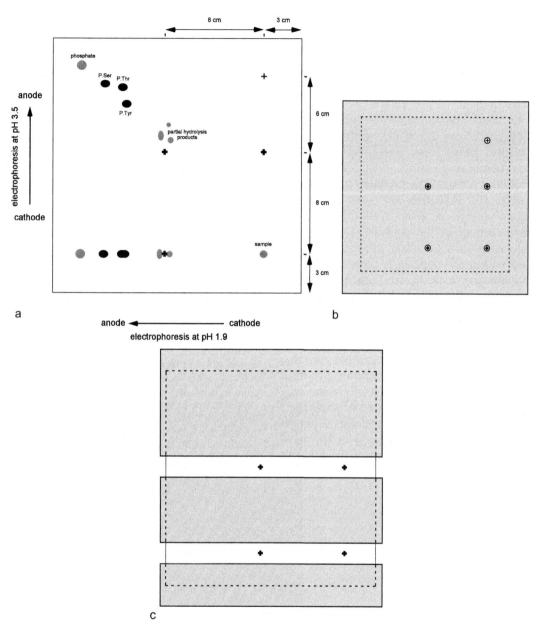

FIGURE 2.3   Separation of phosphoamino acids in two dimensions on a TLC plate. (a) Four samples can be separated on a single TLC plate at the same time. The four sample origins and an origin for marker dye are marked on the TLC plate with a blunt, soft pencil. Samples are first resolved in the horizontal dimension by electrophoresis at pH 1.9 (lower left quadrant). Subsequently the samples are resolved in the vertical dimension by electrophoresis at pH 3.5 (upper left quadrant). (b) After spotting the samples and marker dye, the plate is wetted with a blotter containing five holes, as shown. This will concentrate the samples and dye at their origins. (c) After electrophoresis in the first dimension, the plate is air dried. The plate is wetted with pH 3.5 buffer using three rectangular strips of Whatman 3 M paper as blotters. Buffer will move from each strip as a horizontal line converging with the buffer from the adjacent strips, thereby concentrating the sample on this midpoint. When the plate is completely wet it is turned 90° counterclockwise and placed on the HTLE 7000 apparatus.

If the ultimate goal is to identify phosphorylation sites, then spots/peptides on the phosphopeptide map have to be matched to the amino acid sequence of the protein of interest. This is done best by making a list of possible peptides that can be generated by the protease used to cleave the protein. Now peptides can be eliminated from this list by gathering more information about the phosphopeptides of interest (Gould and Hunter, 1988; van der Geer and Hunter, 1990). The peptides can be isolated from the TLC plate and used to answer additional questions. What is the phosphoamino acid content of the peptide? Does it contain cleavage sites for other proteases? Which residue within the peptide is phosphorylated? Double-labeling experiments can be used to find out whether the peptide contains cysteine or methionine. Information obtained in these experiments can be used to eliminate candidate peptides from the list of possibilities until only one or two are left. The protein can also be analyzed using a peptide mapping program available on the Internet; in this way some candidate peptides can be eliminated because their predicted mobility is substantially different from actual mobility on the map of the peptide in question. As verification of the identity of a phosphorylation site, the mutant protein in which the phosphorylation site has been changed to a nonphosphorylatable amino acid can be subjected to phosphopeptide mapping to confirm that the phosphorylation site of interest/spot(s) on the map has been eliminated.

References

Boyle, W. J., van der Geer, P., and Hunter, T. (1991). Phosphopeptide mapping and phosphoamino acid analysis by two-dimensional separation on thin-layer cellulose plates. *Methods Enzymol.* **201**, 110–148.

Gould, K. L., and Hunter, T. (1988). Platelet-derived growth factor induces multisite phosphorylation of pp60c-src and increases its protein-tyrosine kinase activity. *Mol. Cell. Biol.* **8**, 3345–3356.

Luo, K., Hurley, T. R., and Sefton, B. M. (1990). Transfer of proteins to membranes facilitates both cyanogen bromide cleavage and two-dimensional proteolytic mapping. *Oncogene* **5**, 921–923.

Luo, K., Hurley, T. R., and Sefton, B. M. (1991). Cyanogen bromide cleavage and proteolytic peptide mapping of proteins immobilized to membranes. *Methods Enzymol.* 149–152.

Meisenhelder, J., Hunter, T., and van der Geer, P. (1999). Phosphopeptide mapping and identification of phosphorylation sites. *In "Current Protocols in Protein Science"*, pp. 13.19.11–13.19.28. Wiley, New York.

van der Geer, P., and Hunter, T. (1990). Identification of tyrosine 706 in the kinase insert as the major colony-stimulated factor 1 (CSF-1)-stimulated autophosphorylation site in the CSF-1 receptor in a murine macrophage cell line. *Mol. Cell. Biol.* **10**, 2991–3002.

Radioiodination of Proteins and Peptides

Martin Béhé, Martin Gotthardt, and Thomas M. Behr

I. INTRODUCTION

Radioiodination of biomolecules was established in 1948, when ^{131}I was the first radioisotope of iodine to be used for the labeling of a, at that time, polyclonal antikidney serum, performed and described by Pressmen and Keighley. The technical production of iodine-131 was pioneered in 1938 by Seaborg and Livingood. A variety of other iodine isotopes exist (Table 3.1), from which ^{123}I, ^{125}I, and ^{131}I are the most widely used for the labeling of biomolecules for *in vitro* (i.e., radioimmunoassay) and *in vivo* (i.e., pharmacokinetic and metabolism) applications. ^{123}I, ^{125}I, and ^{131}I are γ emitters that can technically be detected easily, whereas ^{124}I emits positrons, which annihilate with an electron to produce two photons with an energy of

TABLE 3.1   Relevant Isotopes of Iodine

	$t_{1/2}$	γ energy (keV)	β energy (MeV)
[123]I	13.1 h	159 (97.7%)	(Auger/conversion e[-])
[124]I	4.17 days	603, 1691	β^+ 2.1
[125]I	60.1 days	35 (100%)	(Auger/conversion e[-])
[126]I	13.0 days	389, 688	β^+ 1.1; β^- 0.9, 1.3
[127]I	Stable		
[129]I	1.59×10^7 year	40	β^- 0.2
[131]I	8.04 days	364 (83%)	β^- 0.6, 0.8
		637 (6.7%)	
		284 (6.9%)	

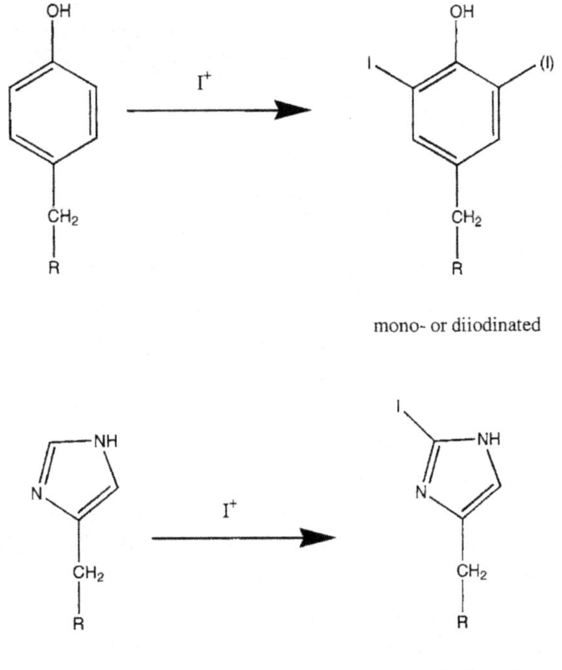

FIGURE 3.1   Schematic representation of the radioiodination procedure.

512 keV each. Due to this property, it can be used for positron emission tomography (PET). All these nuclides can be detected directly *in vitro* or *ex vivo* without a scintillation cocktail, whereas for *in vivo* imaging, the most suitable nuclides are [124]I for PET and [123]I (as well as, under certain conditions, ^{131}I) for single photon emission computer tomography (SPECT). [125]I is not suitable for *in vivo* imaging, due to its low energy of 35 keV, which is absorbed within a very short path length (a few millimeters, at best). However, this makes it very useful, together with the long half-life of 60 days for *ex vivo* (microautoradiography) and *in vitro* studies. [131]I decays by emitting photons with high energies (284, 364, and 637 keV) and electrons (0.6 and 0.8 MeV), the latter of which may cause substantial radiation exposure to tissues, rendering the isotope less optimal for its use in experiments. The main value of this isotope is its clinical application as a therapeutic nuclide in various therapies, including benign and malignant thyroid disorders or radioimmunotherapy.

All isotopes are available as NaI in neutral or basic solution, whereas ^{123}I, ^{125}I, and ^{131}I are commercially produced isotopes; [124]I can be delivered only by specialized cyclotron facilities. Because iodide (I[-]) is a nonreactive form, it must be activated by an oxidizing agent to a reactive cationic species (I[+]), which allows a spontaneous electrophilic substitution on

aromatic rings with a good leaving group such as H[+] in p-kresol (tyrosine) or 4-methylimidazol (histidine) (Fig. 3. 1). The iodination place is pH driven. The tyrosine moiety is labeled mainly at a pH around 7.5, whereas at a pH around 8.5, mainly the histidine is labeled, at a much lower yield though.

The tyrosine moiety can be labeled twice (Fig. 3. 1). The second step, yielding di-iodinated tyrosyl moieties, occurs faster than the monoiodination reaction. Due to the stoichiometry of the reactants, in proteins, usually monoiodotyrosine residues prevail, whereas in peptides, di-iodinated tyrosyl moieties prevail at sufficiently high specific activities (Table 3.2).

The possibility of rapid enzymatic deiodination of the mono- or di-iodinated tyrosine is a disadvantage. Two different kinds of iodination are known: (a) direct iodination as described earlier or (b) indirect iodination with the Bolton

TABLE 3.2 Iodine Molecules per mCi (or per MBq), as well as Antibody Molecules per mg of Protein

	$t_{1/2}$	N (mol)/mCi	N (molecules)/MBq
^{123}I	13.1 h	4.2×10^{-12}	1.1×10^{-13}
^{124}I	4.2 days	3.3×10^{-11}	9.1×10^{-13}
^{125}I	60.1 days	4.7×10^{-10}	1.3×10^{-11}
^{131}I	8.04 days	6.2×10^{-11}	1.7×10^{-12}
	N (mol)/mg		
IgG	6.7×10^{-9}		
(Fab')$_2$	1.0×10^{-8}		
Fab'	2.0×10^{-8}		

and Hunter reagents or similar via free amino groups.

II. MATERIALS AND INSTRUMENTS

Na^{125}I (Cat. No. IMS30) and ^{125}I Bolton and Hunter reagent (Cat. No. IM5861 or IM5862) are from Amersham Bioscience. IM5862 is the di-iodo derivative used to achieve a higher specific activity. Na^{131}I (I-131-S) and Na^{123}I (I-123-S) are delivered by MDS Nordion. All radioisotopes of iodine are purchased in no carrier-added form. Iodo-Gen (T0656), chloramine-T (C9887), sodium metabisulfite (S1516), 3-(4-hydroxyphenyl) propionic acid, N-hydroxysuccinimide ester (H1256, Bolton–Hunter reagent), sodium acetate (S5636), chloroform (C5312), dichloromethane (32,026-0), phosphate-buffered saline (PBS) solution (P5119), sodium dihydrogen phosphate (S9638), disodium hydrogen phosphate (S9390), 1 M hydrochloric acid solution (H9892), boric acid (B0394), 1 M sodium hydroxide solution (S2770), 0.9% saline (8776), trifluoroacetic acid (T8506), HPLC water (27,073-3), and HPLC acetonitrile (57472-4) are from Sigma-Aldrich.

Tyr^3-octreotide and Exendin-3 are synthesized by Bachem. Minigastrin (G0267) is from Sigma-Aldrich.

Rituximab is a mouse–human chimeric anti-CD20 antibody (clone IDEC-C2B8, Mabthera), which is obtained commercially from Roche Pharma (via a pharmacy).

We use 1.5-ml tubes from Eppendorf (0030 120.086). The IB-F silical gel thin-layer strips (4463-02) are from Baker.

Radioactivity is measured with a Cobra II quantum automatic gamma counter from Perkin Elmer. HPLC is performed on a 535pump and 545UV/vis analyzer from Biotek with online flow radioactive analyzer 500TR from Perkin Elmer. PD-10 columns (17-0851-01) are from Amersham Bioscience. The HPLC protein column Bio-Silect SEC 250-5 (125-0476) is from Bio-Rad. The HPLC column is a CC 250/4.6 Nucleosil 120-5 C-18 (721712.46) from Marcherey-Nagel. The Speed-Vac Savant (SPD 101 B) is from Thermo Life Science connected with a MD4C vacuum pump (69 62 92) from Vacubrand.

III. PROCEDURES

A. Iodogen Radioiodination of an Antibody

Solutions

1. *Iodogene solution:* Dissolve 2.5 mg Iodogen in 10 ml of chloroform or dichloromethane.
2. *0.05 M phosphate buffer, pH 7.4:* Prepare by dissolving 2.87 g Na$_2$HPO$_4$ and 0.66 g NaH$_2$PO$_4$ in 500 ml water (pH control must give a pH value of 7.4)
3. *0.1 M NaOH:* Dilute 1 M sodium hydroxide solution (S2770) 10 times

Steps

1. Add 200 µl of Iodogen solution (corresponding to 50 µg of Iodogen) to a

1.5-ml tube. Evaporate the chloroform or dichloromethane under gentle heating in a water bath (40–50°C) with constant and homogeneous rotation, plating the Iodogen homogeneously onto the inner surface of the vial. Batches of 20–30 vials can be produced simultaneously and stored for several months at −20°C.

2. Buffer 67 µg of the antibody Rituximab (1 mg of protein per 555–740 MBq (15–20mCi) of ^{131}I, for molar ratios and other iodine isotopes, see Tables 3.1 and 3.2) in 100 µl 0.05 M phosphate buffer, pH 7.4, and put into the Iodogen vial.
3. Add 37 MBq of ^{131}I in 5 µL 0.1 M NaOH to the reaction vial.
4. Stop the reaction after incubating for 30 min at room temperature by removing the reaction solution from the Iodogen vial to another 1.5-ml tube.
5. For quality control and purification, see special section.

B. Chloramine-T Radioiodination of Tyr^3-Octreotide

Solutions

1. *0.05 M sodium acetate buffer, pH 4.2:* Dissolve 410 mg of sodium acetate in 100 ml water. Adjust the pH to 4.2 by adding 1 M HCl.
2. *Tyr^3-octreotide solution:* Dissolve 5.3 mg tyr^3-octreotide in 10 ml 0.05 M sodium acetate buffer, pH 4.2
3. *0.05 M phosphate buffer:* Prepare 0.05 M phosphate buffer, pH 7.4, by dissolving 2.87 g Na_2HPO_4 and 0.66 g NaH_2PO_4 in 500 ml water (pH control must give a pH value of 7.4)
4. *Phosphate buffered saline (PBS)*
5. *Sodium meta-bisulfite:* Dissolve 20 mg sodium meta-bisulfite in 100 ml PBS
6. *Na^{125}I solution:* Available commercially

Steps

1. Add 3 µl of a 0.51 M tyr^3-octreotide solution in 0.05 M sodium acetate buffer, pH 4.2, to 20 µl of 0.05 M phosphate buffer, pH 7.4, into a 1.5-ml reaction tube.
2. Add 48 MBq of Na^{125}I solution in 2.8 µL 10^{-5}M NaOH.
3. Start the reaction by 1.6 µg chloramine-T in 20 µl 0.05 M phosphate buffer, pH 7.4.
4. Stop the reaction after incubating 1 min at room temperature by adding 20 µg sodium meta-bisulfite in 100 µl PBS.
5. For quality control and purification, see special section.

C. Production of ^{125}I-Labeled Bolton–Hunter Reagent

Radioiodination by the Bolton–Hunter procedure was performed essentially according to the original description.

Solutions

1. *PBS:* Available commercially
2. *Bolton–Hunter reagent solution:* Available commercially
3. *Chloramine-T solution:* Dissolve 2 mg of chloramine-T in 400 µl 0.25 M PBS, pH 7.4
4. *Sodium meta-bisulphite solution:* Solve 1.2 mg sodium meta-bisulphite in 600 µl 0.05 M PBS, pH 7.4

Steps

1. Add 37 MBq of Na^{125}I solution in 2.2 µl 10^{-5}M NaOH to 10 µl of Bolton–Hunter reagent solution in a 1.5-ml vial.
2. Add 40 µl of the chloramine-T solution to the reaction vial to start the reaction.
3. Stop the reaction after 10 s by the addition of the sodium meta-bisulphite solution.
4. After the addition of 200 µl dimethylformamide, extract the

radioiodinated Bolton–Hunter reagent with two 500-µl portions of benzene.

D. Conjugation with the Bolton–Hunter Reagent

Solution

0.1 M sodium borate, pH 8.5: Dissolve 618 mg boric acid in 90 ml water. Adjust the pH with 1 M NaOH to pH 8.5 and make up volume to 100 ml with pure water.

Steps

1. Evaporate the solvent of the ^{125}I-labeled Bolton–Hunter reagent (from Section IIIC or commercially bought) in a hood under a gentle stream of dry nitrogen at room temperature.
2. Add 250 ng of minigastrin in 200 µl ice-cold borate buffer to the Bolton–Hunter tube.
3. Vortex the mixture for 2 h on ice.
4. Stop the reaction by removing the reaction solution from the reaction tube.
5. For quality control and purification, see special section.

E. Radioiodination of Exendin-3 on the Histidine Moiety

Solutions

1. *Iodogene solution:* Dissolve 2.5 mg in 10 ml of chloroform or dichloromethane.
2. *0.05 M Tris buffer, pH 8.5:* Dissolve 0.606 g of tris base (M_r 121.14) in approximately 90 ml of pure water. Titrate to pH 8.7 with 1 M HCl. Make up volume to 100 ml with pure water.

Steps

1. Add 200 µl of Iodogen solution (corresponding to 50 µg of Iodogen) to a 1.5-ml tube. Evaporate the chloroform or dichloromethane under gentle heating in a water bath (40–50°C) and constant and homogeneous rotation, plating the Iodogen homogeneously onto the inner surface of the vial. Batches of 20–30 vials can be produced simultaneously and stored for several months at −20°C.
2. Buffer 10 µg of exendin-3 in 100 µl 0.05 M tris buffer, pH 8.5, and put into the Iodogen vial.
3. Add 37 MBq of ^{125}I in 2.2 µl 0.01 M NaOH to the reaction vial.
4. Stop the reaction after incubating 60 min at room temperature by removing the reaction solution from the Iodogen vial to another Eppendorf tube.
5. For quality control and purification, see special section.

F. Quality Control of Radioiodinated Antibody with Thin-Layer Chromatography

Solution

0.9% saline: Available commercially

Steps

1. Cover the floor of the chromatography tank with 0.9% saline to a depth of 5 mm.
2. Spot a droplet (approximately 1–5 µl) of the final product (radioiodinated antibody or protein) onto an ITLC silicagel IB-F flexible strip with a length of 10 cm at a distance of approximately 1 cm from the bottom.
3. After 20–30 s, place the strip into the chromatography tank.
4. Allow the strip to develop until the solvent front migrates to approximately 1 cm from the top of the strip. At this point, remove the strip from the tank and allow to dry.
5. Cut the strip in half.
6. Count its lower half, containing the radio-labeled protein, as well as its upper half, containing the unincorporated radionuclide ("free iodine"), in a γ counter.

7. Determine the amount of incorporated radioiodine as follows: Incorporated = 100% × (counts or activity of lower strip half)/ (total counts or activity).

G. Quality Control and Purification of Radioiodinated Antibody with Low-Pressure PD-10 Column (Fig. 3.2)

Solution

0.1M phosphate buffer, pH 7.4: Dissolve 12 g of NaH_2PO_4 in approximately 900 ml of pure water. Titrate to pH 7.42 at the laboratory temperature of 20°C with monovalent strong base or acid as needed. Make up volume to 1000 ml with pure water.

Steps

1. Preequilibrate the PD-10 column with 40 ml phosphate buffer, pH 7.4.
2. Apply 370 kBq (for purification: 37 MBq) of the final radiolabeled product onto the PD-10 column.
3. Elute the PD-10 column with 20 ml of phosphate buffer, pH 7.4, collecting 0.5-ml fractions.
4. Count the activity in each fraction.
5. Purification: Combine the three samples of the first peak with the highest activity.

H. Quality Control and Purification of Radioiodinated Antibody with High-Pressure Liquid Chromatography (HPLC)

Solution

0.1M phosphate buffer, pH 7.4: Dissolve 12 g of NaH_2PO_4 in approximately 900 ml of pure water. Titrate to pH 7.42 at the laboratory temperature of 20°C with monovalent strong base or acid as needed. Make up volume to 1000 ml with pure water.

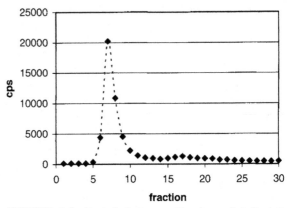

FIGURE 3.2 Typical elution profile of a radioiodinated monoclonal antibody (complete IgG) on a PD-10 column (Sephadex G-25 preequilibrated in a 15×50-mm polypropylene tube). The labeled IgG elutes in fractions 6–9, unbound iodine in fractions 15–22 (fraction size: 0.5 ml).

Steps

1. Preequilibrate the size-exclusion HPLC column with 40 ml phosphate buffer, pH 7.4
2. Apply 37 kBq (for purification: 37 MBq or larger activities) of the final radiolabeled product onto the HPLC column.
3. Elute the substances with phosphate buffer, pH 7.4.
4. Collect the samples with a sample collector (purification or quality control) and measure with an online radioactivity detector (quality control).
5. Purification: Combine the three samples of the first peak with the highest activity.

I. Quality Control and Purification of Radioiodinated Peptides with HPLC

Solutions

1. *0.1% TFA:* Add 1 ml trifluoroacetic acid in 1 liter HPLC water.
2. *Acetonitrile HPLC:* Available commercially

Steps

1. First preequilibrate the C-18 HPLC column by starting an empty run (see Table 3.3).

TABLE 3.3 Gradient for HPLC Quality Control and Purification for Radiolabeled Peptides

	Flow ml/ min	0.1% TFA (%)	Acetonitrile (%)
0 min	1	95	5
5 min	1	95	5
20 min	1	45	55
25 min	1	45	55
40 min	1	0	100
45 min	1	0	100

2. Apply 37 kBq (for purification: 37 MBq) of the final radiolabeled product onto the HPLC column.
3. Elute the substances with the gradient (Table 3.3).
4. Collect samples with a sample collector (purification or quality control) and measure on a γ counter or measure with an online radioactivity detector (quality control).
5. The acetonitrile can be evaporated on a Speed-Vac at a temperature of 60°C.
6. Purification: The radioionated peptide peak can be used for further experiments.

IV. PITFALLS

1. The iodination procedure must be performed within a hood because elementary iodine can be produced during synthesis, which is volatile.
2. The protein or peptide must be in the reaction vial before adding the radioiodine to avoid the formation of elementary iodine.
3. There must not be any additive proteins or substances with tyrosine-like structures in the solutions, as they would be radioiodinated too.
4. The solution with the iodinated protein or peptide should be adjusted to a protein

content of about 2% with albumin after iodination to avoid unspecific deposition on the equipment.

V. COMMENTS

The described procedures can be changed easily between the isotopes and can be applied to proteins and their fragments or peptides without too much difficulty. The quality control part must be modified for any particular substance and can be applied to a purification procedure. The protocols can be transferred between the different iodine isotopes without major problems.

Immunoreactivity testing must be performed for antibodies and their fragments.

For the described Iodogen or chloramine-T radioiodination procedure, incorporation yields were between 60 and 90%, regardless of the antibody isotype or molecular form (IgG or fragments). After purification, the percentage of unbound radionuclide fell below 2% in all cases. Also, tyrosyl residues containing peptides were labeled successfully with yields between 60 and 85% using these procedures. With immunoconjugates, their molecular integrity was maintained at over 90% by using this Iodogen methodology, whereas chloramine-T iodination of F(ab')$_2$ fragments led to an up to 50% degradation to monovalent Fab' fragments. With peptides, oxidation of methionine residues to their respective sulfoxide derivatives was only 15–30% with Iodogen, which stands in contrast to over 90% with the chloramine-T methodology. Finally, the tested immunoreactivities were above 85% in all cases (regardless of whether Iodogen or chloramine-T methodology was used).

Much higher variability in labeling yields was observed while using the Bolton–Hunter procedure (with incorporation yields between 20 and 55%). However, in accordance to the lack of contact to oxidants or reducing agents by using this procedure, no methionine oxidation was observed.

4

Protein Detection in Gels by Silver Staining: A Procedure Compatible with Mass Spectrometry

Irina Gromova and Julio E. Celis

I. INTRODUCTION

Silver staining is one of the procedures, in addition to Coomassie blue, R and G types, (Neuhoff *et al.*, 1988) and fluorescent dyes (Steinberg *et al.*, 1996; Patton, 2002; see also article by Patton in this volume), that are available for detecting proteins separated by gel electrophoresis. Switzer *et al.*, (1979) introduced silver staining in 1979, a technique that today provides a very sensitive tool for protein visualization with a detection level down to the 0.3- to 10-ng level.

The basic mechanisms underlying silver staining of proteins in gels are relatively well understood. Basically, protein detection depends on the binding of silver ions to the amino acid side chains, primary the sulfhydril and carboxyl groups of proteins (Switzer *et al.*, 1979; Oakley *et al.*, 1980; Merril *et al.*, 1981, 1986), followed by reduction to free metallic silver (Rabilloud, 1990, 1999). The protein bands are visualized as spots where the reduction occurs and, as a result, the image of protein distribution within the gel is based on the difference in oxidation–reduction potential between the gel area occupied by the proteins and the free adjacent sites. A number of alterations in the silver-staining procedure can shift the oxidation–reduction equilibrium in a way that gel-separated proteins will be visualized as either positively or negatively stained bands (Merril *et al.*, 1986).

Silver staining protocols can be divided into two general categories: (1) silver amine or alkaline methods and (2) silver nitrate or acidic methods (Merril, 1990). In general, the detection level using the various procedures is determined by how quickly the protein bands develop in relationship to the background (e.g., signal-to-noise ratio). The silver amine or alkaline methods usually have lower background and, as a result, are most sensitive but require extended procedures. Acidic protocols, however, are faster but slightly less sensitive. A comparative analysis of the sensitivity of a number of silver staining procedures has been published (Sorensen *et al.*, 2002; Mortz *et al.*, 2001). Clearly, each protocol has different advantages regarding timing, sensitivity, cost, and compatibility with other analytical methods, especially mass spectrometry (MS), a tool that is being used in combination with gel electrophoresis or chromatographic methods for rapid protein identification (see various articles in this volume). Until recently, most of the silver staining protocols used glutaraldehyde-based sensitizers in the fixing and sensitization step, thus introducing chemical modifications into proteins. The utilization of those chemicals causes the cross-linking of two lysine residues within protein chains, which affects MS analysis by hampering trypsin digestion and highly reduces protein extraction from the gel (Rabilloud, 1990).

Several modifications of the silver nitrate staining procedure have been developed for visualizing proteins that can be subsequently digested, recovered from the gel, and subjected to MS analysis (Shevchenko *et al.*, 1996; Yan *et al.*, 2000). This article describes a procedure that is slightly modified from these.

II. MATERIALS AND INSTRUMENTATION

Ultrapure water ($>18\,\mathrm{M}\Omega/\mathrm{cm}$ resistance) for preparation of all buffers as well as during the washing steps is recommended. Use high-quality laboratory reagents that can be purchased from any chemical company.

III. PROCEDURE

To achieve the best results, i.e., high sensitivity and low background, it is very important to follow closely the incubation time throughout all steps as given in the protocol.

Solutions

1. *Fixation solution*: 50% ethanol (or methanol), 12% acetic acid, 0.05% formalin. To make 1 liter, add 120 ml of glacial acetic acid to 500 ml of 96% ethanol and 500 µl of 35% formaldehyde (commercial formalin is 35% formaldehyde). Complete to final volume with deionized water. Store at room temperature.
2. *Washing solution*: 20% ethanol (or methanol). To make 1 liter, add 200 ml 96% ethanol to 800 ml of deionized water. Store at room temperature.
3. *Sensitizing solution*: 0.02% (w/v) sodium thiosulfate ($Na_2S_2O_3$). To make 1 liter, add 200 mg of sodium thiosulfate anhydrate to a small volume of deionized water, mix well, and bring to the final volume of 1 liter.
4. *Staining solution*: 0.2% (w/v) silver nitrate ($AgNO_3$), 0.076% formalin. Prepare fresh. To make 1 liter, add 2 g of $AgNO_3$ to a small amount of deionized water. Add 760 µl of 35% formaldehyde. Dissolve and bring to final volume with deionized water. Precool the solution at 4°C before using.

5. *Developing solution*: 6% (w/v) sodium carbonate (Na_2CO_3), 0.0004% (w/v) sodium thiosulfate ($Na_2S_2O_3$), 0.05% formalin. To make 1 liter, add 60 g Na_2CO_3 to a small amount of deionized water and dissolve. Add 4 mg of sodium thiosulfate anhydrate to a small volume of deionized water and dissolve. Mix both solutions, add 500 µl of 35% formaldehyde, and bring to the final volume with water. Store at room temperature.

6. *Terminating solution*: 12% acetic acid. To make 1 liter, add 120 ml of glacial acetic acid to 500 ml of deionized water. Mix well and bring to the final volume with water. Store at room temperature.

7. *Drying solution*: 20% ethanol. To make 1 liter, add 200 ml of ethanol to 800 ml of deionized water. Mix well. Store at room temperature.

Steps

Use powder-free rubber gloves throughout the procedure. Wash the gloves with water during the staining procedure. The gel fixation and washing procedure can be carried in a staining try (polypropylene trays are recommended), but make sure that these are only used for silver staining. The size of the container has to be big enough to perform free movement of the gel during the shaking. For each step use sufficient volumes of the solutions to fully immerse the gels. Close the plastic trays or place the trays on top of each other to protect the gels from dust. Perform all steps at room temperature, one gel per tray, placed on a shaker at a very gentle speed. Do not touch the gel with bare hands or metal objects during handling. Plastic or Teflon bars (or ordinary glass pipettes) can be used to handle the gel. The staining procedure can be performed on any type of rotary shakers.

1. After electrophoresis, remove the gel from the cassette and place into a tray containing the appropriate volume of fixing solution. Soak the gel in this solution for approximately 2 h. Fixation will restrict protein movement from the gel matrix and will remove interfering ions and detergent from the gel. Fixation can also be done overnight. It may improve the sensitivity of the staining and decrease the background.

2. Discard the fixative solution and wash the gel in 20% ethanol for 20 min. Change the solution three times to remove remaining detergent ions as well as fixation acid from the gel.
   Note: We recommend using 20% ethanol solution instead of deionized water to prevent swallowing of the gel. The size of the gel can be restored by incubating the gel in 20% ethanol for 20 min if water is used during the washing step

3. Discard the ethanol solution and add enough volume of the sensitizing solution. Incubate for 2 min with gentle rotation. It will increase the sensitivity and the contrast of the staining.

4. Discard the sensitizing solution and wash the gel twice, 1 min each time, in deionized water. Discard the water.

5. Add the cold silver staining solution and shake for 20 min to allow the silver ions to bind to proteins. *Note*: Do not pour the staining solution directly on the gel as it may result in unequal background. Add the solution to the corner of the tray.

6. After staining is complete, pour off the staining solution and rinse the gel with a large volume of deionized water for 20–60 s to remove excess unbound silver ions. Repeat the washing once more. *Note*: Washing the gel for more than 1 min will remove silver ions from the gel, resulting in decreased sensitivity.

7. Rinse the gel shortly with the developing solution. Discard the solution.

8. Add new portion of the developing solution and develop the protein image by incubating the gel in 300 ml of developing solution for 2–5 min. The reaction can be stopped as soon as the desired intensity of the bands is reached.
9. Stop the reduction reaction by adding 50 ml of terminating solution directly to the gel while still immersed in the developing solution. Gently agitate the gel for 10 min. Development is stopped as soon as "bubbling" is over.
10. Moist gels can be kept in 12% acetic acid at 4°C in sealed plastic bags or placed in the drying solution for 2 h prior to vacuum drying.

Figure 4.1 shows silver-stained gels of whole protein extracts from a normal colon tissue biopsy and a breast tumor biopsy separated by two-dimensional gel electrophoresis as described by Celis et al., in this volume. Since the sensitivity of silver staining is in the same range as modern mass spectrometry, it makes this staining one of the most attractive techniques for protein visualization before MS analysis. Protein bands of various intensity can be excised from the gels and identified by MS analysis (see various articles in this volume).

IV. COMMENTS

When choosing the silver staining protocol it is necessary to remember that not all proteins are stained equally by this technique. Thus, several classes of highly negative charged proteins, including proteoglycans and mucins, which contain high levels of sulfated sugar residues, and some very acidic proteins are detected poorly by silver staining (Goldberg et al., 1997). Note that the linear dynamic range of the stain is restricted to approximately the 10-fold range, thus hampering the use of this method for quantitative protein expression analysis.

We replaced methanol for ethanol in all fixative solutions because of methanol toxicity. However, the use of ethanol in combination with acetic acid can result in the formation of ethyl acetate, which may interfere with protein identification by mass spectrometry. Several silver staining kits that offer improved sensitivity and that are compatibility with subsequent mass spectrometric analysis are available commercially, including: Silver Stain PlusOne, Amersham Pharmacia Biotech, Amersham, UK; and SilverQuest silver staining kit, Invitrogen, USA.

V. PITFALLS

1. To increase the sensitivity of the staining, use extended washing after fixation to remove all residual acid. This extra washing will reduce the background during development.
2. Development of the gel for a long period of time can decrease the yield of the peptides for subsequent mass spectrometric analysis. This is due to the fact that mainly unstained peptides from the inner part of the gel are eluted to the solutions following "in gel" tryptic digestion of the proteins.
3. Negative staining can be observed when an excess of protein is applied.
4. In some cases, artificial bands with a molecular mass of around 50–70 kDa, as well as streaking or yellow background, can be observed due to the presence of a high concentration of reducing agents such as 2-mercaptoethanol or dithiothreitol in the sample buffer.
5. It has been reported that the recovery of peptides from the gel for MS analysis could be increased by destaining of the silver-stained protein bands immediately after the staining procedure (Gharahdaghi et al., 1999).

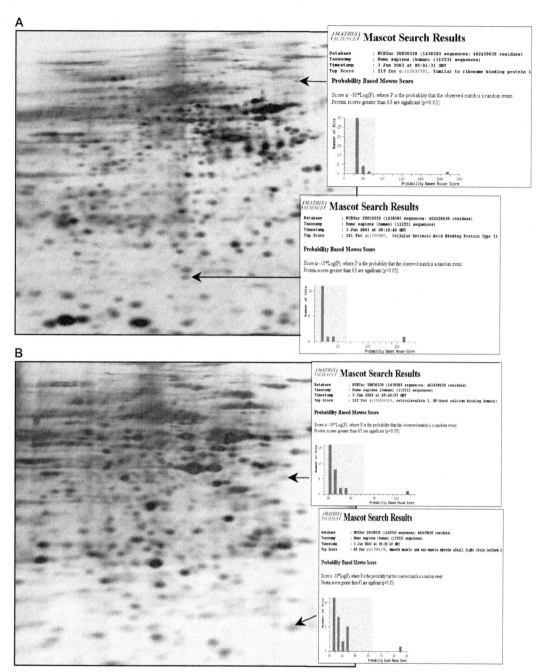

FIGURE 4.1 Gel image of normal human colon, location 7 (A), and human breast tumor biopsy (B) separated by two-dimensional gel IEF electrophoresis. Protein spots labeled on the gel images were identified by matrix-assisted laser desorption/ionization time-of-flight MS analysis. MS analysis of proteins resolved by gel electrophoresis utilizes extraction of the protein spot from the stained gel, followed by trypsin digestion, measurement, and database analysis (see also other articles in this volume).

References

Gharahdaghi, F., Weinberg, C. R., Meagher, D. A., Imai, B. S., and Mische, S. M. (1999). Mass spectrometric identification of proteins from silver-stained polyacrylamide gel: A method for the removal of silver ions to enhance sensitivity. *Electrophoresis* **20**, 601–605.

Goldberg, H. A., and Warner, K. J. (1997). The staining of acidic proteins on polyacrylamide gels: Enhanced sensitivity and stability of "Stains-all" staining in combination with silver nitrate. *Anal. Biochem.* **251**, 227–233.

Merril, C. R. (1990). Silver staining of proteins and DNA. *Nature* **343**, 779–780.

Merril, C. R., Dunau, M. L., and Goldman, D. (1981). A rapid sensitive silver stain for polypeptides in polyacrylamide gels. *Anal. Biochem.* **110**, 201–207.

Merril, C. R., and Pratt, M. E. (1986). A silver stain for the rapid quantitative detection of proteins or nucleic acids on membranes or thin layer plates. *Anal. Biochem.* **156**, 96–110.

Mortz, E., Krogh, T. N., Vorum, H., and Gorg, A. (2001). Improved silver staining protocols for high sensitivity protein identification using matrix-assisted laser desorption/ionization-time of flight analysis. *Proteomics* **1**, 1359–1363.

Neuhoff, V., Arold, N., Taube, D., and Ehrhardt, W. (1988). Improved staining of proteins in polyacrylamide gels including isoelectric focusing gels with clear background at nanogram sensitivity using Coomassie brilliant blue G-250 and R-250. *Electrophoresis* **9**, 255–262.

Oakley, B. R., Kirsch, D. R., and Morris, N. R. (1980). A simplified ultrasensitive silver stain for detecting proteins in polyacrylamide gels. *Anal. Biochem.* **105**, 361–363.

Patton, W. F. (2002). Detection technologies in proteome analysis. *J. Chromatogr. B Anal. Technol. Biomed. Life Sci.* **771**, 3–31.

Rabilloud, T. (1990). Mechanisms of protein silver staining in polyacrylamide gels: A 10-year synthesis. *Electrophoresis* **10**, 785–794.

Rabilloud, T. (1999). Silver staining of 2-D electrophoresis gels. *Methods Mol Biol.* **112**, 297–305.

Shevchenko, A., Wilm, M., Vorm, O., and Mann, M. (1996). Mass spectrometric sequencing of proteins silver-stained polyacrylamide gels. *Anal. Chem.* **68**, 850–858.

Sorensen, B. K, Hojrup, P., Ostergard, E., Jorgensen, C. S., Enghild, J., Ryder, L. R., and Houen, G. (2002). Silver staining of proteins on electroblotting membranes and intensification of silver staining of proteins separated by polyacrylamide gel electrophoresis. *Anal. Biochem.* **304**, 33–41.

Steinberg, T. H., Jones, L. J., Haugland, R. P., and Singer, V. L. (1996). SYPRO orange and SYPRO red protein gel stains: One-step fluorescent staining of denaturing gels for detection of nanogram levels of protein. *Anal. Biochem.* **239**, 223–237.

Switzer, R. C., 3rd, Merril, C. R., and Shifrin, S. (1979). A highly sensitive silver stain for detecting proteins and peptides in polyacrylamide gels. *Anal. Biochem.* **98**, 231–237.

Yan, J. X., Wait, R., Berkelman, T., Harry, R. A., Westbrook, J. A., Wheeler, C. H., and Dunn, M. J. (2000). A modified silver staining protocol for visualization of proteins compatible with matrix-assisted laser desorption/ionization and electrospray ionization-mass spectrometry. *Electrophoresis* **17**, 3666–3672.

5

Fluorescence Detection of Proteins in Gels Using SYPRO Dyes

Wayne F. Patton

I. INTRODUCTION

Operationally, fluorescent, noncovalent staining methods using SYPRO dyes resemble traditional, colorimetric staining procedures such as colloidal Coomassie blue dye staining. After electrophoresis, gels are incubated in a stain solution and proteins are visualized based upon differential dye binding to protein bands relative to the polyacrylamide gel matrix. Because proteins are not covalently modified with dye molecules and staining is performed postelectrophoretically, no alteration in the migration of proteins during electrophoresis occurs. These fluorescence-based staining methods are also highly compatible with downstream microchemical methods, such as Edman-based protein sequencing and peptide mass profiling by matrix-assisted laser desorption time-of-flight mass spectrometry (MALDI-TOF MS).

SYPRO Orange dye, SYPRO Red dye, and SYPRO Tangerine dye bind noncovalently to proteins in gels through interaction with SDS micelles (Steinberg et al., 1996a,b, 1997, 2000). These fluorophores are virtually nonfluorescent in aqueous solution, but they fluoresce in nonpolar solvents or when associated with SDS–protein complexes. Protein quantitation with fluorophores of this type is generally more reliable than that achieved with fluorophores that label primary amines alone (Patton, 2000). The dyes offer detection sensitivities comparable to those of colloidal Coomassie blue staining methods (Patton, 2000).

Because the staining properties of SYPRO Orange and SYPRO Red dyes are similar, the two protocols describing them have been grouped together in this article. SYPRO Orange protein gel stain is slightly brighter, whereas SYPRO Red protein gel stain has somewhat lower background fluorescence in gels. These dyes offer high sensitivity and rapid staining of SDS–polyacrylamide gels using a simple, one-step, 30- to 60-min staining procedure, with no destaining required (Steinberg et al., 1996a,b, 1997, 2000). Staining exhibits low protein-to-protein variability, high selectivity for proteins, and a broad linear detection range extending over three orders of magnitude. The 4- to 10-ng detection sensitivity of these SYPRO dyes is as high as rapid silver staining and colloidal Coomassie blue staining methods (Steinberg et al., 1996a,b, 1997, 2000; Patton, 2000, 2002). Both dyes are efficiently excited by UV and by visible illumination. Thus, stained gels can be viewed and photographed with a standard laboratory UV transilluminator, CCD camera, or any of a variety of laser scanners, in conjunction with the proper filters. SYPRO Orange stain (excitation/emission: ~470/569 nm) is preferable for argon ion or second-harmonic generation (SHG) laser-based instruments, and SYPRO Red stain (excitation/emission: ~547/631 nm) is preferable for green He-Ne or Nd:YAG lasers.

It is possible to stain SDS–PAGE gels during electrophoresis using the SYPRO stains in the running buffer; however, detection sensitivity in this case is four- to eight-fold poorer (Steinberg et al., 1996b). After electrophoresis, gels are briefly washed prior to visualizing proteins. In contrast to many silver staining and reverse staining methods, SYPRO dyes do not stain nucleic acids or bacterial lipopolysaccharides to a significant extent (Steinberg et al., 1996a,b, 1997, 2000). Gels may be completely destained in 30% aqueous methanol (Steinberg et al., 1996b). SYPRO Orange and Red protein gel stains are not suitable for staining proteins on blotting membrane or in isoelectric focusing (IEF) gels and show reduced sensitivity when staining proteins in two-dimensional gels. Both SYPRO Orange and SYPRO Red stain require acetic acid, making them less suitable for applications involving electroblotting, electroelution, or measuring enzyme activity. While acetic acid may be omitted from the SYPRO Orange and SYPRO Red staining solutions, and proteins then recovered from gels, this yields substantially reduced sensitivity of detection and increased protein-to-protein variability in staining (Steinberg et al., 1996b, 2000).

SYPRO Tangerine protein gel stain is a versatile stain for detecting proteins separated by SDS–polyacrylamide gel electrophoresis (Steinberg *et al.*, 2000). Like SYPRO Orange and SYPRO Red stains, it offers high sensitivity, a rapid and simple staining procedure, low protein-to-protein variability, high selectivity for proteins, and a broad linear range of detection. Staining is performed in a nonfixative solution that permits subsequent electroblotting, electroeleution, or detection of enzyme activity. Proteins stained without fixation can be used for further analysis by zymography (in-gel enzyme activity assay), provided SDS does not inactivate the protein of interest. Stained proteins can also be eluted easily from gels and used for further analysis. The stain is fully compatible with Edman-based sequencing and mass spectrometry (Steinberg *et al.*, 2000). In addition, staining does not interfere with the transfer of proteins to blotting membranes, allowing visualization of proteins *before* proceeding with Western blotting or other blotting applications. Small regions of a gel or even individual bands can be excised before blotting. This enables one to use much smaller amounts of transfer membrane and immunodetection reagents. After transfer to membranes, proteins can be visualized using SYPRO Ruby protein blot stain. If protein fixation is preferred, the dye can be used with 7% acetic acid or 12.5% trichloroacetic acid (Steinberg *et al.*, 2000). In this case, however, one should expect slightly higher background staining than with SYPRO Orange and SYPRO Red stains. SYPRO Tangerine protein gel stain is not suitable for staining proteins in IEF gels and shows only moderate sensitivity when staining proteins on two-dimensional gels.

SYPRO Ruby protein gel stain differs from the other SYPRO dyes that all bind through intercalation into SDS micelles (Berggren *et al.*, 2000, 2002; Patton, 2000, 2002). Instead, this stain binds to proteins by a mechanism that is quite similar to Coomassie blue stain, via direct electrostatic interaction with basic amino acid residues (Patton, 2000). SYPRO Ruby dye readily stains glycoproteins, lipoproteins, calcium-binding proteins, fibrillar proteins, and other difficult-to-stain proteins. The dye is used in a simple staining procedure and is ideal for high-throughput gel staining. The stain is as sensitive as the best silver staining methods available and superior to them in terms of ease of use, linear dynamic range, and compatibility with downstream microchemical characterization techniques (Lopez *et al.*, 2001; Nishihara and Champion, 2002; Gerner *et al.*, 2002). SYPRO Ruby protein gel stain is an ultrasensitive dye for detecting proteins separated by SDS–polyacrylamide gels or two-dimensional gels (Berggren *et al.*, 2000, 2002). The background fluorescence is low and the linear dynamic range of the stain extends over three orders of magnitude and shows low protein-to-protein variation. The stain is more sensitive than colloidal Coomassie blue dye and SYPRO Orange, Red, or Tangerine dyes and is comparable in sensitivity to the best available silver stains (Patton, 2000, 2002). The stain is ready to use and gels cannot overstain. Staining protocols are simple, although optimal staining incubation requires about 4h, slower than times required for SYPRO Orange, SYPRO Red, and SYPRO Tangerine stains. Staining times are not critical though and staining can be performed overnight. SYPRO Ruby protein gel stain will not stain extraneous nucleic acids or lipopolysaccharides and is compatible with further downstream microchemical processing. SYPRO Ruby protein gel stain can be used with many types of gels, including two-dimensional gels, Tris–glycine SDS gels, Tris–tricine precast SDS gels, isoelectric focusing gels, and nondenaturing gels. SYPRO Ruby stain is also compatible with gels adhering to plastic backings, although the inherent blue fluorescence of the plastic must be removed with an appropriate emission filter. The stain does not interfere with subsequent analysis of proteins by Edman-based sequencing or mass spectrometry and the stain is especially well suited for peptide mass profiling using MALDI-TOF mass spectrometry (Berggren *et al.*, 2000, 2001, 2002; Lopez *et al.*, 2001). Stained gels can be visualized with a 300-nm UV transilluminator, various laser scanners, or other blue light-emitting sources. The dye maximally emits at about 610 nm. SYPRO Ruby protein gel stain has exceptional photostability, allowing long exposure times for maximum sensitivity.

This article presents protocols for staining proteins in SDS–polyacrylamide and two-dimensional gels using the different SYPRO dyes. The following protocols describe several steps following the preparation and running of SDS–polyacrylamide gels or two-dimensional gels. Some issues regarding further processing of proteins in the gels are also included. In general, gel electrophoresis should be performed according to standard procedures (Laemmli, 1970; O'Farrell, 1975). There are several important considerations to take into account when choosing effective and appropriate stains for an application. Table 5.1 outlines key features of the stains discussed in this article. Important

notes regarding the protocols are included at the end of the article.

II. PROTOCOLS USING SYPRO ORANGE AND SYPRO RED PROTEIN GEL STAINS

A. Materials and Reagents

SYPRO Orange protein gel stain (Molecular Probes, Inc., Cat. No. S-6650, S-6651) and SYPRO Red protein gel stain (Molecular Probes, Inc., Cat. No. S-6653, S-6654) are provided as 5000× concentrated solutions in dimethyl sulfoxide (DMSO), either as a single vial containing 500 μl of stock solution or as a set of 10 vials, each containing 50 μl of stock solution. In each case, enough reagent is supplied to prepare a total of 2.5 liter of working stain solution, which

TABLE 5.1

Protein stain emission	Excitation/ applications	Major	Features
SYPRO Ruby protein gel stain	280 nm, 450 nm/610 nm	Two-dimensional gels, IEF gels, one-dimensional (1D) SDS–PAGE	Highest sensitivity (1–2 ng/band; comparable to best silver staining methods), linear quantitation range over 3 orders of magnitude, compatible with fluorescence-based phosphoprotein and glycoprotein detection (multiplexed proteomics technology)
SYPRO Orange protein gel stain	300 nm, 470 nm/570 nm	1D SDS–PAGE	Good sensitivity (4–10 ng/band; comparable to colloidal Coomassie blue stain), linear quantitation range over 3 orders of magnitude
SYPRO Red protein gel stain	300 nm, 550 nm/630 nm	1D SDS–PAGE	Good sensitivity (4–10 ng/band; comparable to colloidal Coomassie blue stain), linear quantitation range over 3 orders of magnitude
SYPRO Tangerine protein gel stain	300 nm, 490 nm/640 nm	1D SDS–PAGE, staining before blotting, zymography, electroelution	Good sensitivity (4–10 ng/band; comparable to colloidal Coomassie blue stain), linear quantitation range over 3 orders of magnitude, requires no alcohol or acid fixatives, ideal for protein elution from gels

is sufficient to stain ~50 polyacrylamide mini-gels. Before opening the vial, warm it to room temperature to avoid water condensation and subsequent precipitation. After thawing completely, briefly centrifuge the vial in a micro-centrifuge to deposit the DMSO solution at the bottom of the vial. If particles of dye are present, redissolve them by briefly sonicating the tube or vortexing it vigorously after warming. Staining

A

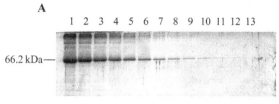

SYPRO Ruby protein gel stain

B

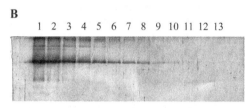

SYPRO Orange protein gel stain

FIGURE 5.1 Comparison of detection sensitivity and brightness of SYPRO Orange protein gel stain and SYPRO Ruby protein gel stain using a laser-based gel-imaging system. Solution-quantified bovine serum albumin standard (Cat. No. P7656, Sigma Chemical Company, Saint Louis, MO) was applied to lanes of 10% SDS–polyacrylamide gels (1000–0.25 ng/lane) and separated by electrophoresis using standard procedures (Laemmli, 1970). Gels were then stained using either SYPRO Ruby protein gel stain or SYPRO Orange protein gel stain as described in the text. Subsequently, gels were imaged using a Fuji FLA 3000 laser-based gel scanner. Gels were scanned using the 473-nm second-harmonic generation (SHG) laser and 580-nm long-pass filter. (A) Gel stained with SYPRO Ruby protein gel stain and imaged using a laser-based gel scanner. (B) Gel stained with SYPRO Orange protein gel stain and imaged using a laser-based gel scanner. SYPRO Orange protein gel stain was capable of detecting 2 ng of bovine serum albumin while SYPRO Ruby protein gel stain was capable of detecting 0.5 ng of the protein. Figure courtesy of Ms. Courtenay Hart, Molecular Probes, Inc.

should be performed in plastic dishes (preferably polypropylene) (*see* Section V,B).

B. Gel Electrophoresis

1. SDS–PAGE

Prepare and run SDS–polyacrylamide gels according to standard protocols (Laemmli, 1970). Originally it was recommended to use 0.05% SDS in the running buffer instead of the usual 0.1% SDS. However, this recommendation was based upon separating molecular weight markers and would not be advisable for biological specimens containing hydrophobic proteins. The use of standard, 0.1% SDS in the running buffer decreases overall detection sensitivity only slightly and improves the resolution of protein bands in gels.

2. Two-Dimensional Gels

Neither SYPRO Orange nor SYPRO Red protein gel stain is recommended for high sensitivity detection of proteins in two-dimensional gels (See Table 5.1). High-quality silver stains and SYPRO Ruby Protein Gel Stain offer better detection sensitivity.

C. Staining Protocol

1. Clean and thoroughly rinse the staining dishes before use. Residual detergent in staining dishes will compromise the detection of proteins.
2. Prepare the staining solution by diluting the stock SYPRO reagent 1:5000 in 7.5% (v/v) acetic acid and mixing vigorously. For 50 μl of stock, dilute into 250 ml acetic acid solution (*see* Section V).
3. Pour the staining solution into a small, clean plastic dish. For one or two standard-size minigels, use ~50 ml to 100 ml of staining solution; for larger gels, use 500 to 750 ml) (*see* Section V).

4. Place the gel into the staining solution. Protect the gel and staining solution from light at all times by covering the container with a lid or with aluminum foil.

5. Gently agitate the gel in stain solution at room temperature (50 rpm on an orbital shaker). Staining times range from 10 to 60 min, depending on the thickness and percentage of polyacrylamide in the gel. For 1-mm-thick 15% polyacrylamide gels, the signal is typically optimal at 40 to 60 min of staining. For standard SDS–PAGE minigels incubate for 40 to 60 min. Large gels sometimes require a preincubation in 7.5% acetic acid for 15 min prior to staining to reduce background fluorescence due to excess SDS in the gel.

6. Rinse briefly (<1 min) with 7.5% acetic acid to remove excess stain from the gel surface and to avoid accumulation of fluorescent dye on the surface of the transilluminator or gel scanner.

D. Viewing, Photographing, and Storing the Gel

View the stained gel on a standard 300-nm UV transilluminator (*see* Section V,E). Gels may be left in staining solution overnight without losing sensitivity. However, photographs should be taken as soon as possible after staining.

Gels may be photographed with a Polaroid or CCD camera. Use Polaroid 667 black-and-white print film and the SYPRO protein gel stain photographic filter (Molecular Probes, Inc., Cat. No. S-6656). Exposure times vary with the intensity of the illumination source; for an f stop of 4.5, use 2 to 5 s for SYPRO Orange stain and 3 to 8 s for SYPRO Red stain. CCD cameras and laser scanners provide high sensitivity detection; contact the manufacturer to determine the optimal filter sets (*see* Section V).

E. Destaining Gels

To destain gels, incubate overnight in 0.1% Tween 20. Alternatively, if thorough destaining

is desired, incubate for prolonged periods in several changes of 7.5% acetic acid or 30% methanol.

III. PROTOCOL FOR SYPRO TANGERINE PROTEIN GEL STAIN

A. Materials and Reagents

SYPRO Tangerine protein gel stain (Molecular Probes, Inc., Cat. No. S-12010) is provided in a 500-μl unit size, as a 5000× concentrated solution in DMSO. One 500-μl unit size prepares a total of 2.5 liter of working stain solution, which is sufficient to stain ~50 polyacrylamide minigels. Before opening the vial, warm it to room temperature to avoid water condensation and subsequent precipitation. After thawing completely, briefly centrifuge the vial in a microcentrifuge to deposit the DMSO solution at the bottom of the vial. If particles of dye are present, redissolve them by briefly sonicating the tube or vortexing it vigorously after warming.

1. Buffers

SYPRO Tangerine staining solution is prepared by diluting the stock reagent 1:5000 in one of a variety of buffers as described in the protocol. Staining should be performed in plastic dishes (preferably polypropylene) (*see* Section V,A).

2. SDS–PAGE

Prepare and run SDS–polyacrylamide gels according to standard protocols (Laemmli, 1970). The use of standard, 0.1% SDS in the running buffer is recommended to ensure complete solubilization of hydrophobic proteins.

3. 2-D Gels

SYPRO Tangerine gel stain is not recommended for high sensitivity detection of proteins in two-dimensional gels (See Table 5.1). High-quality silver stains and SYPRO Ruby protein gel stain offer better detection sensitivity.

B. Staining Procedure

1. Clean and *thoroughly rinse* the staining dishes before use. Residual detergent in staining dishes will compromise the detection of proteins.
2. Prepare the staining solution by diluting the stock SYPRO Tangerine reagent 1:5000 in an appropriate buffer and mixing vigorously.
2a. If the proteins are to be used for electroelution, electroblotting, or zymography, dilute the stock solution into 50 mM phosphate, 150 mM NaCl, pH 7.0. For 50 μl of stock, dilute into 250 ml buffer. *Note: If no fixative is used before or during staining, some diffusion of the protein bands may occur, especially for smaller proteins.*
2b. Alternatively, use one of a wide range of buffers that are compatible with the stain. These include formate, pH 4.0; citrate, pH 4.5; acetate, pH 5.0; MES, pH 6.0; imidazole, pH 7.0; HEPES, pH 7.5; Tris acetate, pH 8.0; Tris–HCl, pH 8.5; Tris borate, 20 mM EDTA, pH 9.0; and bicarbonate, pH 10.0. Buffers should be prepared as 50–100 mM solutions containing 150 mM NaCl. The stock dye solution may also be diluted directly into 150 mM NaCl. For 50 μl of stock, dilute into 250 ml buffer.
2c. For fixative staining to minimize diffusion of the proteins, dilute the SYPRO Tangerine stock solution in 7.5% (v/v) acetic acid. For low percentage gels and for very small proteins, 10% acetic acid will result in better retention of the protein in the gel without compromising sensitivity (Steinberg *et al.*, 2000). However, note that acetic acid and other fixatives will interfere with the transfer of proteins to blotting membranes.
3. Pour the staining solution into a small plastic dish. For one or two standard-size minigels, use ~50 to 100 ml of staining solution; for larger gels, use 500 to 750 ml (*see* Section V,A).
4. Place the gel into the staining solution. Protect the gel and staining solution from

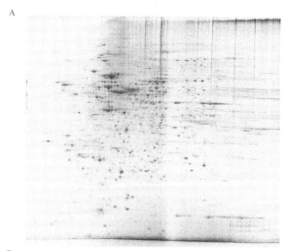

FIGURE 5.2 Comparison of the detection sensitivity of SYPRO Orange protein gel stain and SYPRO Ruby protein gel stain using a laser-based gel-imaging system. Jurkat cell extract (250 μg) was separated on a tube gel as a first dimension followed by 12.5% SDS–PAGE. Two-dimensional gels were stained with either SYPRO Ruby dye or SYPRO Orange dye overnight. Gels were destained for 2–3 h before imaging. (A) Gel stained with SYPRO Ruby protein gel stain. (B) Gel stained with SYPRO Orange protein gel stain. ure courtesy of Dr. Birte Schulenberg.

light at all times by covering the container with a lid or with aluminum foil.

5. Gently agitate the gel in stain solution at room temperature (50 rpm on an orbital shaker). The staining times range from 10 to 60 min, depending on the thickness and percentage of polyacrylamide in the gel. For 1-mm–thick 15% polyacrylamide gels, the signal is typically optimal at 30 to 60 min of staining. For standard SDS–PAGE minigels prepared with 0.1% SDS in the running buffer, incubate for 30 to 60 min. Large gels, including large two-dimensional gels, sometimes require a preincubation in 7.5% acetic acid for 30 min prior to staining to reduce background fluorescence due to excess SDS in the gel. It is important to note, however, that acetic acid interferes with transfer to blots.

6. Use the following step if the proteins are to be transferred to a blot. After staining, incubate the gel in Western blotting buffer containing 0.1% SDS. The SDS is not absolutely required, but it helps in the transfer of some proteins to the blot.

C. Viewing, Photographing, and Storing the Gel

Detect the proteins in the stained gel using a standard 300-nm UV transilluminator, a blue-light transilluminator, 473-nm SHG laser, 488-nm argon ion, or 473-nm He–Ne laser-based imaging system (see Section V,E). Gels may be left in staining solution overnight without loss of sensitivity. However, photographs should be taken as soon as possible after staining. Gels may be documented by a Polaroid or CCD camera. Use Polaroid 667 black-and-white print film and the SYPRO protein gel stain photographic filter (Molecular Probes, Inc., Cat. No. S-6656). Exposure times vary with the intensity of the illumination source; for an f stop of 4.5, use a 2- to 5-s exposure time. CCD cameras and laser

scanners provide high sensitivity; contact the manufacturer to determine the optimal filter sets (see Section V,F).

D. Destaining Gels

SYPRO Tangerine stain is readily destained by incubation in 7% acetic acid or 30% methanol.

E. Notes for SYPRO Tangerine Protein Gel Stain

The SDS front at the bottom of the gel stains very heavily with SYPRO Tangerine stain. Unless the proteins that interest you are comigrating with the SDS front, it will be advantageous to run the SDS front off the gel. Colored stains and marker dyes, as well as commercially prestained protein markers, may interfere with SYPRO Tangerine dye staining and quench fluorescence.

IV. PROTOCOL FOR SYPRO RUBY PROTEIN GEL STAIN

A. Materials and Reagents

SYPRO Ruby protein gel stain (Molecular Probes, Inc., Cat. No. S-12000, S-12001, S-21900) is provided ready to use in 200 ml volume (will stain ~4 minigels), 1 liter volume (~20 minigels or 2–3 large-format gels), or 5 liter volume (~100 minigels or 10–15 large-format gels). Staining should be performed in plastic dishes (preferably polypropylene) (see Section V,B).

B. Protocol

Prepare and run SDS–PAGE or two-dimensional PAGE according to standard protocols (Laemmli, 1970; O'Farrell, 1975). Perform staining with SYPRO Ruby due using continuous, gentle agitation (e.g., on an orbital shaker at 50 rpm).

1. Clean and *thoroughly rinse* the staining dishes before use. Residual detergent in staining dishes will compromise the detection of proteins.

2. A range of fixatives have been validated for use with SYPRO Ruby protein gel stain, including 40% ethanol/10% acetic acid, 10% ethanol/7% acetic, acid 25% ethanol/12.5% trichloroacetic acid, and 10% ethanol/3% phosphoric acid. Harsher fixatives, such as 40% ethanol/10% acetic acid, are recommended as they retain proteins in gels better. Fix the gel in a plastic dish for 30 min. This step improves the sensitivity of the stain in two-dimensional gels, but is optional for one-dimensional SDS–PAGE gels.

3. Pour the staining solution into a small, clean plastic dish. For one or two standard-size minigels, use ~50 to 100 ml of staining solution; for larger gels, use 500 to 750 ml (*see* Section V,A).

4. Place the gel into the staining solution. Protect the gel and staining solution from light at all times by covering the container with a lid or with aluminum foil.

5. Gently agitate the gel in stain solution at room temperature (50 rpm on an orbital shaker). The staining times range from 90 min to 4 h, depending on the thickness and percentage of polyacrylamide in the gel. Specific staining can be seen in as little as 30 min. However, a minimum of 4 h of staining is required for maximum sensitivity and linearity. For convenience, gels may be left in the dye solution overnight or longer without overstaining.

6. *Optional.* After staining, rinse the gel in deionized water for 30–60 min to decrease background fluorescence. To better decrease background fluorescence the gel can be washed in a mixture of 10% methanol and 7% acetic acid for 30 min instead of water. The gel may be monitored periodically using UV illumination to determine the level of background fluorescence.

C. Viewing, Photographing, and Storing the Gel

View the stained gel on a standard 300-nm UV or a blue-light transilluminator (*see* Section V,E). Gels may also be visualized using various laser scanners: 473-nm (SHG) laser, 488-nm argon ion laser, and 532-nm (YAG) laser. Alternatively, use a xenon arc lamp, blue fluorescent light, or blue light-emitting diode (LED) source. Gels may be left in staining solution overnight without losing sensitivity. However, images should be acquired as soon as possible after staining. Gels may be imaged by a Polaroid or CCD camera. Use Polaroid 667 black-and-white print film and the SYPRO protein gel stain photographic filter (Molecular Probes, Inc., Cat. No. S-6656). Exposure times vary with the intensity of the illumination source; for an f stop of 4.5, try 1s. CCD cameras and laser scanners provide high sensitivity; contact the manufacturer to determine the optimal filter sets (*see* Section V,F). To dry the stained gel for permanent storage, incubate the gel in a solution of 2% glycerol for 30 min. Dry the stained gel using a gel dryer. Note that proteins present at very low levels may no longer be detectable after gel drying.

V. NOTES

A. Staining

Minimal staining volumes for typical gel sizes:
50 ml for 8 cm × 10 cm × 0.75 mm gels (minigels)
330 ml for 16 cm × 20 cm × 1 mm gels
500 ml for 20 cm × 20 cm × 1 mm gels
or ~10 times the volume of the gel for other gel sizes

B. Staining Containers

Polypropylene dishes, such as Rubbermaid Servin' Savers, are the optimal containers for staining because the high-density plastic adsorbs only a minimal amount of the dye. Clean and rinse the staining containers well before use, as detergent will interfere with staining. Some rinse the containers with ethanol before use. For small gels, circular staining dishes provide the best fluid dynamics on orbital shakers, resulting in less dye aggregation and better staining. For large-format two-dimensional gels, polyvinyl chloride photographic staining trays, such as Photoquip Cesco-Lite 8 × 10-in. photographic trays (Genomic Solutions, Ann Arbor, MI), also work well. Another convenient staining option for large-format two-dimensional gels uses the Clearview three-drawer organizer (Sterilite, Cat. No. 1790, Townsend, MA). This polypropylene box provides a convenient format for staining three gels per unit and is available at many department stores. Glass dishes are not recommended as they have a tendency to bind dye.

C. Fixing Gels

For low percentage gels and for very small proteins, 10% acetic acid solution will result in better retention of the protein in the gel without compromising sensitivity. Acetic acid will interfere with transfer of the proteins to a blot. Therefore, for applications in which blotting will follow electrophoresis, use SYPRO Tangerine protein gel stain, which does not require acetic acid fixation.

D. Storing Gels

Always store gels in the dark to prevent photobleaching. When gels are stored in the staining solution, the signal decreases somewhat after several days; however, depending on the amount of protein in bands of interest, gels may retain a usable signal for many weeks. Gels may be dried between sheets of cellophane (BioRad Laboratories), although there is sometimes a slight decrease in sensitivity.

E. Viewing Gels

Viewing the Gel with UV Transillumination

Place the gel directly on the transilluminator; do not use plastic wraps or plastic backing. It is important to clean the surface of the transilluminator after each use with deionized water and a soft cloth (such as cheesecloth), as fluorescent dyes, such as SYPRO stains, will accumulate on the glass surface and cause a high background fluorescence.

The polyester backing on some precast gels is highly fluorescent. For maximum sensitivity using a UV transilluminator, the gel should be placed polyacrylamide side down and an emission filter, such as the SYPRO protein gel stain photographic filter (S-6656), used to screen out the blue fluorescence of the plastic. The use of a blue-light transilluminator or laser scanner will reduce the amount of fluorescence from the plastic backing so that the gel may be placed polyester side down.

Noticeable photobleaching can occur after several minutes of exposure to ultraviolet light. If a gel becomes photobleached, it can be restained by simply returning it to the staining solution.

F. Imaging Gels

Because of the low fluorescence, it is possible to take advantage of the integrating capability of photographic or CCD cameras and use long exposure times to increase the sensitivity, often making bands visible that are not visible to the eye. Images are best obtained by digitizing at about 1024 × 1024 pixels resolution with 12- or 16-bit gray scale levels per pixel. Contact the camera manufacturer for recommendations on filter sets. A CCD camera-based image analysis system can gather quantitative information that will allow comparison of fluorescence intensities between different bands or spots.

Acknowledgments

I gratefully acknowledge the many scientists who have contributed to the SYPRO dye development program over the years. These include Tom Steinberg, Kiera Berggren, Birte Schulenberg, Richard Haugland, Vicki Singer, Courtenay Hart, Brad Arnold, Nick Smith, Mary Nunally, and Laurie Jones. SYPRO is a registered trademark of Molecular Probes, Inc.

References

Berggren, K., Chernokalskaya, E., Lopez, M., Beechem, J., and Patton, W. (2001). Comparison of three different fluorescent visualization strategies for detection of *Escherichia coli* ATP synthase subunits after SDS-polyacrylamide gel electrophoresis. *Proteomics* **1**, 54–65.

Berggren, K., Chernokalskaya, E., Steinberg, T., Kemper, C., Lopez, M., Diwu, Z., Haugland, R., and Patton, W. (2000). Background-free, high-sensitivity staining of proteins in one- and two-dimensional sodium dodecyl sulfate-polyacrylamide gels using a luminescent ruthenium complex. *Electrophoresis* **21**, 2509–2521.

Berggren, K., Schulenberg, B., Lopez, M., Steinberg, T., Bogdanova, A., Smejkal, G., Wang, A., and Patton, W. (2002). An improved formulation of SYPRO Ruby protein gel stain: Comparison with the original formulation and with a ruthenium II tris (bathophenanthroline disulfonate) formulation. *Proteomics* **2**, 486–498.

Gerner, C., Vejda, S., Gelbmann, D., Bayer, E., Gotzmann, J., Schulte-Hermann, R., and Mikulits, W. (2002). Concomitant determination of absolute values of cellular protein amounts, synthesis rates, and turnover rates by quantitative proteome profiling. *Mol. Cell Proteomics* **1**, 528–537.

Laemmli, U. (1970). Cleavage of structural proteins during the assembly of the head of bacteriophage T4. *Nature* **227**, 680–685.

Lopez, M., Berggren, K., Chernokalskaya, E., Lazarev, A., Robinson, M., and Patton, W. (2000). A comparison of silver stain and SYPRO Ruby protein gel stain with respect to protein detection in two-dimensional gels and identification by peptide mass profiling. *Electrophoresis* **21**, 3673–3683.

Nishihara, J., and Champion, K. (2002). Quantitative evaluation of proteins in one- and two-dimensional polyacrylamide gels using a fluorescent stain. *Electrophoresis* **23**, 2203–2215.

O'Farrell, P. (1975). High resolution two-dimensional electrophoresis of proteins. *J. Biol. Chem.* **250**, 4007–4021.

Patton, W. (2000). A thousand points of light; the application of fluorescence detection technologies to two-dimensional gel electrophoresis and proteomics. *Electrophoresis* **21**, 1123–1144.

Patton, W. (2002). Detection technologies in proteome analysis. *J. Chromatogr. B Biomed. Appl.* **771**, 3–31.

Patton, W., and Beechem, J. (2002). Rainbow's end: The quest for multiplexed fluorescence quantitative analysis in proteomics. *Curr. Opin. Chem. Biol.* **6**, 63–69.

Steinberg, T., Haugland, R., Singer V., and Jones, L. (1996a). Applications of SYPRO Orange and SYPRO Red protein gel stains. *Anal. Biochem.* **239**, 238–245.

Steinberg, T., Jones, L., Haugland, R., and Singer, V. (1996b). SYPRO Orange and SYPRO Red protein gel stains: One-step fluorescent staining of denaturing gels for detection of nanogram levels of protein. *Anal. Biochem.* **239**, 223–237.

Steinberg, T., Lauber, W., Berggren, K., Kemper, C., Yue, S., and Patton, W. (2000). Fluorescence detection of proteins in SDS-polyacrylamide gels using environmentally benign, non-fixative, saline solution. *Electrophoresis* **21**, 497–508.

Steinberg, T., Martin, K., Berggren, K., Kemper, C., Jones, L., Diwu, Z., Haugland, R., and Patton, W. (2001). Rapid and simple single nanogram detection of glycoproteins in polyacrylamide gels and on electroblots. *Proteomics* **1**, 841–855.

Steinberg, T., White, H., and Singer, V. (1997). Optimal filter combinations for photographing SYPRO Orange or SYPRO Red dye-stained gels. *Anal. Biochem.* **248**, 168–172.

Autoradiography and Fluorography: Film-Based Techniques for Imaging Radioactivity in Flat Samples

Eric Quéméneur

I. INTRODUCTION

Autoradiography (ARG) is the photography-derived technique used to visualize the distribution of a compound labeled radioactively with either a β or γ emitting isotope in a biochemical sample. It has become a fundamental tool, particularly since slab gel electrophoresis has established as an inevitable techniques for separating and analysing complex mixtures of biomolecules. Any other flat samples such as blotting and dotting membranes, thin-layer chromatography plates, or microscopy slides are suitable. It is worth comparing ARG with numeric methods (phosphorimagers) that have popularized in many laboratories for 10 years. Indeed, films appear superior in terms of traceability and resolution. The resolution of a film depends on the size of metal grain in the photographic emulsion. A simple calculation shows that, assuming a grain diameter of 0.2 μm, a film displays a resolution of about 127,000 dpi compared to 600 dpi for a common laser printer. This resolution allows a possible magnification of up to 500-fold until the resolution of human eye (about 100 μm) is surpassed. In other terms, a 10×10-cm^2 autoradiographic picture is equivalent to a 25 million pixel digital image. Furthermore, an autoradiogram is exactly the same size as the sample, making it easy the precise location of a "hot spot" in a complex pattern.

Basically, autoradiographic techniques divide into two modes. The first one, direct ARG, is the direct exposure of film by β particles or γ rays emitted by the sample. In the second mode, the film is sensitized indirectly by the secondary light generated upon excitation of a "fluor" (fluorography) or a "phosphor" (indirect ARG with intensifying screen) by the radioactive emission. Autoradiography and fluorography look simple in appearence, but some fundamental notions should be known and kept in mind for obtaining sensitive, resolutive, and reproducible results. Among them, the notion that sensitivity and resolution are antagonistic concepts that cannot be matched simultaneously. Because of their path lengths, lower energy isotopes such as ^{3}H, ^{35}S, or ^{14}C have better resolution than higher ones such as ^{32}P or ^{125}I. This detection is not sensitive and indirect methods based on the emission of UV/blue photon promote sensitivity but decrease resolution (Fig. 6.1). Other important notions are discussed in Section IV.

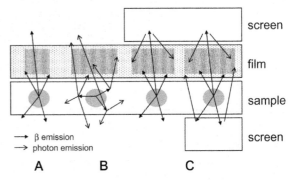

FIGURE 6.1 The underlying principle of the major detection methods and the way to assemble the various components in the cassette correctly. (A) Direct ARG; (B) fluorography; and (C) indirect ARG with either one or two intensifying screens. Note the variation in image size (resolution) depending on the nature and path of radiations.

Labeling the corners of the sample to assist the accurate alignment of the sample to its autoradiogram is useful. Use either a nib and radioactive ink prepared by mixing 100 μl of Indian ink with a few microliters of the diluted radioisotope solution used in the experiment. Alternatively, a pen containing a phosphorescent ink or luminescent stickers can prove more convenient (e.g., Glogos II Autorad Markers, Stratagene).

Microscale standards (Amersham Biosciences) are convenient calibration strips for relative quantitation in all mode of detection. They are available for ^{3}H, ^{14}C, and ^{125}I with activity scales ranging from 3.74 to 4048 Bq/mg, from 3.7 to 31,890 Bq/g, and from 44 to 23,900 Bq/mg, respectively. For films and intensifying screens, see Table 6.1 for the appropriate choice.

II. MATERIALS AND INSTRUMENTATION

For catalog numbers of reagent, products, and equipment, see Table 6.1. Dark room equipped with appropriate lighting systems: the red light could be a 7.5- to 15-W bulb covered with a red Kodak GBX-2 filter.

III. PROCEDURES

Three procedures are reported here. The choice depends practically on the nature of the isotope (energy of emitted radiations) and on the yield of radiolabeling (amount of activity expected in a spot). A compromise has to be found between sensitivity and resolution.

TABLE 6.1 Catalog Numbers of Reagents, Products, and Equipments Used

Darkroom safelight kit (Kodak, Cat. # 852 1429)

GBX-2 filter, 10 × 12in. (Kodak, Cat. # 141 6940)
Pen with phosphorescent ink (Soquelec Cat. # R20105)
Glogos II Autorad Markers, 100 luminescent stickers (Stratagene, Cat. # 420 201)

Microscale standards
^{3}H microscales: 10 strips 0.111–4.07 kBq/mg + 10 strips 3.7–592Bq/mg (Amersham Cat. # RPA510)
^{14}C microscales: 20 strips 1.15–32.7 kBq/g (Amersham Cat. # RPA504L)
^{14}C microscales: 20 strips 3.7–3700Bq/g (Amersham Cat. # RPA511L)
^{125}I microscales: 20 strips 0.046–23.7kBq/mg (Amersham Cat. # RPA523L)

Films
Hyperfilm MP, 20.3 × 25.4cm, 75 sheets/box (Amersham, Cat. # RPN 1678K)
BioMax MS-2, 20.3 × 25.4cm, 50 sheets/box (Kodak, Cat. # 837 7616)
BioMax MR-2, 20.3 × 25.4cm, 50 sheets/box (Kodak, Cat. # 895 2855)
X-Omat AR-2, 20.3 × 25.4cm, 50 sheets/box (Kodak, Cat. # 165 1579)
Cronex 10T, 10 × 12in., 100 sheets/box (Agfa Medical, Cat. # LF5E3)

Intensifying screens
Hyperscreen, 20.3 × 25.4cm (Amersham, Cat. # RPN 1669)
BioMax MS Screen, 20.3 × 25.4cm (Kodak, Cat. # 851 8706)
TransScreen HE, 20.3 × 25.4cm (Kodak, Cat. # 856 3959)
TransScreen LE, 20.3 × 25.4cm (Kodak, Cat. # 162 2034)
Optex HighPlus (MCI Optonix, Cat. # 6102)

Cassettes
BioMax Cassette, 20.3 × 25.4cm (Kodak Cat. # 820 9140)
Hypercassette, 20.3 × 25.4cm (Amersham, Cat. # RPN 11649)
En^{3}Hance spray for fluorography, 2oz. (NEN, Cat. # NEF 9700)

Enlightening rapid autoradiography enhancer, 1 liter (NEN)

Amplify fluorographic reagent, 1 liter (Amersham, Cat. # NAMP 100)

PPO (2,5-diphenyloxazole) scintillation grade, Sigma-Aldrich, Product # D210404

Cellulose paper Grade 3MM Chr (thickness 0.34 mm), 20 × 25cm, 100 sheets/box (Whatman, Cat. # 3030-866)

Cellulose paper grade 1Chr (thickness 0.18mm), 25 × 25cm, 100 sheets/box (Whatman, Cat. # 3001-878)

Mineral oil, "Nujol" (Sigma-Aldrich, Cat. # M1180)

Orange-filtered preflash unit Sensitize (Amersham, Cat. # RPN 2051)

Kodak GBX developer, concentrate to make 3.8 liter (Kodak, Cat. # 190 0943)

Kodak GBX fixer, concentrate to make 3.8 liter (Kodak, Cat. # 190 2485)

Metal hanger for 20.3 × 25.4-cm films (Kodak, Cat. # 150 2764)a

Table 6.3 might help in making the choice and in estimating the necessary exposure time.

A. Direct Autoradiography

This is the method of choice for high resolution whatever the isotope used, although it is rarely suitable for ^{3}H, ^{14}C, or ^{35}S under normal labelling conditions *in vivo*.

Steps

1. Protect the cassette from radioactivity by placing a Saran wrap or plastic sheet under the sample.

2. Put the sample in the cassette. Preferably, the sample should be dry. If not, cover it with Saran wrap in order to avoid sticking of the sample to the film. However, for a sample containing ^{3}H, it is important to remove any barrier such as Saran wrap or a cellophane sheet that may quench the emitted radiations.
3. If necessary, label distinctly two corners of the sample using the radioactive ink. This will help in the future superimposition of the gel and the film. Avoid moisture or glove powder on the surface of the sample.
4. In the dark room, place the film in close contact with the sample. When using single-coated films as recommended for ^{3}H (Fig. 6.2), the sensitive face of the film should face the sample (Fig. 6.1).
5. Close the cassette carefully and let it stand at room temperature for the necessary exposure time. Refer to Table 6.2 for an estimate of this exposure time. An overnight exposure would be sufficient for the detection of most ^{32}P-containing

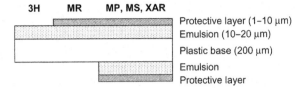

FIGURE 6.2   The structure of some commercially available films. The film types are ordered by increasing sensitivity but decreasing resolution. See Table 6.2 for film names and manufacturers.

samples, whereas a 2- to 3-day exposure might be necessary for ^{35}S.
6. Remove the film from the cassette in the darkroom and process it according to the manufacturer's instructions or as described in Section III,E.

B. Fluorography

Fluorography is particularly recommended for low-energy isotopes, when sensitivity is required and loss of resolution is acceptable. In pratice, it might be the only way to visualise ^{3}H-labeled compounds.

TABLE 6.2   Choosing the Right Combination of Films and Products

Priority	Method	Film	Screen
Speed, sensitivity	Indirect ARG *(high energy isotopes)*	Hyperfilm™ MP[a] BioMax™ MS[b]	Hyperscreen™[a] BioMax™ MS Screen[b]
	Fluorography *(low energy isotopes)*	X-Omat™ AR (XAR)[b] Super RX[c]	TranScreen™ HE[b] *(high energy isotopes)*
	Direct ARG *(high energy isotopes)*	Cronex 10T[d]	TranScreen LE[b] *(low energy isotopes)* Optex HighPlus[e]
Resolution	Direct ARG	BioMax™ MR[b] Hyperfilm™ 3H[a] Hyperfilm™ βmax[a]	

[a] Amersham;
[b] Eastman Kodak;
[c] Fuji;
[d] Agfa;
[e] MCI Optonix.

Solutions

Commercial reagents such as En^{3}Hance (NEN), Enligthning (NEN), or Amplify (Amersham) are efficient and convenient, but expensive.

Steps

1. Stain and destain the sample as usual, avoid TCA for fixing proteins and BET for staining nucleic acids because this would subsequently quench the fluorescence.
2. Work in a fume hood. Soak the sample in the scintillant reagent with gentle agitation for 1 h. The tray should contain enough reagent so that the sample is free floating and does not stick to the walls.
3. Discard the reagent and move to the 4°C cold room. Add slowly 500 ml cold 10% glycerol along the walls of the tray and shake slowly for 1 h. The low temperature prevents the fluor from diffusing out of the gel. The gel will turn white opaque after this step.
4. Put the sample on the top of a thick Whatman paper, cover it with a Saran wrap, and dry the ensemble under vacuum for 2 h at 60°C.
5. Let the sample reach the ambient temperature before putting it in the autoradiography cassette. In the dark room, prepare the cassette in the following order: gel/Saran/film. Exposure should be performed at −70/−80°C. For film processing, see Section III,E.

For seldom use of fluorography if the commercial reagent is not available in the laboratory, or if cost is a limiting factor, an efficient home-made reagent can be made: 20% 2,5-diphenyloxazole (PPO) in undiluted acetic acid (Skinner, 1983). The following procedure works well with most types of samples.

1. Soak sample in undiluted acetic acid for 5 min.
2. Impregnate with the reagent for 1.5 h with gentle agitation.

3. Soak in water for 30 min.
4. Dry and expose to film at −70°C.

C. Indirect Autoradiography with Intensifying Screens

The use of intensifying screens is not efficient for low-energy isotopes such as ^{3}H, ^{14}C, ^{33}P, or ^{35}S because the path of their β emission is too short to cross the film and reach the screen.

Steps

Steps 1–3 are the same as in Section III,A.

4. Place the film over the sample.
5. Place the screen over the film so that the matt white side (the side where the fluorescent material stands) faces the film. Note that the screen must have been kept in the dark for several hours before being introduced in the cassette in order to avoid a glow effect that would increase background on the film.
6. Close the cassette carefully and put it at −70/−80°C for the necessary exposure time (see Table 6.3 for estimation). By default, a 24-h exposure might be fine.

For transparent samples such as polyacrylamide gels, a second screen could be placed under the sample (white face up) to further increase sensitivity. Blots can be rendered translucent by impregnation with mineral oil (Rust, 1987). Kodak has introduced the TranScreen to solve the problem of the film attenuating the β particle before it reaches the screen. In our experience, with ^{3}H, the result has about twofold higher sensitivity compared to the conventional screen.

D. Film Preflashing

The preflashing of films increases their sensitivity for small amounts of radioactivity and expands the range for linear response. It is used only with photon-based detection (Sections III,B and III,C) and has no effect for direct

TABLE 6.3   Choosing the Optimal Detection Method

Isotope	Radiation	E_{max} (MeV)	Path max in air	Minimal activity detected in 24 h in a 0.1-cm^2 band (in Bq)		
				Direct ARG	Fluorography	Indirect ARG
^{3}H	β^-	0.018	6 mm	15,000	15	Inadequate
^{14}C, ^{35}S	β^-	0.156, 0.167	25 cm	10	400	Inadequate
^{32}P	β^-	1.710	8 m	1	Inadequate	0.1
^{125}I	γ	0.035	>10 m	2.5	Inadequate	0.2
	X	0.027				
	Auger e$^-$	0.030				

autoradiography or for large amounts of radioactivity in short exposures. Preflashed films are only valid for a couple of days and must be stored at $-70/-80°C$ if not used immediately.

Steps

1. In the dark room, install the orange-filtered flash, e.g., Sensitize (Amersham), at 50 cm above the film area.
2. Cut six film strips (just large enough to be processed) and cover them with a sheet of Whatman paper No. 1. Keep one of the strips in a closed cassette.
3. Make five consecutive flashes of 1 ms each while removing a strip between each in order to obtain a set of strips ranging from 1- to 5-ms exposure times.
4. Process the six strips and cut them to the format of a spectrophotometer cuvette holder (usually 1×4 cm).
5. Read absorbances at 545 nm on a spectrophotomer, plot the values of exposure time versus absorbance, and select the exposure time that gives an A_{545} in the range of 0.1–0.2.
6. In the dark room, select the film to be used for fluorography or with intensifying screens. Preflash it for the selected exposure time. The preflashed side of the film should face the sample when preparing the cassette (Fig. 6.1).

E. Processing of Films

Most laboratories dealing with a large number of autoradiograms have an automated processor. If this is the case, for optimal results with your sample, make sure that chemicals are fresh and the machine is regularly cleaned up. This article provides a method for the manual processing of films to those who are not equipped with such a machine or seldom use autoradiographic methods. Furthermore, some films are not protected by an antiscratch layer in order to maximize sensitivity to low-energy isotopes, e.g., Hyperfilms ^{3}H and βmax. They must be processed manually. The present method should work well in most cases, but it is obviously preferable to refer to the manufacturer's specific instructions.

Solutions

If only a single film or few films should be processed, prepare three trays or deep tanks for multiple processings containing the following solutions.

1. *Developer*: Kodak GBX developer
2. *Rinser*: Kodak GBX indicator stop or a large volume of tap water
3. *Fixer*: Kodak GBX fixer

Minimizing volumes is an environment-friendly attitude, but good results are obtained with generous volumes of fresh reagents. Collect carefully used solutions 1 and 3 in a dedicated can, which should then be evacuated as a toxic chemical waste.

Steps

1. In the dark room, open the cassette and remove the film. Gloves should not be damp or leave powder on the film. If the cassette comes from the −70°C freezer, let it reach the ambient temperature before opening.
2. Handle the film with suitable pliers or attach it to a metal hanger.
3. Immerse quickly in developer. Dislodge air bubbles by tapping the film against the tank wall. Then do not agitate during development, which will occur within 2–5 min depending on the temperature of the solution.
4. Remove film from the developer and transfer it to the rinser. Wash with continuous, moderate agitation for 30 s.
5. Place the film in the fixer for 10 min. Moderate agitation is recommended. The background should become uniformly transparent.
6. Wash in a large tank filled with running water for 5 min.
7. Hang the film on a line with a suitable clip attached to one corner and let it dry at room temperature in a dust-free area.

IV. COMMENTS

Whenever possible, blotting samples separated previously by electrophoresis from the gel matrix onto the surface of a nitrocellulose membrane should be performed. This has a minor impact on resolution but it improves tremendously the detection of all types of isotopes, particularly that of low-energy isotopes (Quéméneur, 1995). Films should be handled carefully when unpacking to avoid electrostatic artifacts that would print on the film.

V. PITFALLS

When working with high-energy isotopes such as ^{32}P or ^{125}I, do not stack cassettes during exposure because the path of the emission is sufficient to generate phantom images on the neighbouring film. In the same idea, do not store films too close to the working area. Film storage is extremely important for reproducible high-quality results. Ideal storage conditions are 10–20°C and 30–50% hygrometry. Avoid piling up film packages and use an upright position to diminish physical strains on the surface of the films.

The quality of the dark room and its equipment is critical. The inactinic safelight should be at a minimal distance of 1 m. The performance of the red filter decrease with time and it is reasonable to change it every 5 years.

References

Laskey, R. A. (1993). Efficient detection of biomolecules by autoradiography, fluorography or chemiluminescence, Review booklet #23, 2nd Ed.—Amersham Biosciences.

Laskey, R. A. (2002). Radioisotope detection using X-ray film. In "Radioisotopes in Biology" (R. J. Slater, ed.), 2nd Ed., pp. 63–83. Oxford Univ. Press, Oxford.

Perng, G., Rulli, R. D., Wilson, D. L., and Perry, G. W. (1988). A comparison of fluorographic methods for the detection of 35S-labeled proteins in polyacrylamide gels. Anal. Biochem. 175, 387–392.

Quéméneur, E., and Simonnet, F. (1995). Increased sensitivity of autoradiography and fluorography by membrane blotting. Biotechniques 18, 100–103.

Rust, S., Kunke, H., and Assman, G. (1987). Mineral oil enhances the autoradiographic detection of ^{32}P-labeled nucleic acids bound to nitrocellulose membranes. Anal. Biochem. 163, 196–199.

Skinner, M. K., and Griswold, M. D. (1983). Fluorographic detection of radioactivity in polyacrylamide gels with 2,5-diphenyloxazole in acetic acid and its comparison with existing procedures. Biochem. J. 209, 281–284.

PROTEOMIC ANALYSES BASED ON PROTEIN SEPARATION

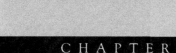

CHAPTER

7

Free-Flow Electrophoresis

Peter J. A. Weber, Gerhard Weber, and Christoph Eckerskorn

I. INTRODUCTION

Without prefractionation and enrichment, none of the existing techniques for the analysis of proteomes, such as two-dimensional electrophoresis (2-DE), or chromatographic methods, such as reversed-phase high-performance liquid chromatography (RP-HPLC), will be able to cope with the enormous complexity of biological samples and the extremely wide dynamic range of the protein concentrations. For example, this means that *low abundant proteins* are very likely to be hidden by highly expressed species.

Free-flow *isoelectric focusing* (Hannig, 1961; Krivankova and Bocek, 1998) of *protein mixtures*

(Bernardo *et al.*, 2000; Hoffmann *et al.*, 2001; Maida *et al.*, 2000; Weber and Bocek, 1996; Weber and Bocek, 1998) is one of the methods that fulfil the prerequisites to meet these *prefractionation* demands, i.e., to increase the amount of low abundant proteins and to dramatically reduce the complexity of protein mixtures. This is based on (1) the continuous operation principle, which allows the processing of large sample amounts; (2) the absence of any kind of gel or matrix that increases the recoveries of the proteins and makes this method highly compatible to virtually all kinds of follow-up analyses; and (3) the gentle procedure, which allows the fractionation of active enzymes and even protein complexes.

The power and resolution of free-flow iso-electric focusing are illustrated with the analysis of pig serum using a Pro Team *free-flow electrophoresis* (FFE) instrument. Detailed information about the instrument can be found in http://www.tecan.com. Please follow the "Proteomics" and "Fractionation" links.

II. MATERIALS AND INSTRUMENTATION

Pro Team HPMC (hydroxypropylmethylcellulose, Cat. No. 5170709) and Pro Team glycerol (Cat. No. 5170708), as well as the Pro Team FFE reagent basic kit (Cat. No. B132001) containing Prolyte 1, Prolyte 2, Prolyte 3, SPADNS (sulfanilic acid azochromotrop), and coloured pI markers, are from Tecan (Grödig, Austria). $1M$ NaOH (Cat. No. 35256), $1M$ H_2SO_4 (Cat. No. 35276), and petroleum benzene (Cat. No. 32248) are from Riedel-de Haën (Seelze, Germany). Urea (analytical grade, Cat. No. 24524) is from Serva (Heidelberg, Germany). Isopropanol (Cat. No. 9866.4) is from Karl Roth (Karlsruhe, Germany).

The Pro Team FFE apparatus is from Tecan (Grödig, Austria). It is equipped with seven 0.64-mm (i.d.) media tubes, one 1.42-mm (i.d.) counterflow tube, one 0.51-mm (i.d.) sample tube, a 0.4-mm spacer, and 0.65-mm filter paper strips. The water cooler IC006 ($P \geq 350W$) is from Huber (Offenburg, Germany).

III. PROCEDURES

A. Preparation and Running of a Denaturing Free-Flow Isoelectric Focusing Experiment

Solutions

1. *Anode stabilisation media (inlet 1)*: 14.5% (w/w) glycerol, 0.12% (w/w) HPMC, 100 mM H_2SO_4, and $8M$ urea. To make 60 ml (~68.9 g), add 6.0 g of $1M$ H_2SO_4, 14.0 g distilled water, 10.0 g 0.8% (w/w) HPMC, 10.0 g glycerol, and 28.9 g urea. Mix thoroughly by stirring. The solution should not be heated. Actual consumption: ~10 g/h.
2. *Separation media 1 (inlet 2)*: 14.5% (w/w) glycerol, 0.12% (w/w) HPMC, 11.6% (w/w) Prolyte 1, and $8M$ urea. To make 60 ml (~68.9 g), add 8.0 g Prolyte 1, 12.0 g distilled water, 10.0 g 0.8% (w/w) HPMC, 10.0 g glycerol, and 28.9 g urea. Mix thoroughly by stirring. The solution should not be heated. Actual consumption: ~10 g/h.
3. *Separation media 2 (inlet 3 + 4)*: 14.5% (w/w) glycerol, 0.12% (w/w) HPMC, 19.4% (w/w) Prolyte 2, and $8M$ urea. To make 120 ml (~137.8 g), add 26.7 g Prolyte 2, 13.3 g distilled water, 20.0 g 0.8% (w/w) HPMC, 20.0 g glycerol, and 57.8 g urea. Mix thoroughly by stirring. The solution should not be heated. Actual consumption: ~20 g/h.
4. *Separation media 3 (inlet 5)*: 14.5% (w/w) glycerol, 0.12% (w/w) HPMC, 14.5% (w/w) Prolyte 3, and $8M$ urea. To make 60 ml (~68.9 g), add 10.0 g Prolyte 3, 10.0 g distilled water, 10.0 g 0.8% (w/w) HPMC, 10.0 g glycerol, and 28.9 g urea. Mix thoroughly by stirring. The solution should not be heated. Actual consumption: ~10 g/h.
5. *Cathode stabilisation media (inlet 6 + 7)*: 14.5% (w/w) glycerol, 0.12% (w/w) HPMC, 100 mM NaOH, and $8M$ urea. To make 120 ml (~137.8 g), add 12.0 g $1M$ NaOH, 28.0 g distilled water, 20.0 g 0.8% (w/w) HPMC, 20.0 g glycerol, and 57.8 g urea. Mix thoroughly by stirring. The solution should not be heated. Actual consumption: ~20 g/h.
6. *Counterflow media (inlet 8)*: 14.5% (w/w) glycerol, 0.12% (w/w) HPMC, and $8M$ urea. To make 288 ml (~330.5 g), add 96.0 g distilled water, 48.0 g 0.8% (w/w) HPMC, 48.0 g glycerol, and 138.5 g urea. Mix

thoroughly by stirring. The solution should not be heated. Actual consumption: ~46 g/h.

7. *Electrolyte anode circuit (+ve):* 100 mM H_2SO_4. To make 400 g, add 40.0 g $1M$ H_2SO_4 and 360.0 g of distilled water. Mix thoroughly by stirring. Actual consumption: none, amount lasts for one working day.

8. *Electrolyte cathode circuit (−ve):* 100 mM NaOH. To make 400 g, add 40.0 g $1M$ NaOH and 360.0 g of distilled water. Mix thoroughly by stirring. Actual consumption: none, amount lasts for one working day.

1. Disassembly, Cleaning, Reassembly, and Filling of the Instrument

Steps

1. Switch on the cooler first and set it to 10°C.
2. Attach the fractionation plate to the separation chamber front part via the magnetic holder and move the separation chamber to the vertical position.
3. Reduce the force of all separation chamber clamps (turn counterclockwise), open the clamps pairwise from the outside to the inside, and open the separation chamber by carefully pulling the front part.
4. Put the electrode membranes into a 1:1 mixture of glycerol:isopropanol and put the paper strips in distilled water.
5. Make sure that the sample tube is connected to the middle sample inlet and that the two other sample inlets are closed.
6. Clean the interior surfaces of the separation chamber with lint-free paper towels in the following order: distilled water—isopropanol—petroleum benzene—isopropanol—high-purity water. Use a separate paper towel for each cleaning operation.
7. Place the wet 0.4-mm spacer on the front plate of the separation chamber with even distance to the electrode seals. Take care that the separation media inlets do not get covered.

8. Place membranes on electrodes starting from the top to the bottom. The smooth side of the membrane should face towards the electrode seal. The membrane must not protrude over the electrode seal. Subsequently, place the paper strips congruently on the membranes in the same fashion.

9. Quickly move the separation chamber front part towards the back part, ensuring that the separation chamber front part is parallel with the back part. Then, close the middle pair of clamps simultaneously. Afterwards, close adjacent clamps pairwise. Finally, increase the clamping force pairwise starting from the centre by opening the pair of clamps, turning the clamps clockwise, and closing the clamps (the water drops underneath the spacer should get displaced, the membranes and filters should not be visible).

10. Place the fractionation plate on the fraction collector housing.

11. To fill the instrument with water, tilt the separation chamber 45° and turn the bubble trap in the filling position.

12. Place all media and counterflow tubes in a bottle with distilled water. Place the sample tube in an empty 2-ml reaction tube without tightening the screw.

13. Open the three-way stopcock on the counterflow tube in all directions and open the Luer lock closure on the cock. Place the upper counterflow tube outlet in the fractionation tray and the three-way stopcock in the bottle grid.

14. Close the media pump tube cassettes, start the pump (direction "IN") with a delivery rate of 50 rpm, and wait until the separation chamber is filled halfway. Reverse pumping direction and empty the separation chamber until the inlet areas of the spacer are reached and all air bubbles are gone. Reverse the pumping direction again and fill the chamber bubble free until

the counterflow reservoir is nearly full. Then reduce the pump speed to 20 rpm and fill the remaining part of the counterflow reservoir.

15. Fill the counterflow tube, including the bubble trap at maximum speed, reduce the flow rate to 20 rpm, and connect the Luer lock closure of the three-way stopcock. Finally, close the remaining opening of the three-way stopcock. The fractionation tubes will start dripping.

16. Tilt the chamber to the horizontal position and tap the fractionation plate on the fractionation collector housing. If a fractionation tube fails to deliver, connect a syringe to the corresponding fractionation tube and suck until liquid is coming out of the tube.

17. Dry the outside of the separation chamber and control for any kind of leakage.

18. Start the sample pump with 4 rpm (direction "IN") and tighten the adjusting screw until the tube starts to deliver. Subsequently, change the pumping direction until no air remains in the tube. Then stop the sample pump.

2. Quality Control and Calibration of the Pumps

Steps

1. To check the flow profile and the proper assembly of the instrument, dilute 500 μl of Pro Team SPADNS with 50 ml of distilled water. Then switch off the media pump, place the media tubes of inlets 2, 4, and 6 in the SPADNS solution, and leave the remaining media tubes of inlet 1, 3, 5, and 7 as well as the counterflow tube of inlet 8 in distilled water.

2. Run the media pump with 40 rpm (direction "IN") for at least 6 min: In the separation chamber, red and colourless stripes should appear that flow in parallel, with identical width, and sharp boundaries (see Fig. 7.1). The equivalent pattern should appear in a 96-well plate. If this is not the case, try to readjust the clamps and make sure that the sample pump screw is closed. Other reasons might be surface contamination of the separation chamber; incorrectly installed spacer, seals, paper strips, or membranes; and partially clogged or leaking media tubes.

3. To continue, switch off the media pump, return the media tubes of inlets 2, 4, and 6 to distilled water, and run the media pump at 40 rpm (direction "IN") to rinse the separation chamber for at least 6 min.

4. To calibrate the sample pump, fill a 2-ml reaction tube with approximately 1.5 ml of distilled water, weigh it accurately to a milligram, and immerse the sample tube into it.

5. Call up the dialogue "Calibrate sample pump" and follow the instructions.

6. Afterwards, weigh the reaction tube again and save the calibration.

FIGURE 7.1   Stripe test to check the flow profile and the proper assembly of the Pro Team FFE instrument.

7. To calibrate the media pump, fill an appropriate bottle with approximately 200 ml of distilled water, weigh it accurately to 10 mg, and immerse the media tubes (inlets 1–7 without counterflow tube 8) in the calibration solution.

8. Call up the dialogue "Calibrate media pump" and follow the instructions.

9. Afterwards, weigh the bottle again and save the calibration.

10. To fill the separation chamber with separation media, make sure that the media pump is switched off.

11. Immerse the liquid circuit tubes in the appropriate media (+ve: anode circuit, −ve: cathode circuit), close the safety cover of the electrode circuit bottles and the media pump, and turn on the electrode pump (flowing air bubbles in the electrode ducts indicate the proper function of the pump).

12. Immerse the media tubes into the corresponding separation media and the counterflow tube into the counterflow media.

13. Turn on the media pump with 20 rpm for 3 s (direction "OUT"). Tap the media bottles against the bottle grid. Reverse the direction of the media pump and rinse the separation chamber with 20 rpm for 15 min. Then adjust the rate of delivery to ~57 ml/h.

14. Set the voltage to 1200 V and the current to 50 mA, switch on the high voltage, and wait approximately 15 min until the current reached a stable minimum.

15. To check the performance of the instrument (Fig. 7.2), dilute the Pro Team pI markers 1:10 with separation media 2 and apply them via the middle sample inlet at a sample flow rate of 0.5 ml/h. As soon as coloured drops are visible at the fractionation tubes, wait 10 more minutes. Then collect the separated pI marker fractions in a 96-well plate by placing it

on the drawer and moving it under the fractionation tubing outlets. You can avoid cross-contaminations during the collection by tapping the fractionation plate on the fractionation collector housing just before introducing and removing the drawer. The width of the red and the six yellow pI markers should be no more than 3–4 wells, otherwise the separation media have not been prepared properly or the separation chamber has not been assembled properly.

3. Fractionation of the Protein Sample by Free-Flow Isoelectric Focusing

Steps

1. Prepare the sample solution in accordance with the information mentioned in Section IV.

2. To avoid carryover when applying a new sample, make sure that the separation chamber and the sample tube get rinsed

FIGURE 7.2 Performance test to check the functionality of the Pro Team FFE instrument. Markers are separated according to their different pI values and mimic the actual separation of the proteins.

for 30 min (media pump direction "IN", ~57 ml/h; sample pump direction "OUT", 2 ml/h), that the voltage is on during this time, and that the current has reached its stable minimum.

3. Stop the sample pump, snip the end of the sample tube to create a tiny bubble at its end, immerse the tube into the sample vial, and turn on the sample pump (direction "IN") with a high flow rate until the bubble reaches the sample inlet. Then reduce the flow rate to ~1 ml/h.

4. As soon as the sample reaches the fractionation manifold (indicated by the red Pro Team SPADNS that you added to the sample), start collecting the sample fractions in a 96-well plate by placing it on the drawer and moving it under the fractionation tubing outlets. As soon as the red dye leaves the separation chamber you can stop the sample collection.

4. Shutting Down the Instrument

Steps

1. For active rinsing of the instrument you have to inactivate the instrument first, i.e., make sure that the high voltage is switched off as well as the media pump and the electrode pump.

2. Move the separation chamber to its horizontal position, place the counterflow tube, as well as the media tubes, in a bottle containing at least 1 liter of distilled water, and place the sample tube in an empty 2-ml reaction tube.

3. Operate the media pump for 30 min with 50 rpm (direction "IN") and the sample pump for 30 min with 4 rpm (direction "OUT").

4. To rinse the electrodes, remove the two "longer" electrode tubes signed with "+" or "−" from the electrode media and operate the electrode pump until the electrode ducts are almost free of electrode

media. Then switch off the electrode pump. Remove the bottles with electrode media, place all four electrode tubes in a vessel containing approximately 500 ml of distilled water, and operate the electrode pump for approximately 10 min. Remove the two "longer" electrode tubes signed with "+" or "−" from the distilled water and operate the electrode pump until the electrode ducts are almost free of water. Finally, switch off electrode pump and remove the water vessel.

5. For passive rinsing of the instrument, exchange the 96-well plate drawer for a rinse tray with approximately 2 liters of distilled water. While operating the media pump (direction "IN"), dunk the fractionation plate into the rinse tray and stop the media pump. Place the counterflow and media tubes onto the bottle grid and remove the water vessel. Turn the bubble trap in the draining position. Tilt the chamber 45° upwards. Open every pair of clamps and relax clamping force for three complete revolutions (turn counterclockwise) and close it again. Remove the media pump safety cover and release the tube cassettes by pushing on the lower right side. Release the adjusting screw of the sample pump and place the sample tube on the bottle grid. Finally, switch off the FFE and the cooler. If the FFE is used within the next 24 h it remains in this state. If the FFE is not used for a longer period of time, open the chamber on the next day. Rinse the spacer and filter paper with distilled water and store them dry. Store the membranes in glycerol:isopropanol (1:1).

B. Gel Examples

Figure 7. 3 shows representative two-dimensional gels of individual FFE fractions after separation by IEF-FFE. For more information, the reader is encouraged to visit Tecan's Web site at http://www.tecan.com.

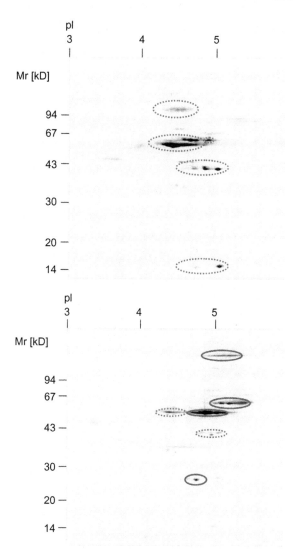

FIGURE 7.3 Silver-stained 2-DE analysis of two individual fractions of FFE–IEF-separated pig serum. Top: fraction 31; bottom: fraction 33. Ellipses indicate the minor overlap of the two protein patterns.

IV. PITFALLS

1. Use only high-purity water as well as chemicals of the highest grade available.
2. Never use acetone, powders (gloves!), silicon (latex gloves!), oil, grease, or adhesive in proximity of the separation chamber.
3. Air bubbles must not enter the separation chamber, as this would disturb the laminar flow profile and the resolution of the fractionation.
4. Make sure to prepare all solutions very carefully and to mix them thoroughly by stirring. The temperature of media should not exceed the working temperature by more than 15°C.
5. For optimal results, prepare the media freshly every day.
6. Prepare all solutions through weighing, as the high viscosity of glycerol and HPMC prevents accurate volumetric measurements.
7. Before fractionating a real sample, make sure to run two quality control experiments (see Figs. 7.1 and 7.2) to check the proper assembly and the functionality of the instrument.
8. The chemical and physical properties (density, conductivity, and viscosity) of the sample and the separation media should be similar, i.e., it is useful to dilute a concentrated sample with separation media.
9. To visualize the actual position of the colourless samples within the separation chamber, mix it with 1% of Pro Team SPADNS.
10. The total salt concentration in the sample should be less than 50 mM.
11. Turbid protein samples are an indication that protein precipitation has already occurred or that insoluble cell components are still present. You have to clarify turbid sample solutions by filtration or centrifugation before use.
12. You can add nonionic detergents such as CHAPS, octylglucoside, Triton X-100, and Triton X-114 (0.1–1%), as well as reducing agents such as dithiothreitol (up to 50 mM) or uncharged phosphines, to the samples to increase solubility.

13. Precipitation of proteins will appear as white lines in the separation chamber. You can tolerate precipitation as long as no immobile "islands" are formed. Otherwise you have to reduce the sample delivery rate or the concentration of the sample.

References

Bernardo, K., Krut, O., Wiegmann, K., Kreder, D., Micheli, M., Schafer, R., Sickman, A., Schmidt, W. E., Schroder, J. M., Meyer, H. E., Sandhoff, K., and Kroenke, M. (2000). Purification and characterization of a magnesium-dependent neutral sphingomyelinase from bovine brain. *J. Biol. Chem.* **275**, 7641–7647.

Hannig, K. (1961). Die trägerfreie kontinuierliche Elektrophorese und ihre Anwendung. *Fresenius Zeitschrift Anal. Chem.* **181**, 244–274.

Hoffmann, P., Ji, H., Moritz, R. L., Connolly, L. M., Frecklington, D. F., Layton, M. J., Eddes, J. S., and Simpson, R. J. (2001). Continuous free-flow electrophoresis separation of cytosolic proteins from the human colon carcinoma cell line LIM 1215: A non two-dimensional gel electrophoresis-based proteome analysis strategy. *Proteomics* **1**, 807–818.

Krivankova, L., and Bocek, P. (1998). Continuous free-flow electrophoresis. *Electrophoresis* **19**, 1064–1074.

Maida, R., Krieger, J., Gebauer, T., Lange, U., and Ziegelberger, G. (2000). Three pheromone-binding proteins in olfactory sensilla of the two silkmoth species *Antheraea polyphemus* and *Antheraea pernyi*. *Eur. J. Biochem.* **267**, 2899–2908.

Weber, G., and Bocek, P. (1996). Optimized continuous flow electrophoresis. *Electrophoresis* **17**, 1906–1910.

Weber, G., and Bocek, P. (1998). Recent developments in preparative free flow isoelectric focusing. *Electrophoresis* **19**, 1649–1653.

Gel-Based Proteomics: High-Resolution Two-Dimensional Gel Electrophoresis of Proteins. Isoelectric Focusing and Nonequilibrium pH Gradient Electrophoresis

Julio E. Celis, Signe Trentemølle, and Pavel Gromov

I. INTRODUCTION

The sequencing of the human and other important genomes is only the beginning of the quest to understand the functionality of cells, tissues, and organs in both health and disease. Together with advances in bioinformatics, this development has paved the way to the revolution in biology and medicine that we are experiencing today. We are rapidly moving from the study of single molecules to the analysis of complex biological systems, and the current explosion of emerging technologies within proteomics and functional genomics (see other articles in this volume) promises to elicit major advances in medicine in the near future. In particular, proteomic technologies are expected to play a key role in the study and treatment of diseases as they provide invaluable resources to define and characterize regulatory and functional networks, investigate the precise molecular defect in diseased tissues and biological fluids, and for developing specific reagents to precisely pinpoint a particular disease or stage of a disease. For drug discovery, proteomics assist with powerful tools for identifying new clinically relevant drug targets and provide functional insight for drug development.

High-resolution two-dimensional (2D) polyacrylamide gel electrophoresis (PAGE), often referred as gel-based proteomics, multidimensional chromatography, and protein biochips in combination with mass spectrometry (McDonald and Yates, 2002; Yip and Lomas, 2002; Wu and McCoss, 2002 and references therein), are among the proteomic tools that are available for biomarker and drug target discovery. Considerable work is currently underway to explore applications of nongel-based proteomics in various areas of biology as this technology has much to offer.

2D PAGE is often considered the method of choice to separate complex protein mixtures present in cells, tissues, and fluids (Cash and Kroll, 2003; Ong and Pandey, 2001; Celis and Gromov, 1999). The technique separates proteins in terms of both their isoelectric points (pI) and molecular weights (M_r), and it is essentially a stepwise separation tool that combines isoelectric focusing and SDS–polyacrylamide gel electrophoresis. Using the current 2D PAGE technologies it is possible to (i) separate complex protein mixtures into their individual polypeptide components, (ii) compare the protein expression profiles of sample pairs (normal versus transformed

cells, cells at different stages of growth or differentiation, etc.), and (iii) choose a condition of interest, e.g., the addition of a cytokine or a drug to a given cell type or tissue, and allow the cell or tissue to reveal the global protein behavioral response under conditions in which all of the detected proteins can be analyzed, both qualitatively and quantitatively in relation to each other (Celis and Olsen, 1994). Protein profiles can be scanned and quantitated to search for protein differences (changes in the levels of preexisting proteins, induction of new products, coregulated polypeptides), and interesting targets or molecular signatures can be identified using additional proteomic technologies, such as mass spectrometry and Western blotting. Furthermore, by carrying out studies in a systematic manner, it is possible to store the information in comprehensive 2D PAGE databases that record how genes are regulated in health and disease (see, e.g., http://proteomics.cancer.dk; Celis *et al.*, 1995, 1998; Gromov *et al.*, 2002).

Today, 2D PAGE can be carried out using two separation modes in the first dimension: (i) conventional isoelectic focusing (IEF) gels and (ii) immobilised pH gradient (IPG) gels (see article by Görg and Weiss in this volume). In the first case, the pH gradient is generated and maintained by special amphoteric compounds, carrier ampholytes, that migrate and stack according to their pI when an electric field is applied. In contrast, IPGs are an integral part of the polyacrylamide matrix, a fact that is achieved by copolymerization of several nonamphoteric buffering species with various pK values (Immobilines), within the fibres of a gel (Bellqwist, 1982). This article presents protocols to resolve proteins based on conventional carrier ampholytes.

II. MATERIALS AND INSTRUMENTATION

Ampholines are from Pharmacia Biotech (pH 3.5–10, Cat. No. 80-1125-87; pH 7–9, Cat. No.

80-1125-94, and pH 8–9.5, Cat. No. 80-1125-95; the pH 8–9.5 ampholyte can be replaced by SERVALYT 9-11, Cat. No. 42909) and Serva (pH 5–7, Cat. No. 42905. Acrylamide (Cat. No. 161-0100), *N,N'*-methylenebisacrylamide (Cat. No. 161-0200), *N,N,N'N'*-tetramethylenediamine (TEMED, Cat. No. 161-0800), agarose (Cat. No. 162-0100), and ammonium persulfate (Cat. No. 161-0700) are from Bio-Rad. Dithiothreitol (DTT, Cat. No. D-0632) and bromphenol blue (Cat. No. B-6131) are from Sigma. Glycine (Cat. No. 808822) and urea (Cat. No. 821527) are from ICN Biomedical. Tris base (Cat. No.6483111) and Nonidet P-40 (NP-40, Cat. No. 492015) are from Calbiochem. SDS (Cat. No. 20763) is from Serva. Acrylamide (Cat. No. 10674) from Serva has also been used for the second dimension with essentially the same results. Filter-Count is from Packard (Cat. No. 6013149).

Dulbecco's modified Eagle medium (DMEM, Cat. No. 31966-021) is from GIBCO. Penicillin/ streptomycin (Cat. No. A2213) are from Biocrom KG. Fetal calf serum (FCS, Cat. No. 04-001-1A) is from Biological Industries. The [35S]methionine (Cat. No. SJ 204) and Amplify fluorographic reagent (Cat. No. NAMP100) are from Amersham. The 96-well plates (Cat. No. 655 180) and 50-ml culture flasks (Cat. No. 690 160) are from Greiner.

First-dimension glass tubes (14 cm in length and 2 mm inside diameter) are from Euro-GLAS. Prior to use they are washed with a solution containing 60 ml of alcohol and 40 ml of HCl. The glass tubes should be immersed in this solution for at least 30 min. Afterward they are washed thoroughly with glass-distilled water. Spacers are cut from 1-mm-thick polystyrene plates (Metzoplast SB/Hk). First- and second-dimension chambers, as well as the rack to hold the first dimension tubes, are homemade. X-ray films (X-OMAT UV; 18 × 24 cm, Cat. No. 524 9792) are from Kodak. The scalpels (Paragon No. 11) are from Paragon, and the long (Cat. No. V2A 1415 LL-10) and short (Cat. No. V2A 1406 LL-7) needles are from Acufirm.

The aspiration pump (recirculates water) for drying gels is from Holm & Halby (HETO, Cat No. SUE 300Q).

Power supplies are from Pharmacia Biotech (EPS 500/400) or similar. The orbital shaker (Red Rotor PR70) is from Pharmacia.

III. PROCEDURES

A. Sample Preparation

1. Labeling of Cultured Cells with [35S]Methionine

Solutions

1. *Complete Dulbecco's modified Eagle's medium*: To make 500 ml, mix 445 ml of DMEM medium, 5 ml of 10× stock penicillin/ streptomycin, and 50 ml of FCS.
2. *Methionine-free solution*: Supplement MEM lacking methionine with antibiotics (100 U/ ml penicillin–100 μg/ml streptomycin) and 2% dialyzed (against 0.9% NaCl) fetal calf serum (FCS). Dispense in 1-ml aliquots and keep at −80°C.
3. *[35S]Methionine (SJ 204, Amersham)*: Aliquot in 100-μCi portions in sterile 1-ml cryotubes. Keep at −20°C. Freeze dry just before use.
4. *Labeling medium*: Add 0.1 ml of MEM lacking methionine to each ampoule containing 100 μCi of [35S]methionine.

Steps

1. Grow the cells in complete DMEM and seed in a 96-well microtiter plate. Leave in a 37°C humidified, 5% CO_2 incubator until they reach the desired density (3000–4000 cells per well).
2. Freeze dry the [35S]methionine and resuspend in labeling medium at a concentration of 1 mCi/ml. For one well, one needs 100 μCi of [35S]methionine in 0.1 ml of labeling medium.

3. Remove the medium from the well with the aid of a sterile, drawn-out (under a flame) Pasteur pipette. Wash once with labeling medium. Add the labeling medium containing the radioactivity.

4. Wrap the plate in Saran wrap and place in the 37°C humidified, 5% CO_2 incubator for 16h or shorter period (if necessary).

5. At the end of the labeling period, remove the medium with the aid of a drawn-out Pasteur pipette. Keep the medium if you want to analyze secreted or externalised proteins. Place the 96-well plate at an angle to facilitate removal of the liquid. Dispose of the radioactive medium according to the regulations enforced in your laboratory.

6. Resuspend the cells in 0.1 ml of O'Farrell's lysis solution. Pipette up and down (avoid foaming). Keep at −20°C until use.

7. Apply about 10^6 cpm to the first-dimension gels (IEF and NEPHGE) as described in Section III,B steps 6 and 7.

2. Labeling of Tissue Samples with [^{35}S]Methionine

1. Place the tissue sample on ice immediately after dissection and transport to the laboratory as fast as possible.

2. Remove clots and contaminating tissue with the aid of a scalpel. Rinse the piece two to three times in Hank's solution.

3. Mince the tissue sample in small specimens (about $1\,mm^3$) with the aid of a scalpel.

4. Place a tissue specimen in a 10-ml sterile plastic conical tube containing 0.2 ml of MEM lacking cold methionine and containing 2% dialyzed FCS and 200 μCi of [^{35}S]methionine and incubate for 16 h at 37°C in a humidified 5% CO_2 incubator.

5. Following incubation, carefully aspirate the medium and resuspend the tissue specimen in 0.2–0.3 ml of lysis solution. Homogenize using a small glass homogenizer.

6. Apply 20–50 μl to the first-dimension gels (IEF and/or NEPHGE) as described in Section III,B steps 6 and 7.

3. Preparation of Cell Extracts from Cultured Cells for Silver Staining

1. Plate cells in 50-ml culture flasks and grow until they reach 80% confluence.

2. Wash the monolayer three times with Hank's buffered saline. Carefully aspirate the fluid with the aid of an extended Pasteur pipette.

3. Add 0.6 ml of lysis solution. Rock at room temperature for a couple of min.

4. Aspirate and keep the extract at −20°C until use. Usually 20–30 μl of the sample can be applied to the first-dimension gel.

4. Preparation of Tissue Extracts for Silver Staining

1. Place the tissue sample on ice immediately after dissection and proceed further as described in Section III,A,2, steps 1–3.

2. Place four to six small tissue pieces in a glass homogenizer and add 1–2 ml of lysis solution. Homogenize at room temperature until the suspension clears up.

3. Keep at −20 °C until use. Usually 20–30 μl of the sample can be applied to the first-dimension gel.

B. Preparation and Running of First-Dimension Gels (IEF, NEPHGE)

This procedure is modified from those of O'Farrell (1975), O'Farrell et al. (1977), and Bravo (1984).

Solutions

1. *Lysis solution:* 9.8M urea, 2% (w/v) NP-40, 2% ampholytes, pH 7–9, and 100 mM DTT.

To make 50 ml, add 29.42 g of urea, 10 ml of a 10% stock solution of NP-40, 1 ml of ampholytes, pH 7–9, and 0.771 g of DTT. After dissolving, complete to 50 ml with distilled water. The solution should not be heated. Aliquot in 2-ml portions and keep at −20°C.

2. *Overlay solution:* 8M urea, 1% ampholytes, pH 7–9, 5% (w/v) NP-40, and 100 mM DTT. To make 25 ml, add 12.012 g of urea, 0.25 ml of ampholytes, pH 7–9, 12.5 ml of a 10% stock solution of NP-40, and 0.386 g of DTT. After dissolving, complete to 25 ml with distilled water. The solution should not be heated. Aliquot in 2-ml portions and keep at −20°C.

3. *Equilibration solution:* 0.06M Tris–HCl, pH 6.8, 2% SDS, 100 mM DTT, and 10% glycerol. To make 250 ml, add 15 ml of a 1M stock solution of Tris–HCl, pH 6.8, 50 ml of a 10% stock solution of SDS, 3.857 g of DTT, and 28.73 ml of glycerol (87% concentration). After dissolving, complete to 250 ml with distilled water. Store at room temperature.

4. *Acrylamide solution:* 28.38% (w/v) acrylamide and 1.62% (w/v) N,N'-methylenebisacrylamide. To make 100 ml, add 28.38 g of acrylamide and 1.62 g of bisacrylamide. After dissolving, complete to 100 ml with distilled water. Filter if necessary. Store at 4°C and use within 3 to 4 weeks.

5. *NP-40:* 10% (w/v) NP-40 in H_2O. To make 100 ml, weigh 10 g of NP-40 and complete to 100 ml with distilled water. Dissolve carefully. Store at room temperature.

6. *Agarose solution:* 0.06M Tris–HCl, pH 6.8, 2% SDS, 100 mM DTT, 10% glycerol, 1% agarose, and 0.002% bromphenol blue. To make 250 ml, add 15 ml of a 1M stock solution of Tris–HCl, pH 6.8, 50 ml of a 10% stock solution of SDS, 3.857 g of DTT, 28.73 ml of glycerol (87% concentration), 2.5 g of agarose, and 2.5 ml of a 0.2% stock solution of bromphenol blue. Add distilled water

and heat in a microwave oven. Complete to 250 ml with distilled water and aliquot in 20-ml portions while the solution is still warm. Keep at 4°C.

7. *1M NaOH stock:* Weigh 4 g of NaOH and complete to 100 ml with distilled water. Keep at 4°C for no more than 2 weeks.

8. *1M H_3PO_4 stock:* To make 100 ml, take 6.74 ml of H_3PO_4 (87%) and complete to 100 ml with distilled water. Keep at room temperature.

9. *20 mM NaOH:* To make 500 ml, take 10 ml of 1M NaOH and complete to 500 ml with distilled water. Prepare fresh.

10. *10 mM H_3PO_4:* To make 500 ml, take 5 ml of 1M H_3PO_4 and complete to 500 ml with distilled water. Prepare fresh.

Steps

1. Mark the glass tubes with a line (use Easy Marker from Engraver or a diamond-tipped pencil) 12.5 cm from the bottom (Fig. 8.1A, f). Seal the bottom end of the tube by wrapping with Parafilm and place it standing up in a rack (Fig. 8.1B).

2a. To make 12 first-dimensional IEF gels, use 4.12 g urea; 0.975 ml of acrylamide solution; 1.5 ml of 10% NP-40, 1.5 ml of H_2O; 0.30 ml of carrier ampholytes, pH range 5–7; 0.10 ml of carrier ampholytes, pH range 3.5–10; 15 μl of 10% ammonium persulfate; and 10 μl of TEMED.

2b To make 12 first-dimensional NEPHGE gels, use 4.12 g urea; 0.975 ml of acrylamide solution; 1.5 ml of 10% NP-40; 1.69 ml of H_2O; 0.170 ml of carrier ampholytes, pH range 7–9; 0.020 ml of carrier ampholytes, pH range 9–11; 15 μl of 10% ammonium persulfate; and 10.5 μl of TEMED.

2c Mix the urea, H_2O, acrylamide, NP-40, and ampholytes (kept at −20°C in 1-ml aliquots) in a tube containing a vacuum outlet (Fig. 8.1A, g). Swirl the solution gently until the urea is dissolved. The solution should not be heated. Add

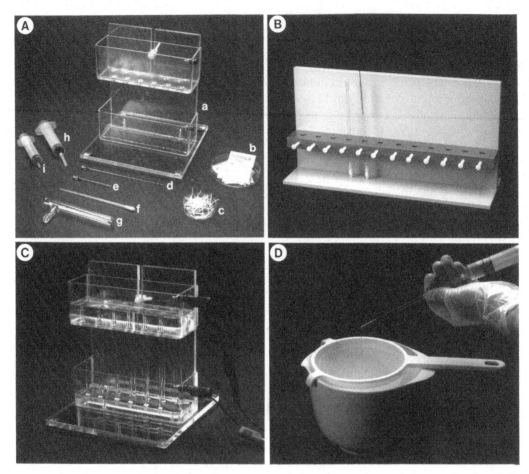

FIGURE 8.1 (A) First-dimension chamber and accessories for first-dimension. (a) First-dimension chamber, (b) Parafilm, (c) paper strips, (d) long needle for filling the tubes, (e) short needle for extruding the gel, (f) first-dimension tubes, (g) vacuum tube, (h) syringe connecting to a piece of rubber tubing, and (i) syringe. (B) Filling first-dimensional glass tubes with gel solution. (C) First-dimension chamber filled with tubes. (D) Extruding the first-dimension gel.

ammonium persulfate and TEMED, mix gently, and degas using a vacuum pump. Use a clean rubber stopper to control the vacuum.

3. Pour the solution into a 55-mm culture dish. Aspirate the liquid with a 10-ml syringe and add to the thin glass tubes (Fig. 8.1A, f) using a long needle (Fig. 8.1A, d). Insert the tip of the needle into the bottom of the tube and slowly fill to the mark to avoid air bubbles while moving up the needle (Fig. 8.1B).

4. Overlay the gel mix with 10 µl of glass-distilled water and leave to polymerise for 45 min. In the meantime, fill the lower chamber of the first dimension (Fig. 8.1C) with 250 ml of 10 mM H_3PO_4 (+; IEF gels) or 250 ml of 20 mM NaOH (−; NEPHGE gels).

5. Take the tubes from the rack and remove the Parafilm using a scalpel. Remove excess liquid from the upper part of the tube by shaking and dry using a thin strip of Whatman 3 MM paper (Fig. 8.1A, c). Insert

the tubes into the chamber, which holds up to 12 tubes (Fig. 8.1C). Tap the bottom of the tubes to remove trapped air bubbles.

6. Prerun IEF gels before adding the sample. First add 10 µl of lysis solution and then 10 µl of overlay solution. Use a Gilson microman pipette to apply the solutions. Fill the tubes as well as the upper chamber (−) with 20 mM NaOH. Prerun gels at room temperature for 15 min at 200 V, 30 min at 300 V, and 60 min at 400 V. After prerunning, disconnect the power supply and discard the upper and bottom solutions. Remove the tubes and wash the top of the gels with distilled water. Dry with a thin strip of Whatmann 3 MM paper and apply the sample (up to 50 µl in lysis solution). Add 10 µl of overlay solution and fill the tubes with 20 mM NaOH. Fill the upper chamber with 20 mM NaOH (−) and the bottom one with 20 mM H_3PO_4. Run for 19 h at 400 V at room temperature.

7. NEPHGE gels are not prerun. Add the sample in lysis solution (up to 50 µl). Cover with 10 µl of overlay solution. Add 250 ml of 20 mM NaOH to the bottom chamber (−) and fill the tubes and the upper chamber with 10 mM H_3PO_4 (+). Run the gels for 4.5 h at 400 V at room temperature.

8. Before stopping the run, add 3.5 ml of equilibration solution to 35-mm tissue culture dishes marked with the gel number in both the bottom part and the lid. Turn off the power supply and take the gels out with the aid of a syringe (Fig. 8.1A, h) filled with glass-distilled water. First, use a short needle (Fig. 8.1A, e) to loosen the gel at both ends of the tube. Then extrude the gel with the aid of pressure applied by a 20-ml syringe (Fig. 8.1D). Collect the gel in a sieve and place in the 35-mm culture dish containing the equilibration solution. Leave 2–5 min at room temperature and store at −20°C until use. Samples can be stored for at least 2 months under these conditions.

C. Second Dimension: SDS–Polyacrylamide (15%) Gel Electrophoresis

This procedure is performed essentially according to Laemmli (1970).

Solutions

1. *Solution A:* To make 500 ml, add 150 g of acrylamide and 0.75 g of bisacrylamide. After dissolving, complete to 500 ml with distilled water. Filter if necessary. Aliquot 100-ml portions and store at 4°C.
2. *Solution B:* To make 1 liter of 1.5M Tris–HCl, pH 8.7, add 181.6 g of Tris base and titrate with HCl. Complete to 1 liter with distilled water. Aliquot 200-ml portions and store at 4°C.
3. *Solution C:* To make 1 liter of 1M Tris–HCl, pH 6.8, add 121.1 g of Tris base and titrate with HCl. Complete to 1 liter with distilled water. Aliquot in 200-ml portions and store at 4°C.
4. *Solution D:* To make 100 ml, add 10 g of acrylamide and 0.5 g of bisacrylamide. Complete to 100 ml with distilled water. Filter if necessary. Aliquot in 200-ml portions and store at 4°C.
5. *10% SDS:* To make 1 liter, weigh 100 g of SDS and complete to 1 liter with distilled water. Filter if necessary. Store at room temperature.
6. *10% ammonium persulfate:* To make 10 ml, weigh 1 g of ammonium persulfate and complete to 10 ml with distilled water. This solution should be prepared just before use.
7. *Electrode buffer:* To make 1 liter of a 5× solution, add 30.3 g of Tris base, 144 g of glycine, and add 50 ml of 10% SDS solution. Complete to 1 liter with distilled water. Store at room temperature.

Steps

1. Store clean glass plates in dust-free boxes. One of the plates is 16.5 cm wide and 20 cm

high and has a notch 2 cm deep and 13 cm wide. Cover the edges of the plate with a thin line of Vaseline. Use a plastic 10-ml syringe filled with vaseline and fitted with a Gilson tip (Fig. 8.2A). Place 1-mm-thick polystyrene spacers at the edges of the plate and cover with Vaseline (Fig. 8.2A). Place a small piece of paper without lines containing the gel number (written with pencil) at the corner of the plate (Fig. 8.2A).

2. Assemble the rectangular glass place 16.5 cm wide and 20 cm high together with the notched plate and spacers. Make sure that the vertical spacers are in contact with the horizontal one at the bottom. Hold the assembled plates together with the aid of fold-back clamps. Mark a line 2.5 cm from the top of the notched plate.

3. To make six 15% separation gels, mix the following solutions in a 250-ml filter flask containing a magnetic stirrer: solution A (acrylamide:bisacrylamide, 30:0.15), 75.0 ml; 10% SDS, 1.5 ml; solution B (1.5M Tris–HCl, pH 8.8), 37.5 ml; H_2O, 35.22 ml; 10% ammonium persulfate, 750 μl; and TEMED, 30 μl.

4. Add ammonium persulfate and TEMED to the separation gel solution just before degassing using a vacuum pump. Pour the solution into the assembled plates until the marked line and overlay with distilled

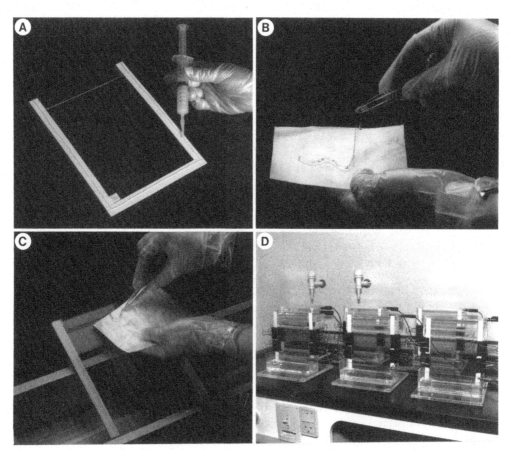

FIGURE 8.2   (A) Covering the spacers with Vaseline. (B) First-dimension gel prior to application to second-dimension gel. (C) Application of first-dimension gel to second-dimension gel. (D) Second-dimension chambers.

water. Leave the gels to polymerize for approximately 1 h.

5. Remove excess liquid and dry the top of the gel with a strip of Whatman 3 MM paper (2 × 9 cm).

6. To make 5% stacking solution for six gels, mix the following solutions: solution D (acrylamide:bisacrylamide, 10:0.5), 15.0 ml; 10% SDS, 0.3 ml; solution C (1.0 M Tris–HCl, pH 6.8), 3.6 ml; H_2O, 10.8 ml; 10% ammonium persulfate, 0.24 ml; and TEMED, 12 μl.

7. After degassing, add the stacking gel solution and insert a polystyrene spacer a few millimetres into the assembled plates. Keep in place with the aid of a fold-out clamp. Leave the gels to polymerize for approximately 1 h.

8. When the gel has polymerised, remove the top spacer and clean the top of the gel with a strip of Whatman 3 MM paper. Remove the clamps as well as the horizontal spacer at the bottom with the aid of a spatula. Clean the space between the two glass plates using a thin spatula covered with tissue paper.

9. Lay the gels at an angle in order to facilitate application of the first dimension (Fig. 8.2C). Take out the culture dishes containing the first-dimension gels from the freezer 20 min before application. Once they are defrosted, melt the agarose solution in a microwave oven and immediately cover the top of the stacking gel with a small amount of agarose to fill the space left by the spacer.

10. Collect the first-dimension gel into a sieve (Fig. 8.1D) and place it on a piece of Parafilm (Fig. 8.2B). Place the gel carefully on top of the second dimension with the aid of plastic tweezers (Fig. 8.2C). Do not stretch the gel. Cover the gels with 2–3 ml of melted agarose. Eliminate air bubbles by pushing them out with the same pipette.

11. Clamp the gel plates to the electrophoresis chambers, which have been prefilled with 1× electrode buffer. Fill the upper chamber with enough electrode buffer to cover the agarose. Remove air bubbles at the bottom of the gel with electrode buffer using a 10-ml syringe joined to a bent needle.

12 Connect the electrodes (upper, −; lower, +) to the power supply. Run the gels at 10 mA for 4 h and at 3 mA overnight at room temperature (until the tracking dye has reached 1 cm from the bottom) (Fig. 8.2D). At the end of the run turn off the power supply, disassemble the plates, and remove the stacking gel with the aid of a scalpel. Process the separation gel for autoradiography, fluorography (see later), or for staining with either silver or Coomassie brilliant blue. Gels can be used directly for blotting.

Figure 8.3 shows several representative auto-radiographs and silver-stained gels of normal (Fig. 8.3A) and malignant tissues (Figs. 8.3B–3D) run under the conditions (IEF and NEPHGE) described in this article. For more information, the reader is encouraged to visit the group's Web site at http://proteomics.cancer.dk.

D. Other Procedures

1. Fluorography

This protocol is essentially from Amersham. The procedure increases the detection efficiency 1000-fold for ^3H and 15-fold for ^{35}S.

Solutions

1. *Fixation solution:* To make 1 liter, add 450 ml of methanol and 75 ml of acetic acid. Complete to 1 liter with distilled water.

2. *Amplify fluorographic reagent:* Available commercially from Amersham (Cat. No. NAMP100).

Steps

1. Place the gel in a rectangular glass pie dish (24 × 19 cm) and fix for 60 min at room

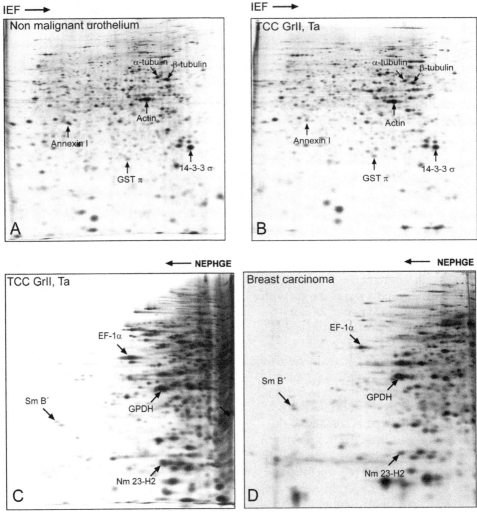

FIGURE 8.3 Two-dimensional patterns of whole protein extracts from human normal urothelium (A), a transitional cell carcinoma (B and C), and a breast tumor (D). Proteins were separated by 2D PAGE—IEF (A and B) and 2D PAGE—NEPHGE (C and D) and were visualized by autoradiography (A–C) and silver staining (D). The identity of a few proteins is indicated for reference.

temperature in fixative solution. Shake while fixing.

2. Place gel in 120 ml of amplifying fluorographic reagent and agitate for 30 min.
3. Dry gels.
4. Sensitize X-ray films by preflashing in the dark and expose the gels at −80°C in cassettes.

2. Quantitation of [35S]Methionine-Labeled Protein Spots Excised from Two-Dimensional Gels

Steps

1. Localize protein spots with the aid of the X-ray film (Fig. 8.3). Before exposing, make four crosses at the corner of the gels using

Transitional Cell Carcinomas-IEF database

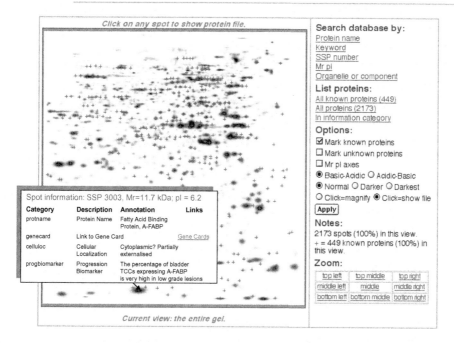

FIGURE 8.4 Master synthetic image of human bladder TCC proteins separated by IEF 2D PAGE as depicted in the Internet (http://proteomics.cancer.dk). Proteins flagged with a cross correspond to known proteins. By clicking on any spot, it is possible to open a file that contains protein information available in the database as well as links to other related Web sites. Only part of the file for A-FABP is shown.

radioactive ink. Excise the proteins from dry gels using a scalpel.

2. Place gel pieces in counting vials containing 4 ml of Filter-Count, leave for 1 h, and count for 5 min in a scintillation counter.

IV. COMMENTS

Using the protocols described in this article, it is possible to resolve proteins having apparent molecular masses between 8.5 and 230 kDa and pI values from 4 to 12 (see also http://proteomics.cancer.dk). Autoradiographs can be quantitated using phosphorimaging autoradiography (BioRad; Amersham; Fuji). Several softwares for the analysing of 2D protein images are available. These include PDQUEST (BioRad), Z3 (Compugen), Phoretix (Nonlinear Dynamics), GelFox (Imaxia), ProteinMine (Scimagix), and Melanie, as well as several others. By carrying out systematic studies it is possible to store protein information in comprehensive 2D PAGE databases that record how genes are regulated in health and disease (see, e.g., http://proteomics.cancer.dk) (Fig. 8.4). As these databases achieve critical mass of data, they will become valuables sources of information for expediting the identification of signaling pathways and components that are affected in diseases (Celis *et al.*, 1991a,b, 1998; Gromov *et al.*, 2002).

V. PITFALLS

1. Do not heat the sample when dissolving in lysis solution or when defrosting.
2. For optimal results, titrate the ampholytes. Mix in various proportions, maintaining the final percentage fixed, and run first and second dimensions. Select the combination that gives the best separation (well-focused spots, no streaking, etc.). Store them at $-20°C$ in 1- or 2-ml aliquots. Ampholytes can be stored for many years at $-20°C$.
3. Make sure that all glassware is washed properly. Rinse with glass-distilled water before drying.
4. The pH of the $1.5M$ Tris–HCl, pH 8.8, solution should be checked carefully and regularly.
5. Use sterile, disposable pipettes to dispense the solutions. Do not allow other colleagues to use your solutions.

References

Anderson, N. L., and Anderson, N. G. (1978). Analytical technique for cell fractions. XXII. Two-dimensional analysis of serum and tissue proteins: Multiple gradient-slab gel electronporesis. *Anal. Biochem.* **85**, 341–354.

Bravo, R. (1984). Two-dimensional gel electrophoresis: A guide for the beginner. *In* "Two-Dimensional Gel Electrophoresis of Proteins: Methods and Applications" (J. E. Celis *et al.*, eds.), pp. 3–36. Academic Press, New York.

Celis, J. E. (ed.) (1996). Electrophoresis in cell biology. *Electrophoresis* **11**.

Celis, J. E., and Bravo, R. (eds.) (1984). "Two-Dimensional Gel Electrophoresis of Proteins: Methods and Applications. " Academic Press, New York.

Celis, J. E., Gromov, P., Østergaard, M., Madsen, P., Honoré, B., Dejgaard, K., Olsen, E. Vorum, H., Kristensen, D. B., Gromova, I., Haunsø, A., Van Damme, J., Puype, M., Vandekerckhove, J., and Rasmussen, H. H. (1996). *FEBS Lett.* **398**, 129–134.

Celis, J. E., Leffers, H., Rasmussen, H. H., Madsen, P., Honore, B., Gesser, B., Dejgaard, K., Olsen, E., Ratz, G. P., Lauridsen, J. B., *et al.* (1991a). The master two-dimensional gel database of human AMA cell proteins: Towards linking protein and genome sequence and mapping information (update 1991). *Electrophoresis* **12**, 765–801.

Celis, J. E., Østergaard, M., Jensen, N. A., Gromova, I., Rasmussen, H. H., and Gromov, P. (1998). Human and mouse proteomic databases: Novel resources in the protein universe. *FEBS Lett.* **430**, 64–72.

Celis, J. E., Rasmussen, H. H., Leffers, H., Madsen, P., Honore, B., Gesser, B., Dejgaard, K., and Vandekerckhove, J. (1991b). Human cellular protein patterns and their link to genome DNA sequence data: Usefulness of two-dimensional gel electrophoresis and microsequencing. *FASEB J.* **5**, 2200–2208.

Celis, J. E., Rasmussen, H. H., Olsen, E., Gromov, P., Madsen, P., Leffers, H., Honoré, B., Dejgaard, K., Vorum, H., Kristensen, D. B., Østergaard, M., Haunsø, A., Jensen, N. A., Celis, A., Basse, B., Lauridsen, J. B., Ratz, G. P., Andersen, A. H., Walbum, E., Kjærgaard, I., Andersen, I., Puype, M., Van Damme, J., and Vandekerckhove, J. (1995). The human keratinocyte two-dimensional gel protein database (update 1995): Mapping components of signal transduction pathways. *Electrophoresis* **12**, 2217–2184.

Gromov. P., Østergaard, M., Gromova, I., and Celis, J. E. (2002). Human proteomic databases: A powerful resource for functional genomics in health and disease. *Prog. Biophys. Mol. Biol.* **80**, 3–22.

Klose, J. (1975). Protein mapping by combined isoelectric focusing and electrophoresis of mouse tissues: A novel approach to testing for induced point mutations in mammals. *Humangenetik* **26**, 231–243.

Laemmli, U. K. (1970). Cleavage of structural proteins during the assembly of the head of bacteriophage T4. *Nature (London)* **227**, 680–685.

Laskey, R. A., and Mills, A. D. (1975). Quantitative film detection of ^3H and ^{14}C in polyacrylamide gels by fluorography. *Eur. J. Biochem.* **56**, 335–341.

McDonald, W. H., and Yates, J. R., 3rd (2002). Shotgun proteomics and biomarker discovery. *Dis. Mark.* **18**, 99–105.

O'Farrell, P. H. (1975). High resolution two-dimensional electrophoresis of proteins. *J. Biol. Chem.* **250**, 4007–4021.

O'Farrell, P. Z., Goodman, H. M., and O'Farrell, P. H. (1977). High resolution two dimensional electrophoresis of basic as well as acidic proteins. *Cell* **12**, 1133–1142.

Ong, S. E., and Pandey, A. (2001) An evaluation of the use of two-dimensional gel electrophoresis in proteomics. *Biomol. Eng.* **18**, 195–205.

Wu, C. C., and McCoss, M. J. (2002). Shotgun proteomics: Tools for the analysis of complex biological systems. *Curr. Opin. Mol. Ther.* **4**, 242–250.

Yip, T. T., and Lomas, L. (2002). SELDI ProteinChip array in oncoproteomic research. *Cancer Res. Treatment* **1**, 273–274.

High-Resolution Two-Dimensional Electrophoresis with Immobilized pH Gradients for Proteome Analysis

Angelika Görg and Walter Weiss

I. INTRODUCTION

Although promising progress has been made in the development of alternative protein separation techniques for proteome analysis, such as multidimensional chromatography–tandem mass spectrometry, there is still no generally applicable method that can replace two-dimensional gel electrophoresis (2D PAGE) in its ability to simultaneously separate and display thousands of proteins from complex biological samples. 2D PAGE separates proteins according to two independent parameters, i.e., isoelectric point (pI) in the first dimension and molecular mass (M_r) in the second dimension, by coupling isoelectric focusing (IEF) and sodium dodecyl sulfate polyacrylamide gel electrophoresis (SDS–PAGE). In comparison with the classical O'Farrell method (1975) of 2D PAGE based on the use of carrier ampholyte-generated pH gradients, 2D PAGE using immobilized pH gradients (IPG) in the first dimension (IPG-Dalt) (Görg et al., 1988) has proved to be extremely flexible with respect to the requirements of proteome analysis.

For proteome analysis it is essential to generate reproducible, high-resolution protein separations. Using the classical 2D PAGE approach of O'Farrell, it is, however, often difficult to obtain reproducible results even within a single laboratory, let alone between different laboratories. The problem of limited reproducibility is largely due to the synthetic carrier ampholytes (CA) used to generate the pH gradient required for IEF, for reasons such as batch-to-batch variability of CAs, pH gradient instability over time, cathodic drift etc. with resultant loss of alkaline proteins. These problems have been largely overcome by the development of immobilized pH gradients (IPG) (Bjellqvist et al., 1982) based on the use of the bifunctional Immobiline reagents, a series of 10 chemically well-defined acrylamide derivatives with the general structure $CH_2=CH–CO–NH–R$, where R contains either a carboxyl or an amino group. These form a series of buffers with different pK values between pK 1 and >12. Because the reactive end is copolymerized with the acrylamide matrix, extremely stable pH gradients are generated, allowing true steady-state IEF with increased reproducibility, as has been demonstrated in several interlaboratory comparisons (Corbett et al., 1994; Blomberg et al., 1995).

More than a decade ago, a basic protocol of IPG-Dalt was described by Görg et al. (1988), summarizing the critical parameters inherent to isoelectric focusing with IPGs and a number of experimental conditions that were not part of the classical 2D electrophoresis repertoire with CAs. In principle, this protocol is still valid today: The first dimension of IPG-Dalt, IEF, is performed in individual 3-mm-wide IPG gel strips cast on GelBond PAGfilm (either ready-made Immobiline DryStrips or laboratory made). Samples are applied either by cup loading or by in-gel rehydration. After IEF, the IPG strips are equilibrated with SDS buffer in the presence of urea, glycerol, dithiothreitol (DTT), and iodoacetamide and are applied onto horizontal or vertical SDS gels in the second dimension.

II. MATERIALS AND INSTRUMENTATION

Immobiline chemicals (Cat. Nos. 80125570–80125575), Pharmalyte pH 3–10 (Cat. No. 17-0456-01), Immobiline DryStrips pH 4–7, 3–10 and 3–10NL (Cat. Nos. 17-1233-01, 17-1234-01, and 17-1235-01), narrow-range Immobiline DryStrips covering the pH range between pH 3.5 and pH 6.7 (Cat. Nos. 17-6001-83–17-6001-87), ExcelGel SDS and buffer strips (Cat. No. 17-1236-01 and Cat. No. 80-1129-42), IEF electrode strips (Cat. No. 18-1004-40), CHAPS (Cat. No. 17-1314-01), acrylamide (Cat. No. 17-1302-01), N, N'-methylenebisacrylamide (Cat. No. 17-1304-01), TEMED (Cat. No. 17-1312-01), GelBond PAGfilm (Cat. No. 80-1129-37), and DryStrip cover fluid (Cat. No. 17-1335-01) are from Amersham Biosciences. Sodium dodecyl sulfate (SDS) (Cat. No. 20763), ammonium persulfate (Cat. No. 13375), glycine (Cat. No. 23390), kerosene (Cat. No. 26940), and ion exchanger (Serdolite MB-1, Cat. No. 40701) are from Serva. Trizma base (Cat. No. T-1503), DTT (Cat. No. D-0632), acrylamido buffers (pK 1.0, 10.3, and 12.0; Cat. No. 84885), thiourea (Cat. No. T8656), and iodoacetamide (Cat. No. I-6125) are from Sigma-Aldrich. Glycerol (Cat. No. 4093), urea (Cat. No. 84879), and a serine protease inhibitor (Pefabloc, Cat. No. 24839) are from VWR. Dimethylchlorosilane (Cat. No. 40140) is from Fluka. Water is deionized using the Milli-Q system of Millipore.

Apparatus for isoelectric focusing and horizontal SDS electrophoresis (Multiphor II, Cat. No. 18-1018-06), multiple vertical SDS electrophoresis (Ettan DALT Vertical System, Cat. No. 80-6466-46), thermostatic circulator (Multitemp III, Cat. No. 18-1102-78), power supply (Multidrive XL, Cat. No. 18-1013-68), IPGphor (Cat. No. 80-6469-88), IPGphor strip holders (Cat. No. 80-6416-68), cup-loading strip holders (Cat. No. 80-6459-43), IPG DryStrip reswelling tray (Cat. No. 80-6465-32), IPG DryStrip kit (Cat. No. 18-1004-30), and gradient gel kit (including glass plates with a 0.5-mm-thick U frame and gradient maker, Cat. No. 80-1013-74) are from Amersham Biosciences. Prior to use the glass plates are washed with a mild detergent, rinsed with deionized water, and air dried. If new glass plates are used, pipette 1–2 ml of dimethylchlorosilane (diluted 1 + 9 in trichlorethane) on the glass plate that bears the U frame and distribute it evenly with a fuzz-free filter paper. Let it dry for a few minutes, rinse again with water, and let it air dry. Repeat this procedure occasionally in order to prevent the gels from sticking to the glass plates.

III. PROCEDURES

A. Preparation of First-Dimensional IPG Gel Strips

Solutions

1. *Acrylamide/bisacrylamide solution (30% T, 3% C)*: 29.1% (w/v) acrylamide, 0.9% (w/v) N, N'-methylenebisacrylamide. To make 100 ml of the solution, dissolve 29.1 g of acrylamide and 0.9 g of bisacrylamide in deionized water and fill up to 100 ml. Add 1 g of Serdolite MB-1, stir for 10 min, and filter. This solution can be stored for 1 week at 4°C; however, for optimum results it is advisable to prepare it freshly the day you use it.

2. *Ammonium persulfate solution*: 40% (w/v) in deionized water. To prepare 1 ml of the solution, dissolve 0.4 g of ammonium persulfate in 1 ml of deionized water. This solution should be prepared freshly just before use.

3. *Solutions for casting immobiline gels*: To prepare 15 ml each of acidic and basic solutions, mix chemicals and reagent solutions as described in Table 9.1. A huge selection of recipes for many types of narrow or broad pH gradient have been calculated (Righetti, 1990). Table 9.1 describes several of our favourite pH gradients for 2D electrophoresis.

Steps

IPG slab gels for 180-mm separation distance $(250 \times 180 \times 0.5 \, \text{mm}^3)$ are cast on GelBond PAGfilm. After polymerization, the IPG gels are washed extensively with deionized water, air dried, and stored frozen until used.

1. To assemble the polymerisation cassette, wet the plain glass plate (size $260 \times 200 \, \text{mm}^2$) with a few drops of water. Place the Gelbond PAGfilm, hydrophilic side upward, on the wetted surface of the plain glass plate. The GelBond PAGfilm should overlap the upper edge of the glass plate for 1–2 mm to facilitate filling of the cassette. Expel excess water with a roller. Place the glass plate that bears the U frame (0.5 mm thick) on top of the Gelbond PAGfilm and clamp the cassette together (Fig. 9.1A). Put it in the refrigerator for 30 min.

2. To cast the IPG gel, pipette 12.0 ml of the acidic, dense solution into the mixing chamber of the gradient mixer. The outlet and connecting line between the mixing chamber and reservoir have to be closed! Add 7.5 μl of TEMED and 12 μl of ammonium persulfate and mix. Open the connecting line between the

TABLE 9.1   Recipes for Casting Immobiline Gels with pH Gradients[a]

Linear pH gradient	pH 4–7		pH 4–9		pH 6–12		pH 4–12		pH 3–12	
	Acidic solution (pH 4)	Basic solution (pH 7)	Acidic solution (pH 4)	Basic solution (pH 9)	Acidic solution (pH 6)	Basic solution (pH 12)	Acidic solution (pH 4)	Basic solution (pH 12)	Acidic solution (pH 3)	Basic solution (pH 12)
Immobiline pK 1.0	—	—	—	—	—	—	—	—	1287 µl	—
Immobiline pK 3.6	578 l	302 µl	829 µl	147 µl	1367 µl	—	950 µl	—	306 µl	—
Immobiline pK 4.6	110 µl	738 µl	235 µl	424 µl	—	—	352 µl	74 µl	414 µl	—
Immobiline pK 6.2	450 µl	151 µl	232 µl	360 µl	188 µl	251 µl	319 µl	206 µl	558 µl	336 µl
Immobiline pK 7.0	—	269 µl	22 µl	296 µl	323 µl	125 µl	294 µl	103 µl	496 µl	168 µl
Immobiline pK 8.5	—	—	250 µl	71 µl	365 µl	84 µl	48 µl	522 µl	112 µl	699 µl
Immobiline pK 9.3	—	876 µl	221 µl	663 µl	497 µl	32 µl	52 µl	219 µl	84 µl	157 µl
Immobiline pK 10.0	—	—	—	—	50 µl	485 µl	41 µl	325 µl	25 µl	342 µl
Immobiline pK > 13	—	—	—	—	—	345 µl	—	531 µl	—	258 µl
Acrylamide/ bisacrylamide	2.0 ml	2.0 ml	2.0 ml	2.0 ml	2.25 ml	2.25 ml	2.5 ml	2.5 ml	2.25 ml	2.25 ml
deionized water	8.9 ml	10.7 ml	8.3 ml	11.1 ml	7.0 ml	10.5 ml	7.4 ml	10.5 ml	6.45 ml	10.8 ml
Glycerol (100%)	3.75 g	—	3.75 g	—	3.75 g	—	3.75 g	—	3.75 g	—
TEMED (100%)	10.0 µl	10.0 µl	10.0 µl	10.0 µl	10.0 µl	10.0 µl	10.0 µl	10.0 µl	10.0 µl	10.0 µl
Persulfate (40%)	15.0 µl	15.0 µl	15.0 µl	15.0 µl	15.0 µl	15.0 µl	15.0 µl	15.0 µl	15.0 µl	15.0 µl
Final volume	15.0 ml	15.0 ml	15.0 ml	15.0 ml	15.0 ml	15.0 ml	15.0 ml	15.0 ml	15.0 ml	15.0 ml

[a]For effective polymerization, acidic and basic solutions are adjusted to pH 7 with 4*N* sodium hydroxide and 4*N* acetic acid, respectively, before adding the poymerization catalysts (TEMED and ammonium persulfate).

chambers for a second to release any air bubbles.

3. Pipette 12.0 ml of the basic, light solution into the reservoir of the gradient mixer. Add 7.5 µl of TEMED and 12 µl of ammonium persulfate and mix with a spatula.

4. Switch on the magnetic stirrer at a reproducible and rapid rate; however, avoid excessive vortex. Remove the polymerisation cassette from the refrigerator and put it underneath the outlet of the gradient mixer. Open the valve connecting the chambers and, immediately afterwards, the pinchcock on the outlet tubing so that the gradient mixture is applied centrally into the cassette from a height of about 5 cm just by gravity flow. Formation of the gradient is completed in 2–3 min (Fig. 9.1B).

5. Keep the mold at room temperature for 15 min to allow adequate leveling of the density gradient. Then polymerize the gel for 1 h at 50 °C in a heating cabinet.

6. After polymerization, wash the IPG gel for 1 h with 10-min changes of deionized water (500 ml each) in a glass tray on a rocking platform. Then equilibrate the gel in 2% (w/v) glycerol for 30 min and dry it

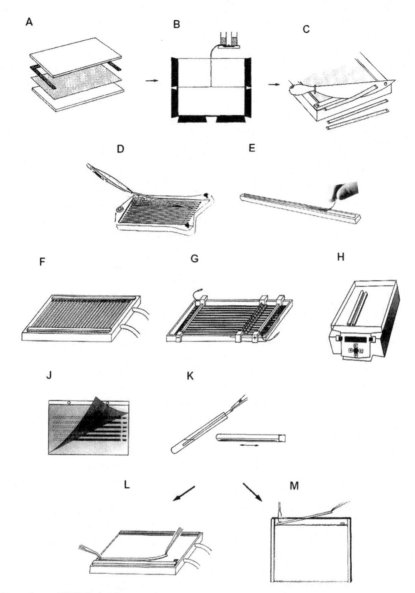

FIGURE 9.1 Procedure of IPG-Dalt (Görg *et al.*, 2000). (A) Assembly of the polymerization cassette for the preparation of IPG and SDS gels on plastic backings (glass plates, GelBond PAGfilm and 0.5-mm-thick U frame). (B) Casting of IPG gels. (C) Cutting of washed and dried IPG slab gels (or IPG DryPlates) into individual IPG strips. (D) Rehydration of individual IPG strips in the IPG DryStrip reswelling tray. (E) Rehydration of individual IPG strips in the IPGphor strip holder. (F) IEF of individual IPG gel strips directly on the cooling plate of the IEF chamber. (G) IEF of individual IPG gel strips in the IPG DryStrip kit. (H) IEF of individual IPG gel strips on the IPGphor. (J) Storage of IPG strips after IEF. (K) Equilibration of IPG strips prior to SDS–PAGE. (L) Transfer of the equilibrated IPG strip onto onto the surface of a ready-made horizontal SDS gel along the cathodic buffer strip. (M) Loading of the equilibrated IPG strip onto onto the surface of a vertical SDS gel.

overnight at room temperature, using a fan, in a dust-free cabinet. Afterward, protect the surface of the dry gel with a sheet of plastic film. The dried IPG gel can be stored in a sealed plastic bag at −20°C for at least several months without loss of function. Dried IPG gels in several pH ranges are also available commercially (Immobiline DryPlate).

7. For IEF in individual IPG gel strips, cut the dried IPG gels or the ready-made Immobiline DryPlates into 3-mm-wide strips with the help of a paper cutter (Fig. 9.1C). Alternatively, ready-cut IPG strips (Immobiline DryStrip) can also be used.

B. Running of First-Dimensional IPG Strips

The first dimension of IPG-Dalt, isoelectric focusing, is performed in individual 3-mm-wide IPG gel strips cast on GelBond PAGfilm. Instead of laboratory-made gels, ready-made gels (Immobiline DryStrip) can be used. Samples are applied either by in-gel rehydration or by cup loading after IPG strip rehydration (*see* Table 9.2 and Fig. 9.2). IPG-IEF can be simplified by using an integrated system, the IPGphor. This instrument features strip holders that provide rehydration of individual IPG strips with or without sample, optional separate sample cup loading, and subsequent IEF, all without handling the strip after it is placed in a ceramic strip holder.

Solutions

1. *Lysis solution*: 9.5M urea, 2% (w/v) CHAPS, 2% (v/v) Pharmalyte pH 3–10, 1% (w/v) DTT. To prepare 50 ml of lysis solution, dissolve 30.0 g of urea in deionized water and make up to 50 ml. Add 0.5 g of Serdolite MB-1, stir for 10 min, and filter. Add 1.0 g of CHAPS, 0.5 g DTT, 1.0 ml of Pharmalyte pH

TABLE 9.2 Sample Application

In-gel rehydration	Wide pH range IPGs between pH 3 and pH 12 (analytical and micropreparative runs)
Cup loading at the anode	Narrow pH range IPGs at the basic extreme (e.g., IPG 9–12)
Cup loading at the cathode	Narrow pH range IPGs at the acidic extreme (e.g., IPG 2.5–5)

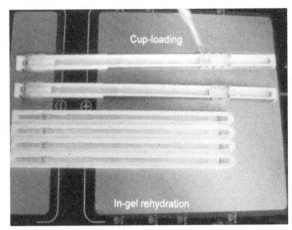

FIGURE 9.2 IPGphor: Sample application by in-gel rehydration and by cup loading.

3–10, and 50 mg of Pefabloc to 48 ml of the urea solution.

2. *IPG strip rehydration solution*: 8M urea, 0.5% (w/v) CHAPS, 15 mM DTT, 0.5% (v/v) Pharmalyte pH 3–10. To prepare 50 ml of the solution, dissolve 25.0 g of urea in deionized water and complete to 50 ml. Add 0.5 g of Serdolite MB-1, stir for 10 min, and filter. To 48 ml of this solution add 0.25 g of CHAPS, 0.25 ml Pharmalyte pH 3–10 (40% w/v), and 100 mg of DTT and complete to 50 ml with deionized water.

Notes

a. For solubilization of the more hydrophobic proteins, *thiourea/urea lysis solution* [2M

thiourea, 5–7M urea, 2–4% (w/v) CHAPS, and/or sulfobetaine detergents, 1% DTT, 2% (v/v) carrier ampholytes] and *rehydration solution* consisting of a mixture of urea/thiourea [6M urea, 2M thiourea, 1% (w/v) CHAPS, 15 mM DTT, 0.5% (v/v) Pharmalyte pH 3–10] are recommended (Rabilloud *et al.*, 1997).

b. Lysis solution and IPG strip rehydration solution should always be prepared freshly. Alternatively, small aliquots (1 ml) can be stored at −80°C. Lysis and rehydration solution thawn once should not be refrozen again. Never heat urea solutions above 37°C! Otherwise, protein carbamylation may occur.

c. It is important that the urea solution is deionized with an ion exchanger prior to adding the other chemicals because urea in aqueous solution exists in equilibrium with ammonium cyanate, which can react with the NH_3^+ of protein side chains (e.g., lysine) and introduce charge artifacts (= additional spots) on the 2D gel.

1. Rehydration of IPG DryStrips

Prior to IEF, the dried IPG strips must be rehydrated to their original thickness of 0.5 mm. IPG DryStrips are rehydrated either with sample already dissolved in rehydration buffer ("sample in-gel rehydration") or with rehydration buffer only, followed by sample application by "cup loading." Sample in-gel rehydration is generally not recommended for samples containing very high molecular weight, very alkaline, and/or very hydrophobic proteins, as these are taken up into the gel with difficulties only. In these cases, cup loading is preferred.

a. Rehydration of IPG DryStrips Using the IPG DryStrip Reswelling Tray

1. For *sample in-gel* rehydration (Rabilloud *et al.*, 1994), directly solubilize a cell lysate

or tissue sample (1–10 mg protein/ml) in an appropriate quantity of rehydration buffer. For 180-mm-long and 3-mm-wide IPG dry strips, pipette 350 μl of this solution into the grooves of the IPG DryStrip reswelling tray (Fig. 9.1D). For longer or shorter IPG strips, the rehydration volume has to be adjusted accordingly.

2. Remove the protective covers from the surface of the IPG DryStrips and apply the IPG strips, gel side down, into the grooves without trapping air bubbles. Then cover the IPG strip, which must still be moveable and not stick to the tray, with silicone oil (which prevents drying out during reswelling) or IPG cover fluid and rehydrate the strips overnight at approximately 20°C. Higher temperatures (>37°C) hold the risk of protein carbamylation, whereas lower temperatures (<10°C) should be avoided to prevent urea crystallization on the IPG gel.

3. For *cup loading*, rehydrate the IPG dry strips overnight in rehydration buffer in the reswelling tray as described earlier, however without sample.

b. Sample In-Gel Rehydration of IPG DryStrips Using IPGphor Strip Holders

1. Solubilize proteins in sample solubilization buffer and dilute the extract (protein concentration ≈10 mg/ml) with IPG strip rehydration solution (dilution: 1 + 1 for micropreparative runs, and 1 + 19 for analytical runs, respectively) to a final volume of 350–400 μl for 180-mm-long IPG strips.

2. Apply the required number of IPGphor strip holders (Fig. 9.1E) onto the cooling plate/electrode contact area of the IPGphor.

3. Pipette 350 μl of sample-containing rehydration solution (for 180-mm-long IPG strips) into the strip holder base. For shorter (e.g., 110 mm) IPG strips in shorter strip holders, use correspondingly less liquid.

4. Peel off the protective cover sheets from the IPG strip and slowly lower the IPG strip (gel side down) onto the rehydration solution. Avoid trapping air bubbles. The IPG strip must still be moveable and not stick to the tray. Cover the IPG strips with 1–2 ml of cover fluid and apply the plastic cover. Pressure blocks on the underside of the cover assure that the IPG strip keeps in good contact with the electrodes as the gel swells.

2. IEF on the Multiphor Flat-Bed Electrophoresis Apparatus

1. After IPG DryStrip rehydration, rinse the rehydrated IPG strips with distilled water for a few seconds and then blot them between two sheets of moist filter paper to remove excess rehydration buffer in order to avoid urea crystallization on the gel surface, which is held responsible for prolonged IEF and "empty" vertical lanes in the 2D pattern.
2. Apply up to 40 IPG strips (*rehydrated with or without sample solution*) side by side and 1–2 mm apart onto the surface of the kerosene-wetted cooling plate (Fig. 9.1F) of the horizontal flat-bed electrophoresis apparatus (e.g., Multiphor, Amersham Biosciences). It is of utmost importance that the acidic ends of the IPG gel strips face towards the anode.
3. Soak electrode paper strips (cut from 1-mm-thick filter paper, e.g., MN 440, Macherey & Nagel) with deionized water, blot against filter paper to remove excess liquid, and place them on top of the aligned IPG gel strips at the cathodic and anodic ends. If samples have already been applied by sample in-gel rehydration, proceed as described in step 5).
4. In case of sample application by cup loading, apply the samples (20 μl; protein concentration 5–10 mg/ml) into silicon rubber frames (size: 2 × 5 mm²) or special

sample cups placed at either the anodic or the cathodic end of the IPG strips (see Fig. 9.1F). For analytical purposes, load 50–100 μg of protein onto a single, 180-mm-long IPG gel strip. For micropreparative purposes, up to several milligrams of protein may be applied with the help of a special strip tray, the IPG DryStrip kit (Fig. 9.1G) (*see* **Notes**).
5. Position the electrodes and press them down on top of the IEF electrode paper strips. Place the lid on the electrofocusing chamber, connect the cables to the power supply, and start IEF. For maximal sample entry, the initial voltage should be limited to 150 V for 60 min and then increased progressively until 3500 V is attained. Current and power settings should be limited to 0.05 mA and 0.2 W per IPG strip, respectively. The optimum focusing temperature is 20°C (Görg *et al.*, 1991). The time required for the run depends on several factors, including the type of sample, the amount of protein applied, the length of the IPG strips, and the pH gradient being used. Some typical running conditions are given in Table 9.3 (Görg *et al.*, 2000).
6. After termination of IEF, the IPG strips can be used immediately for the second dimension. Alternatively, strips can be stored between two sheets of plastic film at −70°C for several months.

Notes

a. When basic IPG gradients are used for the first dimension (e.g., IPG 6–10), horizontal streaking can often be observed at the basic end of 2D protein profiles. This problem may be resolved by applying an extra electrode strip soaked in 20 mM DTT on the surface of the IPG strip alongside the cathodic electrode strip (Görg *et al.*, 1995). This has the advantage that the DTT within the gel, which migrates towards

TABLE 9.3   Running Conditions Using the Multiphor[a]

Gel length	180 mm	
Temperature	20°C	
Current max.	0.05 mA per IPG strip	
Power max.	0.2 W per IPG strip	
Voltage max.	3500 V	

I. Analytical IEF

  Initial IEF

   Cup loading (20–50 μl)

    150 V, 1 h

    300 V, 1–3 h

    600 V, 1 h

   In-gel rehydration (350 μl)

    150 V, 1 h

    300 V, 1–3 h

  IEF to the steady state at 3500 V

1–1.5 pH units	**4 pH units**	**7 pH units**
e.g., IPG 5–6 24 h	IPG 4–8 10 h	IPG 3–10L 6 h
e.g., IPG 4–5.5 20 h	IPG 6–10 10 h	IPG 3–10NL 6 h
3 pH units	**5–6 pH units**	**8–9 pH units**
IPG 4–7 12 h	IPG 4–9 8 h	IPG 3–12 6 h
IPG 6–9 12 h	IPG 6–12 8 h	IPG 4–12 8 h

II. Extended Separation Distances (240 mm)

  IEF to the steady state at 3500 V

    IPG 3–12 8 h

    IPG 4–12 12 h

    IPG 5–6 40 h

III. Micropreparative IEF

  Initial IEF

   Cup loading (100 μl)

    50 V, 12–16 h

    300 V, 1 h

   In-gel rehydration (350 μl)

    50 V, 12–16 h

    300 V, 1 h

  IEF to the steady state at 3500 V

  Focusing time of analytical IEF plus approximately 50%

[a]From Görg *et al.* (2000).

the anode during IEF, is replenished by the DTT released from the strip at the cathode. Alternative approaches are (i) to use the noncharged reducing agent, tributyl phosphine, which does not migrate during IEF (Herbert, 1999), (ii) to substitute DTT in the rehydration buffer of the IPG strip by a disulfide such as hydroxyethyl disulfide (Olsson et al., 2002), or (iii) to apply high voltages (8000V) for short running times (Wildgruber et al., 2002).

b. After sample entry, the filter papers beneath the anode and cathode should be replaced with fresh ones. This is because salt contaminants have quickly moved through the gel and have now collected in the electrode papers. For high salt and/or protein concentrations, it is recommended to change the filter paper strips several times. In case of IEF with very alkaline, narrow range IPGs, such as IPG 10–12, this procedure should be repeated once an hour.

c. For micropreparative IEF or for running very alkaline IPGs exceeding pH 10, a special strip tray (IPG DryStrip kit) is recommeded that has a frame that fits on the cooling plate of the Multiphor (Fig. 9.1G). This tray is equipped with a corrugated plastic plate that contains grooves allowing easy alignment of the IPG strips. In addition, the tray is fitted with bars carrying the electrodes and a bar with sample cups allowing application of samples at any desired point on the gel surface. The advantages are that the cups can handle a larger quantity of sample solution (100μl can be applied at a time, but it is possible to apply a total of up to 200μl portion by portion, onto a single IPG gel), and the frame allows one to cover the IPG strips with a layer of silicone oil that protects the gel from the effects of the atmosphere during IEF. In case of very basic pH gradients exceeding pH 10 (e.g., IPG 3–12, 4–12, 6–12, 9–12, 10–12) or narrow-range

("zoom" gels) pH gradients with 1.0 or 1.5 pH units with extended running time (>24h), the IPG strips *must* be covered by a layer of silicone oil.

d. In case of cup loading, it is not recommended to apply proteins at (or proximate to) the pH area that corresponds with their pI in order to avoid that they are uncharged and poorly soluble and thus prone to precipitation at the sample application site.

e. The IEF run should be performed at 20°C, as at lower temperatures there is a risk of urea crystallization and at higher temperatures carbamylation might occur. Precise temperature control is also important because it has been found that alterations in the relative positions of some proteins on the final 2D patterns may happen (Görg, 1991).

3. IEF on the IPGphor

Using the IPGphor, sample in-gel rehydration and IEF can be carried out automatically according to the programmed settings, preferably overnight. Alternatively, the IPGphor can also be used with a cup-loading procedure, which allows the application of quantities up to 100μl. The instrument can accommodate up to 12 individual strip holders and incorporates Peltier cooling with precise temperature control and a programmable 8000-V power supply. The IPGphor saves about a day's worth of work by combining sample application and rehydration, as well as by starting the run at preprogrammed times, and by running the IEF at rather high voltages (Fig. 9.3).

a. IEF following In-Gel Rehydration of Sample

1. Apply the required number of strip holders onto the cooling plate/electrode contact area of the IPGphor. Pipette the appropriate amount of sample-containing rehydration solution into the strip holders, lower the

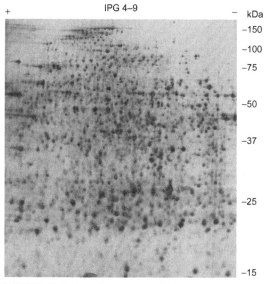

IPG 4–9

kDa
−150
−100
−75
−50
−37
−25
−15

FIGURE 9.3 IPG-Dalt of mouse liver proteins. First dimension: IEF with IPG 4–9 run on the IPGphor. Separation distance: 24 cm. Sample application by cup loading near the anode. Second dimension: vertical SDS–PAGE, 13% T. Silver stain.

IPG dry strips gel side down into the rehydration solution, and overlay with cover fluid (Fig. 9.1E).

2. Program the instrument (Fig. 9.1H) (desired rehydration time, volthours, voltage gradient). Apply low voltage (30–50 V) during IPG strip rehydration for improved entry of high M_r proteins into the polyacrylamide gel matrix (Görg *et al.*, 1999, 2000).
3. After the IPG gel strips have been rehydrated (which requires 6 h at least, but typically overnight), IEF starts according to the programmed parameters (*see* Table 9.4).
4. After completion of IEF, store those IPG gel strips that are not used immediately for a second-dimension run between two sheets of plastic film at −70°C (Fig. 9.1J).

b. IEF after Sample Application by Cup Loading

When protein separation is performed in basic pH ranges, much better separations are obtained by applying sample *via* cup loading on separately rehydrated IPG strips than by sample in-gel rehydration. Sample cup loading is accomplished with a special cup-loading ("universal") IPGphor strip holder.

1. Rehydrate IPG dry strips with rehydration buffer, without sample, in an IPG DryStrip reswelling tray (Fig. 9.1D) as described earlier for the Multiphor instrument.
2. Apply the required number of the cup-loading IPGphor strip holders onto the cooling plate/electrode contact area of the IPGphor instrument (Fig. 9.1H), and apply the rehydrated IPG gel strips into the cup-loading strip holders, gel side upwards and acidic ends facing towards the anode.
3. Moisten two filter paper electrode pads (size: $4 \times 10 \, mm^2$) with deionized water, remove excess liquid by blotting with a filter paper, and apply the moistened filter paper pads at the anodic and cathodic ends of the IPG strip between the IPG gel and the electrodes. This is particularly important when analyzing samples containing high salt amounts.
4. Position the movable electrodes above the electrode filter paper pads and gently press the electrodes on the filter papers. Filter papers should be replaced after 2–3 h.
5. Position the movable sample cup near either the anode or the cathode and gently press the sample cup onto the surface of the IPG gel strip. Overlay the IPG strip with 5 ml of cover fluid (do not use silicone oil or kerosene!). If cover fluid leaks into the sample cup, rearrange the cup and remove the cover fluid from the cup with the help of tissue paper. Check again for leakage before pipetting the sample (20–100 μl) into the cup.
6. Program the instrument (desired volt hours, voltage gradient, temperature, etc.) and run IEF according to the settings recommended in Table 9.4.

TABLE 9.4   IPGphor Running Conditions[a]

Gel length	180 mm		
Temperature	20°C		
Current max.	0.05 mA per IPG strip		
Voltage max.	8000 V		
I. Analytical IEF			
Reswelling	30 V, 12–16 h		
Initial IEF:	200 V, 1 h		
	500 V, 1 h		
	1000 V, 1 h		
IEF to the steady state:	Gradient from 1000 to 8000 V within 30 min 8000 V to the steady state, depending on the pH used used		
1–1.5 pH units	**4 pH units**		**7 pH units**
e.g., IPG 5–6 8 h	IPG 4–8 4 h		IPG 3–10L 3 h
e.g., IPG 4–5.5 8 h			IPG 3–10NL 3 h
3 pH units	**5–6 pH units**		**8–9 pH units**
IPG 4–7 4 h	IPG 4–9 4 h		IPG 3–12 3 h
			IPG 4–12 3 h
II. Micropreparative IEF			
Reswelling	30 V, 12–16 h		
IEF to the steady state	Focusing time of analytical IEF + additional 50% (approximate)		

[a]From Görg *et al.* (2000).

Note

When IPG strips with separation distances <11 cm are used, voltage should be limited to 5000 V.

C. Second Dimension: Horizontal SDS Gel Electrophoresis

Solutions

1. *Tris–HCl buffer*: 1.5M Tris–HCl, pH 8.8, 0.4% (w/v) SDS. To make 250 ml, dissolve 45.5 g of Trizma base and 1 g of SDS in about 200 ml of deionized water. Adjust the pH of the solution with 4N HCl and fill up to 250 ml with deionized water. Add 25 mg of sodium azide and filter. The buffer can be stored at 4°C up to 2 weeks.

2. *Equilibration buffer*: 6M urea, 30% (w/v) glycerol, 2% (w/v) SDS in 0.05M Tris–HCl buffer, pH 8.8. To make 500 ml, add 180 g of urea, 150 g of glycerol, 10 g of SDS, and 16.7 ml of gel buffer. Dissolve in deionized water and fill up to 500 ml. The buffer can be stored at room temperature up to 2 weeks.

3. *Bromphenol blue solution*: 0.25% (w/v) of bromphenol blue in stacking gel buffer. To make 10 ml, dissolve 25 mg of bromphenol blue in 10 ml of gel buffer. Store at 4°C.

Steps

For the second dimension, an SDS pore gradient gel (0.5 mm thick on GelBond PAGfilm) is applied on the cooling plate of the horizontal electrophoresis unit. Then the equilibrated IPG gel strip is simply placed gel side down onto the surface of the SDS gel alongside the cathodic electrode buffer strip without any embedding procedure. Horizontal setups are particularly suited for the use of ready-made gels on film supports. This section describes the procedure for ready-made ExcelGels. For casting and running of laboratory-made horizontal SDS gels, *see* Görg and Weiss (2000).

1. Equilibration of IPG Gel Strips

The IPG gel strips are equilibrated twice, 15 min each (*see* Table 9.5). During the second equilibration, 260 mM iodoacetamide is added to the equilibration buffer in order to remove excess DTT (responsible for the "point streaking" in silver-stained patterns) and to alkylate sulfhydryl groups for subsequent MALDI-MS.

1. Dissolve 100 mg of DTT in 10 ml of equilibration buffer (= equilibration buffer I). Take out the focused IPG gel strips from the freezer and place them into individual test tubes (200 mm long, 20 mm i.d.). Add 10 ml of equilibration buffer I and 50 µl of the bromphenol blue solution. Seal the test tubes with Parafilm, rock them for 15 min on a shaker (Fig. 9.1K), and then pour off the equilibration buffer.
2. Dissolve 480 mg of iodoacetamide in 10 ml of equilibration buffer (= equilibration buffer II). Add equilibration buffer II and 50 µl of bromophenol blue solution to the test tube as just described and equilibrate for another 15 min on a rocker.
3. After the second equilibration, place the IPG gel strip on a piece of moist filter paper to remove excess equilibration buffer. The

strip should be turned up at one edge for a few minutes to drain off excess equilibration buffer.

2. Horizontal SDS–PAGE with Ready-Made Gels

Ready-made SDS gels (ExcelGel, 250 × 200 × 0.5 mm³, on plastic backing; Amersham Biosciences) in combination with polyacrylamide buffer strips are used.

1. Equilibrate the IPG gel strips as described earlier (Table 9.5).
2. While the strips are being equilibrated, begin the assembly of the SDS ExcelGel for the second dimension: Remove the ExcelGel from its foil package. Pipette 2–3 ml of kerosene on the cooling plate of the horizontal electrophoresis unit (15°C). Remove the protective cover from the top of the ExcelGel and place the gel on the cooling plate, with cut-off edge toward the anode. Avoid trapping air bubbles between the gel and the cooling block.
3. Peel back the protective foil of the cathodic SDS buffer strip. Wet your gloves with a few drops of deionized water and place the buffer strip on the cathodic end of the gel. Avoid trapping air bubbles between gel surface and buffer strip.
4. Repeat this procedure with the anodic buffer strip.
5. Place the equilibrated IPG gel strips gel side down on the surface of the ExcelGel, 2–3 mm apart from the cathodic buffer strip (Fig. 9.1 L).
6. Press gently on top of the IPG gel strips with forceps to remove any trapped air bubbles.
7. Align the electrodes with the buffer strips and lower the electrode holder carefully onto the buffer strips.
8. Start SDS–PAGE at 100 V for about 60 min with a limit of 20 mA. When the bromphenol blue tracking dye has moved 4–5 mm from the IPG gel strip, interrupt the run, remove

TABLE 9.5 Equilibration of IPG Strips

Reagent	Purpose
50 mM Tris–HCl, pH 8.8	
+2% SDS	Improved protein transfer onto SDS gel
+6M urea	Improved protein transfer onto SDS gel
+30% glycerol	Improved protein transfer onto SDS gel
+1% DTT	Complete reduction of disulphide bonds
+4.8% iodoacetamide	Removal of excess DTT (point streaking) Alkylation of SH groups for MALDI-MS

	DTT	Iodoacetamide
First step 15min	+	−
Second step 15min	−	+

the IPG gel strip, and move the cathodic buffer strip forward so that it just covers the former contact area of the IPG gel strip. Readjust the electrodes and continue with electrophoresis at 800 V and 35 mA for about 180min until the bromphenol blue dye front has reached the anodic buffer strip.

9. Proceed with protein staining with Coomassie blue, silver nitrate, or fluorescent dyes or with blotting.

D. Second Dimension: Multiple Vertical SDS–Polyacrylamide Gel Electrophoresis

First-dimension IEF and the equilibration step are performed as described previously, no matter whether the second dimension is run horizontally or vertically. After equilibration, the IPG gel strip is placed on top of the vertical SDS gel, with or without embedding in agarose.

Vertical SDS–PAGE is performed as described by Görg & Weiss (2000). A stacking gel is usually not necessary. For multiple runs, the Ettan DALT II (Amersham Biosciences) vertical electrophoresis apparatus is recommended because this system allows a large batch of SDS slab gels (up to 12) to be run under identical conditions. Ready-made gels on plastic backing are also available (Amersham Biosciences).

Solutions

1. *Gel buffer*: 1.5M Tris–HCl, pH 8.6, 0.4% (w/v) SDS (Laemmli, 1970).

   To make 500 ml, dissolve 90.85 g of Trizma base and 2.0 g of SDS in about 400 ml of deionized water. Adjust to pH 8.6 with 4N HCl and fill up to 500 ml with deionized water. Add 50 mg of sodium azide and filter. The buffer can be stored at 4°C up to 2 weeks.

2. *Acrylamide/bisacrylamide solution (30.8% T, 2.6% C)*: 30% (w/v) acrylamide, 0.8% (w/v) methylenebisacrylamide in deionized water. To make 1000 ml, dissolve 300.0 g of acrylamide and 8.0 g of methylenebisacrylamide in deionized water and fill up to 1000 ml. Add 1 g of Serdolit MB-1, stir for 10min, and filter. The solution can be stored up to 2 weeks in a refrigerator.

3. *Ammonium persulfate solution*: 10% (w/v) of ammonium persulfate in deionized water. To prepare 10 ml of the solution, dissolve 1.0 g of ammonium persulfate in 10 ml of

deionized water. This solution should be prepared freshly just before use.

4. *Displacing solution*: 50% (v/v) glycerol in deionized water, 0.01% (w/v) bromphenol blue. To make 500 ml, mix 250 ml of glycerol (100%) with 250 ml of deionized water, add 50 mg of bromphenol blue, and stir for a few minutes.

5. *Overlay buffer*: Buffer-saturated 2-butanol. To make 30 ml, mix 20 ml of gel buffer with 30 ml of 2-butanol, wait for a few minutes, and pipette off the butanol layer.

6. *Electrode buffer*: To make 5 liter of electrode buffer stock solution, dissolve 58.0 g of Trizma base, 299.6 g of glycine, and 19.9 g of SDS in deionized water and complete to 5.0 liter.

7. *Equilibration buffer*: 6M urea, 30% (w/v) glycerol, 2% (w/v) SDS in 0.05M Tris–HCl buffer, pH 8.6. To make 500 ml, add 180 g of urea, 150 g of glycerol, 10 g of SDS, and 16.7 ml of gel buffer. Dissolve in deionized water and fill up to 500 ml. The buffer can be stored at room temperature up to 2 weeks.

8. *Agarose solution*: Suspend 0.5% (w/v) agarose in electrode buffer and melt it in a boiling water bath or in a microwave oven.

1. Casting of Vertical SDS Gels

Steps

1. The polymerization cassettes (200 × 250 mm^2) are made in the shape of books consisting of two glass plates connected by a hinge strip with two 1.0-mm-thick spacers in between them.

2. Stack 14 cassettes vertically into the gel-casting box with the hinge strips to the right, interspersed with plastic sheets (e.g., 0.05-mm-thick polyester sheets).

3. Put the front plate of the casting box in place and screw on the nuts (hand tight).

4. Connect a polyethylene tube (*i.d.* 5 mm) to a funnel held in a ring stand at a level

of about 30 cm above the top of the casting box. Place the other end of the tube in the grommet in the casting box side chamber.

5. Fill the side chamber with 100 ml of displacing solution.

6. Immediately before gel casting, add TEMED and ammonium persulfate solutions to the gel solution (Table 9.6). To cast the gels, pour the gel solution (about 830 ml) into the funnel. Avoid introduction of any air bubbles into the tube!

7. When pouring is complete, remove the tube from the side chamber grommet so that the level of the displacing solution in the side chamber falls.

8. Very carefully pipette about 1 ml of overlay buffer onto the top of each gel in order to obtain a smooth, flat gel top surface.

9. Allow the gels to polymerize for at least 3 h (better: overnight) at approximately 20°C.

10. Remove the front of the casting box and carefully unload the gel cassettes from the box using a razor blade to separate the cassettes. Remove the polyester sheets that had been placed between the individual cassettes.

11. Wash each cassette with water to remove any acrylamide adhered to the outer surface and drain excess liquid off the top surface. Discard unsatisfactory gels; in general the gels opposite to the front and rear plate of the gel casting box (due to the uneven thickness of these gels).

12. Gels not needed at the moment can be wrapped in plastic wrap and stored in a refrigerator (4°C) for 1–2 days.

2. Multiple Vertical SDS–PAGE Using Ettan Dalt II

Steps

1. Pour 1875 ml of electrode buffer stock solution and 5625 ml of deionized water in

TABLE 9.6   Recipes for Casting Vertical SDS Gels

	7.5% T 2.6% C	10% T 2.6% C	12.5% T 2.6% C	15% T 2.6% C
Acrylamide/ bisacrylamide (30.8%T, 2.6%C)	202 ml	270 ml	337 ml	404 ml
gel buffer	208 ml	208 ml	208 ml	208 ml
Glycerol (100%)	41 g	41 g	41 g	41 g
Deionized water	383 ml	315 ml	248 ml	181 ml
TEMED (100%)	42 µl	42 µl	42 µl	42 µl
Ammonium persulfate (10%)	6.0 ml	6.0 ml	6.0 ml	6.0 ml
Final volume	830 ml	830 ml	830 ml	830 ml

the lower electrophoresis buffer tank and turn on cooling (20°C).

2. Support the SDS gel cassettes in a vertical position to facilitate application of the first-dimension IPG gel strips.

3. Equilibrate the IPG gel strip as described earlier for horizontal SDS gels. Immerse it in SDS electrode buffer for a few seconds to facilitate insertion of the IPG strip between the two glass plates of the gel cassette.

4. Place the IPG gel strip on top of an SDS gel and overlay it with 2 ml of hot agarose solution (75°C). Carefully press the IPG strip with a spatula onto the surface of the SDS gel to achieve complete contact (Fig. 9.1M). If it is desired to coelectrophorese M_r marker proteins, soak a piece of filter paper (2 × 4 mm^2) with 5 µl of SDS marker proteins dissolved in electrophoresis buffer, let it dry (!), and apply it to the left or right of the IPG strip. Allow the agarose to solidify for at least 5 min and then place the slab gel into the electrophoresis apparatus (see later). Repeat this procedure for the remaining IPG strips. Note: Embedding in agarose is not absolutely necessary, but it ensures better

contact between the IPG gel strip and the top of the SDS gel.

5. Wet the gel cassettes by dipping then into electrode buffer for a few seconds and then insert them in the electrophoresis apparatus. Pour 1250 ml of electrophoresis buffer stock solution and 1250 ml of deionized water in the upper electrophoresis tank, mix, and start SDS electrophoresis.

6. Run the SDS–PAGE gels with 50 mA (100 V maximum setting) for about 2 h. Then continue with 175 mA (200 V maximum setting) for about 16 h. Note: In contrast to the procedure of horizontal SDS–PAGE it is not necessary to remove the IPG gel strips from the surface of the vertical SDS gel once the proteins have migrated out of the IPG gel strip.

7. Terminate the run when the bromphenol blue tracking dye has migrated off the lower end of the gel.

8. Open the cassettes carefully with a plastic spatula. Use the spatula to separate the agarose overlay from the polyacrylamide gel. Peel the gel from the glass plate carefully, lifting it by the lower edge, and

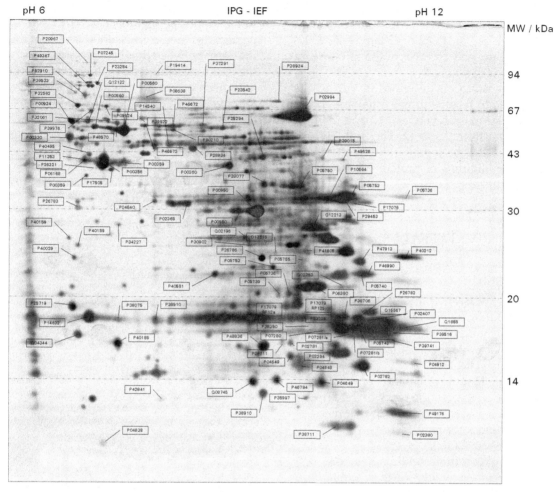

FIGURE 9.4 Web-based 2D reference map of alkaline *Saccharomyces cerevisiae* (strain FY1679) proteins showing 106 mapped and identified spots annotated by SWISS-PROT accession numbers. First dimension: IPG 6–12. Second dimension: SDS–PAGE, 15% T. Spot identification by MALDI-TOF-MS (Ettan z^2 MALDI-TOF, Amersham Biosciences) (with permission from Wildgruber *et al.*, 2002) (http://www.wzw.tum.de/proteomik).

place it in a box of stain solution or transfer buffer, respectively. Then continue with protein staining or blotting.

IV. COMMENTS

Since the early 1990s, IPG-Dalt has constantly been refined to meet the requirements of proteome analysis. In particular, (i) the development of basic IPGs up to pH 12 has facilitated the analysis of very alkaline proteins (Görg *et al.*, 1997, 1998, 1999; Wildgruber *et al.* 2002); (ii) the introduction of overlapping narrow-range IPGs to stretch the first dimension permits increased resolution (Δ pI = 0.01/cm) (Görg *et al.*, 1988), as well as higher loading capacity for the analysis of less abundant proteins

(Wildgruber *et al.*, 2000); and (iii) the availability of ready-made IPG strips and integrated devices such as the IPGphor (Islam *et al.*, 1998) have paved the way towards automation (Görg *et al.*, 1999, 2000) (Fig. 9.4).

Using the protocols described in this article, it is possible to resolve proteins having apparent molecular masses between 10 and 200 kDa and pI values from 2.5 to 12. The protein-loading capacity for micropreparative runs is up to several milligrams of protein if narrow pH range IPGs (0.5–1 pH units wide) are used.

V. PITFALLS

1. Blot the rehydrated IPG gel strips to remove excess rehydration solution; as urea crystallization on the surface of the IPG gel strips might occur, which will disturb IEF patterns or prolong focusing time to reach the steady state.
2. Make sure that the orientation of the IPG gel strips on the cooling block of the IEF chamber is correct (acidic end towards anode). Check the temperature of the cooling block (20°C).
3. The sample solution should not be too concentrated to avoid protein precipitation at the sample application point. If you are in doubt, dilute the sample with lysis solution and apply a larger volume (>20 μl) instead.
4. For improved sample entry, start IEF at low voltage (150 V for 30 min, followed by 300 V for 60 min). Extend these times when the sample contains high amounts of salts or when large sample amounts (>50 μl) are applied *via* cup loading.
5. Sample in-gel rehydration: Use IPGphor and apply low voltage (30–50 V) during IPG strip rehydration for improved entry of high M_r proteins into the polyacrylamide gel matrix.

6. When analyzing samples containing high salt amounts, apply moistened filter paper pads at the anodic and cathodic ends of the IPG strip between the IPG gel surface and the electrodes and replace these filter papers with fresh ones after several hours.
7. Remove the IPG gel strip from the surface of the *horizontal* SDS–PAGE gel as soon as the bromphenol blue dye front has migrated 4–5 mm off the IPG gel strip. Then move the cathodic electrode wick (or buffer strip) forward so that it overlaps the area the IPG gel strip once covered.

References

Bjellqvist, B., Ek, K., Righetti, P. G., Gianazza, E., Görg, A., Westermeier, R., and Postel, W. (1982). Isoelectric focusing in immobilized pH gradients: Principle, methodology and some applications. *J. Biochem. Biophys. Methods* **6**, 317–339.

Blomberg, A., Blomberg, L., Norbeck, J., Fey, S., Mose Larsen, P., Larsen, M., Roepstorff, P., Degand, H., Boutry, M., Posch, A., and Görg, A. (1995). Interlaboratory reproducibility of yeast protein patterns analyzed by immobilized pH gradient two-dimensional gel electrophoresis. *Electrophoresis* **16**, 1935–1945.

Bossi, A., Righetti, P. G., Vecchio, G., and Severinsen, S. (1994). Focusing of alkaline proteases in pH 10–12 immobilized gradients. *Electrophoresis* **15**, 1535–1540.

Corbett, J. M., Dunn, M. J., Posch, A., and Gorg, A. (1994). Positional reproducibility of protein spots in two-dimensional polyacrylamide gel electrophoresis using immobilised pH gradient isoelectric focusing in the first dimension: An interlaboratory comparison. *Electrophoresis* **15**, 1205–1211.

Görg, A., Boguth, G., Obermaier, C., Posch, A., and Weiss, W. (1995). Two-dimensional polyacrylamide gel electrophoresis with immobilized pH gradients in the first dimension (IPG-Dalt): The state of the art and the controversy of vertical versus horizontal systems. *Electrophoresis* **16**, 1079–1086.

Görg, A., Boguth, G., Obermaier, C., and Weiss, W. (1998). Two-dimensional electrophoresis of proteins in an immobilized pH 4–12 gradient. *Electrophoresis* **19**, 1516–1519.

Görg, A., Obermaier, C., Boguth, G., Csordas, A., Diaz, J. J., and Madjar, J. J. (1997). Very alkaline immobilized pH gradients for two-dimensional electrophoresis of ribosomal and nuclear proteins. *Electrophoresis* **18**, 328–337.

Görg, A., Obermaier, C., Boguth, G., Harder, A., Scheibe, B., Wildgruber, R., and Weiss, W. (2000). The current state of two-dimensional electrophoresis with immobilized pH gradients. *Electrophoresis* **21**, 1037–1053.

Görg, A., Obermaier, C., Boguth, G., and Weiss, W. (1999). Recent developments in 2-D gel electrophoresis with immobilized pH gradients: Wide pH gradients up to pH 12, longer separation distances and simplified procedures. *Electrophoresis* **20**, 712–717.

Görg, A., Postel, W., Friedrich, C., Kuick, R., Strahler, J. R., and Hanash, S. M. (1991). *Electrophoresis* **12**, 653–658.

Görg, A., Postel, W., and Günther, S. (1988). The current state of two-dimensional electrophoresis with immobilized pH gradients. *Electrophoresis* **9**, 531–546.

Görg, A., and Weiss. (2000). 2D electrophoresis with immobilized pH gradients. *In "Proteome Research: Two Dimensional Electrophoresis and Identification Methods"* (T. Rabilloud, ed.), pp. 57–106. Springer, Berlin.

Herbert, B. (1999). Advances in protein solubilisation for two-dimensional electrophoresis. *Electrophoresis* **20**, 660–663.

Islam, R., Ko, C., and Landers, T. (1998). A new approach to rapid immobilised pH gradient IEF for 2-D electrophoresis. *Sci. Tools* **3**, 14–15.

Laemmli, U. K. (1970). Cleavage of structural proteins during the assembly of the head of bacteriophage T4. *Nature* **227**, 680–685.

O'Farrell, P. H. (1975). High resolution two-dimensional electrophoresis of proteins. *J. Biol. Chem.* **250**, 4007–4021.

O'Farrell, P. Z., Goodman, H. M., and O'Farrell, P. H. (1977). High resolution two-dimensional electrophoresis of basic as well as acidic proteins. *Cell* **12**, 1133–1142.

Olsson, I., Larsson, K., Palmgren, R., and Bjellqvist, B. (2002). Organic disulfides as a means to generate streak-free two-dimensional maps with narrow range IPG strips as first dimension. *Proteomics* **2**, 1630–1632.

Rabilloud, T., Adessi, C., Giraudel, A., and Lunardi, J. (1997). Improvement of the solubilization of proteins in two-dimensional electrophoresis with immobilized pH gradients. *Electrophoresis* **18**, 307–316.

Rabilloud, T., Valette, C., and Lawrence, J. J. (1994). Sample application by in-gel rehydration improves the resolution of two-dimensional electrophoresis with immobilized pH-gradients in the first dimension. *Electrophoresis* **15**, 1552–1558.

Righetti, P. G. (1990). "Immobilized pH Gradients: Theory and Methodology." Elsevier, Amsterdam.

Wildgruber, R., Harder, A., Obermaier, C., Boguth, G., Weiss, W., Larsen, P. M., Fey, S., and Görg, A. (2000). Towards higher resolution: Two-dimensional electrophoresis of *Saccharomyces cerevisiae* proteins using overlapping narrow immobilized pH gradients. *Electrophoresis* **21**, 2610–2616.

Wildgruber, R., Reil, G., Drews, O., Parlar, H., and Görg, A. (2002). Web-based two-dimensional database of *Saccharomyces cerevisiae* proteins using immobilized pH gradients from pH 6 to pH 12 and matrix-assisted laser desorption/ionization-time of flight mass spectrometry. *Proteomics* **2**, 727–732.

10

Two-Dimensional Difference Gel Electrophoresis: Application for the Analysis of Differential Protein Expression in Multiple Biological Samples

John F. Timms

I. INTRODUCTION

Fluorescence two-dimensional difference gel electrophoresis (2D-DIGE) is a recently developed 2D gel-based proteomics technique that provides a sensitive, rapid, and quantitative analysis of differential protein expression between biological samples. Developed by Minden and co-workers (Unlu *et al.*, 1997), the technique utilizes charge- and mass-matched chemical derivatives of spectrally distinct fluorescent cyanine dyes (Cy3 and Cy5), which are used to covalently label lysine residues in different samples prior to mixing and separation on the same 2D gel. Resolved, labeled proteins are then detected at appropriate excitation and emission wavelengths using a multiwavelength fluorescence detection device and the signals compared. As well as reducing the number of gels that need to be run, differential labeling and mixing mean that samples are subjected to the same handling procedures and microenvironments during 2D separation, raising the confidence with which protein changes can be detected and quantified. Because fluorescence detection also provides a superior linear dynamic range of detection and sensitivity to other methods (Patton, 2000), this technology is suited to the analysis of biological samples with their large dynamic ranges of protein abundance.

The 2D-DIGE methodology is now commercialized as the Ettan DIGE proteomics system (GE Healthcare), with a third dye (Cy2) introduced, allowing simultaneous analysis of three samples on a single gel. In expression profiling experiments, one dye is used to label an internal standard to be run on all gels against pairs of test samples labeled with the other two dyes. This allows the direct comparison of ratios of expression across multiple samples and gels, improving the ability to distinguish biological variation from gel-to-gel variation. Because the labeling strategy employed is compatible with downstream identification of gel spots by mass spectrometry (MS) (Tonge *et al.*, 2001; Gharbi *et al.*, 2002), 2D-DIGE is of particular use as a reproducible, high-throughput proteomic technology. This article describes the necessary materials and instrumentation, experimental design, and work flow for the preparation and labeling of samples for 2D-DIGE analysis. Image capture, analysis, and spot picking for MS identification are also described.

II. MATERIALS AND INSTRUMENTATION

The CyDye DIGE fluors *N*-hydroxysuccinimidyl (NHS) esters of Cy2 (Prod. Code RPK0272), Cy3 (Prod. No. RPK0273), and Cy5 (Prod. No. RPK0275) are from GE Healthcare. NHS-Cy3 and NHS-Cy5 are also synthesized in-house following the original published protocol (Unlu *et al.*, 1997), with modifications (unpublished data). Anhydrous 99.8% *N,N*-dimethylformamide (DMF, Cat No. 22,705-6) is from Aldrich, urea (Cat. No. U-0631), dithiothreitol (DTT, Cat. No. D-9163), thiourea (Cat. No. T-8656), and l-lysine (Cat. No. L-5626) are from Sigma. 1.5M Tris solution, pH 8.8 (Prod. No. 20-79000-10), 1M Tris solution, pH 6.8 (Prod. No. 20-7901-10), and 10× Tris–glycine SDS electrophoresis buffer (Prod. No. 20-64) are from Severn Biotech Ltd. Phosphate-buffered saline (PBS, Cat No. 14190-094) is from GIBCO. Nonidet P-40 (NP-40, Prod. No. 56009D2L), bromphenol blue (Prod. No. 200173J), methanol (Prod. No. 10158BG), and glacial acetic acid (Prod. No. 27013BV) are from VWR. 3-[(3-Cholamidopropyl)-dimethylammonio]-1-propanesulfonate (CHAPS, Cat No. B2006) and sodium dodecyl sulphate (SDS, Cat. No. B2008) are from Melford Laboratories Ltd. Coomassie protein assay reagent (Prod. No. 1856210) is from Pierce, S&S weighing papers (Cat. No. Z13411-2) are from Aldrich, and low-gelling temperature agarose (Cat. No. 05075) is from Fluka.

Ampholines (pH 3.5–10, Cat. No. 80-1125-87), Pharmalyte (pH 3–10, Cat. No. 17-0456-01), Immobiline DryStrip reswelling tray (Cat. No. 80-6465-32), Immobiline DryStrip IEF gels (IPG strips, Cat. No. 17-6002-45), Multiphor II electrophoresis unit (Cat. No. 18-1018-06), Ettan DALT low fluorescence glass plates (27 × 22 cm, Cat. No. 80-6475-58), Plus One Repel Silane (Cat. No. 17-1332-01), Plus One Bind Silane (Cat. No. 17-1330-01), reference markers (Cat. No. 18-1143-34), Ettan DALT 12-gel caster, separation unit, and power supply (Cat. Nos. 80-6467-22 and 80-6466-27), Typhoon 9400 imager (Cat. No. 60-0038-54), Ettan Spot Picker (Cat. No. 18-1145-28), and DeCyder differential analysis software (Cat. No. 56-3202-70) are from GE Healthcare. SYPRO Ruby protein gel stain (Cat. No. S-12000) is from Molecular Probes, and colloidal Coomassie brilliant blue G-250 tablets (CBB, Cat No. K26283182) are from Merck. The Immobilon-P polyvinylidene fluoride transfer membrane (PVDF, Cat. No. IPVH00010) is from Millipore.

III. PROCEDURES

A. Experimental Design

The following steps are guidelines for the design of 2D-DIGE expression profiling experiments. Several experiments are outlined to illustrate how samples can be fluorescently labeled and mixed for 2D gel separation and image analysis so that statistically meaningful data can be acquired. The throughput of the technique is, however, dependent upon the 2D gel running, image capture, and analysis capabilities of the laboratory. We routinely run 12-gel experiments providing accurate differential expression data for 24 samples, including an internal standard run on all gels for accurate spot matching and quantitation. The same gels are then poststained and proteins of interest are picked for MS identification. 2D-DIGE is applicable for the analysis of total cell lysates and complex protein mixtures from cultured cells, whole tissues, sorted or fractionated cells, whole organisms (*E. coli*, *S. pombe*, *C. elegans*, etc.), cellular subfractions (membrane, nuclear, cytoplasmic, etc.), or affinity-purified protein fractions.

Steps

1. Three spectrally distinct fluorescent CyDyes (Cy2, Cy3, and Cy5) can be used for differential labeling of protein samples.

In the simplest expression profiling experiment, two individual samples are labeled with two different dyes, mixed, and resolved on a single 2D gel (see Section III,B). Because the same protein isoforms in each sample will comigrate, one can accurately measure differential expression as the ratio of the fluorescence intensities of comigrating spots. Thus, the problem of gel-to-gel variation is avoided. This type of single gel experiment is useful where only limited sample quantities are available, e.g., laser capture dissection-procured normal and cancer cells (Zhou *et al.*, 2002).

2. To obtain statistically meaningful expression changes, at least triplicate samples should be labeled and analysed on separate gels. These may be biological replicates, e.g., three separately grown cell cultures or tissue from three individual animals, or may be experimental replicates, where three aliquots of the same sample are compared across different gels. Differential expression can then be taken as an average fold change (e.g., the average spot intensity ratio between differentially labeled spots matching across all three gels) with statistical confidence provided by applying a *t* test. Of note we have found considerable interanimal variation in liver lysates from mice (unpublished data), and it is therefore advisable to analyse samples from at least five individual animals for each treatment or condition to provide statistically meaningful data.

3. 2D-DIGE analysis is further improved by running a Cy-labeled internal standard on all gels against pairs of test samples, labeled with the other two CyDyes. This increases the ability to distinguish biological variation from gel-to-gel variation by increasing the confidence with which spots can be matched across gels and by allowing the direct comparison of expression ratios across samples. An equal pool of all samples (including biological replicates) is best

employed as an internal standard as it will contain proteins present in all samples. It is created simply by mixing and labeling equal amounts of protein from every sample and should provide sufficient material for the number of gels to be run. Equal amounts of standard and test samples are then resolved on each gel.

4. In an experiment comparing 100 μg of protein from cell A and cell B, lysates from triplicate cultures are prepared and protein concentrations determined (six samples). For simplicity, samples are adjusted to the same protein concentration by adding lysis buffer. Then 100 μg of each is labeled with Cy3 or Cy5 as shown in Table 10.1. A pool consisting of a mixture of 50 μg of each of all the replicate samples (300 μg total) is labeled with Cy2. Following labeling, the samples are mixed appropriately for separation on three 2D gels as shown in Table 10.1. This scheme controls for dye bias, although labeling combinations are interchangeable so long as each gel is loaded with samples labeled with distinct dyes. This experiment generates nine images for matching, cross-comparison, and statistical analysis in the biological variance analysis (BVA) module of DeCyder software.

5. For complex comparisons we recommend running 12 gels at once. This allows imaging within a day and fits with our downstream laboratory work flow. Although more gels can be run in an individual experiment, consistency may be compromised by running gels at different times or on different electrophoresis units, or may be impractical depending on man power, resources, and automation. Still, it is possible to compare 24 different samples in a single 12-gel 2D-DIGE experiment. For example, our laboratory was able to analyse lysates from two cell lines subjected to growth factor stimulation for four different time periods (8 conditions) using triplicate cultures (Gharbi *et al.*, 2002). The 24 lysates generated were labeled with Cy3 or Cy5 and run in pairs against the internal standard (a pool of all samples) labeled with Cy2. This generated 36 images for DeCyder BVA analysis, with an image acquisition time of ~8 h using a Typhoon 9400 imager.

B. Preparation of CyDye-Labeled Samples for 2D Electrophoresis

The following procedure can be applied for the preparation and labeling of multiple samples for 2D-DIGE comparative protein expression analysis. The procedure is based on the Ettan DIGE System (GE Healthcare), with some modifications. The protocol is designed to generate differentially labeled samples that are compatible with all systems for 2D gel electrophoretic separation. The principles and applications of 2D gel electrophoresis are discussed in more detail elsewhere in this volume. For brevity, the sample preparation is outlined for 24-cm pH 3–10 NL-immobilized pH gradient (IPG) gels. Accordingly, final volumes, protein loads, and IPG buffers may differ depending on the size and pH range of the first-dimension gels.

Solutions

1. *CyDyes (NHS-Cy2, -Cy3, -Cy5)*: From lyophilized powder (stored at −20°C),

TABLE 10.1 Example of Differential Labeling, Mixing, and Loading for Statistical Comparison of Protein Expression in Two Cell Lines Using 2D-DIGE

	Cy3	Cy5	Cy2
Gel 1	100 μg A, replicate 1	100 μg B, replicate 1	100 μg pool
Gel 2	100 μg B, replicate 2	100 μg A, replicate 2	100 μg pool
Gel 3	100 μg A, replicate 3	100 μg B, replicate 3	100 μg pool

reconstitute to 1 mM stock by dissolving in the appropriate volume of anhydrous DMF. Keep stock solutions in the dark at −20°C; they are stable for up to 4 months.

2. *Lysis buffer*: 8M urea, 2M thiourea, 4% (w/v) CHAPS, 0.5% NP-40 (w/v), 10 mM Tris–HCl, pH 8.3. To make 100 ml, dissolve 48 g of urea and 15.2 g of thiourea in 50 ml of distilled H$_2$O. Add 4 g CHAPS, 0.5 g NP-40, and 0.67 ml of 1.5M Tris, pH 8.8, solution. This should give a final pH of 8.3. Make up to final volume, aliquot, and store at −20°C. Do not heat.

3. *40% (w/v) CHAPS*: To make 50 ml, dissolve 20 g CHAPS in distilled H$_2$O and complete to 50 ml. Store at room temperature.

4. *10% (w/v) NP-40*: To make 50 ml, dilute 5 g of 100% NP-40 in distilled H$_2$O and complete to 50 ml. Store at room temperature.

5. *L-lysine solution*: 10 mM L-lysine in H$_2$O. Dissolve 9.1 mg in 5 ml distilled H$_2$O. Aliquot and store at −20°C.

6. *DTT solution*: 1.3M DTT in H$_2$O. To make 10 ml, dissolve 2 g DTT in distilled H$_2$O and complete to 10 ml. Aliquot and store at −20°C. Do not heat.

7. *Ampholines/Pharmalyte mix*: Mix equal volumes of ampholines (pH 3.5–10) and Pharmalyte (pH 3–10). Store at 4°C. These broad pH range IPG buffers can be replaced with narrow range buffers depending on the first-dimension pH range.

8. *Bromphenol blue*: 0.2% (w/v) bromphenol blue in H$_2$O. To make 10 ml, weigh 20 mg bromphenol blue and complete to 10 ml with distilled H$_2$O. Filter and store at room temperature.

Steps

1. Wash cultured or fractionated cells, whole organisms, or tissues in PBS or, if possible, a low-salt buffer that does not compromise cellular integrity. Salts should be kept to a minimum so drain well. Subcellular

or affinity-purified fractions should be prepared at high protein concentration (>2.5 mg/ml) in a low-salt buffer (<10 mM) or dialysed against a low-salt buffer.

2. Lyse cells in lysis buffer using appropriate physical disruption (sonnication, grinding, homogenization, repeated passage through a 25-gauge needle). Do not let samples heat up. A volume of buffer should be used to give a final protein concentration of at least 1 mg/ml. For cellular fractions in a known volume of low-salt buffer and at >2.5 mg/ml, add urea, thiourea, 10% NP-40, 40% CHAPS, and 1.5M Tris, pH 8.8, solution to give final concentrations of 8M urea, 2M thiourea, 4% (w/v) CHAPS, 0.5% (w/v) NP-40, and 10 mM Tris (same as lysis buffer). Rotate on a wheel at room temperature until reagents have dissolved. Because the volume is increased substantially upon reagent addition, amounts should be calculated for 2.5 times the original sample volume. Thus, for a 1-ml sample, add 1.2 g urea, 0.38 g thiourea, 250 μl 40% CHAPS, 125 μl 10% NP-40, 16.67 μl 1.5M Tris, pH 8.8, and make to 2.5 ml with lysis buffer. Use weighing papers to avoid static during weighing. The final pH should be ~8.3, the optimum for NHS-CyDye labeling.

3. Determine protein concentrations using Pierce Coomassie protein assay reagent according to the manufacturer's instructions, using BSA in lysis buffer to generate a standard curve. It is recommended that at least four replicate assays are performed for each sample for accurate protein determination. Dilute concentrated samples with lysis buffer if necessary. For ease, samples should be adjusted to the same protein concentration at this point using lysis buffer.

4. Aliquot desired amount of sample into tubes for CyDye labeling. Typically we label 100 μg of protein in triplicate using a random combination of Cy3 and Cy5 across

the sample set (See Experimental Design). Mix equal amounts of protein from each sample to create an internal standard. This is labeled with Cy2 and should provide enough material for the number of gels to be run (100 μg/gel).

5. Label samples by the addition of 4 pmol of the appropriate CyDye per microgram of protein (400 pmol/100 μg, equivalent to 4 μM for a 1-mg/ml sample). CyDye stocks can be diluted with anhydrous DMF to avoid pipetting submicroliter volumes. Incubate samples on ice in the dark for 30 min. Note that protein lysates are viscous so ensure thorough mixing at all steps to avoid non-uniform labeling.

6. Quench reactions by adding a 20-fold molar excess of L-lysine. For 400 pmol CyDye, add 0.8 μl of 10 mM L-lysine solution. Incubate on ice in the dark for 10 min.

7. Mix Cy3- and Cy5-labeled samples appropriately and add a 100-μg aliquot of the Cy2-labeled pool (to give 300 μg total protein). Note that the final volume should be less than that required for reswelling of IPG strips (450 μl for 24-cm strips). This reswelling volume dictates the practical lower limit for sample protein concentrations.

8. Reduce samples by adding 1.3M DTT to a final concentration of 65 mM (22 μl). Add carrier ampholines/Pharmalyte mix to a final concentration of 2% (v/v) (9 μl) and 1 μl of 0.2% bromphenol blue. Adjust volume to 450 μl with lysis buffer. Spin samples briefly.

9. Rehydrate Immobiline DryStrip pH 3–10 NL gels with samples overnight in the dark at room temperature in a reswelling tray according to the manufacturer's instructions (passive rehydration method). Other methods of sample loading (cup loading, rehydration under voltage) can also be applied depending on user preference.

10. Perform 2D electrophoresis following guidelines for the type of system employed, but see Section III,C for recommended modifications.

Comments

1. Primary amines and reducing agents should be avoided as they interfere with CyDye labeling. These include carrier ampholines/ Pharmalytes and DTT, which are therefore added after labeling but prior to 2D-PAGE.

2. It is often necessary to use protease, kinase, and phosphatase inhibitors for the preparation of lysates and cellular fractions. We have found that aprotinin (17 μg/ml), pepstatin A (1 μg/ml), leupeptin (1 μg/ml), EDTA (1 mM), okadaic acid (1 μM), fenvalerate (5 μM), BpVphen (5 μM), and sodium orthovanadate (2 mM), at the final concentrations shown, do not interfere with CyDye labeling.

3. The quantity of CyDye used for labeling is limiting in the reaction and only ~3% of protein molecules are labeled on an average of one lysine residue (minimal labeling). This minimal labeling approach maintains sample solubility and prevents heterogeneous labeling that would lead to vertical spot trains. However, because 436 Da (Cy2), 467 Da (Cy3), or 465 Da (Cy5) is added to the 3% of labeled molecules, a slight shift in migration is observed between CyDye and poststained images (Gharbi *et al.*, 2002). This is more noticeable in the lower molecular weight range and necessitates poststaining with a general protein stain to attain accurate picking of the majority (97%) of unlabeled protein (see Section III,D).

C. Preparation of 2D Gels, Imaging, and Image Analysis

Isoelectric focusing and second-dimension polyacrylamide gel electrophoresis of CyDye-labeled samples can be performed on any system following the manufacturer's instructions. However, inclusion of the following steps is recommended for high sensitivity, reproducibility, accuracy in the determination of differential expression, and precise excision of protein features for MS identification. The steps are detailed for use with the Multiphor II IEF and Ettan DALT 12 PAGE separation systems for 24 × 20-cm 2D gels, but are generally applicable to other systems. All gel preparation and running steps should ideally be performed in a dedicated clean room to avoid contamination with particulates and nonsample proteins, such as skin and hair keratins. Image analysis and statistical analysis can be performed using various 2D gel analysis softwares (e.g., Melanie, Phoretix, ImageMaster), although DeCyder software is tailored specifically for use with the 2D-DIGE system and is relatively simple to use. Instructions for analysis using DeCyder software are found in the DeCyder software user manual.

Solutions

1. *Bind saline solution*: For twelve 24 × 20-cm plates, mix 16 μl of Plus One bind saline, 400 μl glacial acetic acid, 16 ml ethanol, and 3.6 ml distilled H_2O.
2. *Equilibration buffer*: 6M urea, 30% (v/v) glycerol, 50 mM Tris–HCl, pH 6.8, 2% (w/v) SDS. To make 200 ml, dissolve 72 g urea in 100 ml distilled H_2O. Add 60 ml of 100% glycerol, 10 ml of 1M Tris, pH 6.8, solution, and 4 g SDS. Dissolve all powders and adjust volume to 200 ml with distilled H_2O. Aliquot and store at −20°C.
3. *Agarose overlay*: 0.5% (w/v) low-melting point agarose in 1× SDS electrophoresis buffer. To make 200 ml, melt 1 g of agarose in 200 ml of 1× SDS electrophoresis buffer in a microwave on low heat. Add bromphenol blue solution to give a pale blue colour.

Steps

1. Prior to gel casting, treat low-fluorescence glass plates for gel bonding by applying 1.5 ml of fresh bind saline solution per plate and wiping over one surface with a lint-free tissue. Leave plates to dry for a minimum of 1 h. Note that only one plate in each set should be treated; treat the smaller, nonspacer "front plate" if using Ettan DALT 24-cm gel plates (Fig. 10.1A). Bonding allows easier handling of gels during scanning, protein staining, storage, and, importantly, automated spot excision.
2. Treat the inner surface of clean and dry "spacer plates" with Repel Silane to ensure

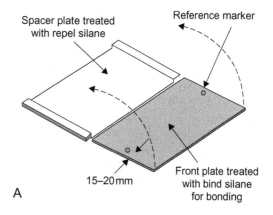

A

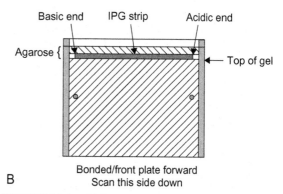

B

FIGURE 10.1 (A) Treatment of plates for bonding and reference marker positioning. (B) Casting and loading of second-dimension gels. Based on the Ettan DALT 24-cm strip format.

easy separation after running (Fig. 10.1A). Apply PlusOne Repel Silane solution to a lint-free tissue and wipe over the surface. Leave to dry for 10 min. Use in a well-ventilated area. Remove excess Repel Saline by wiping with a clean tissue, rinse with ethanol, then with distilled H_2O, and dry with a tissue.

3. Stick two reference markers to the bonded surface of the plates. These should be placed half-way down the plates and 15–20 mm in from each edge (Fig. 10.1A). These markers are used as references for determining cutting coordinates for automated spot picking using the Ettan Spot Picker.

4. Assemble plates in casting chamber and cast gels according to the manufacturer's guidelines.

5. Perform IEF in the dark according to the manufacturer's guidelines.

6. Equilibrate strips for 15 min in equilibration buffer containing 65 mM DTT (reduction) and then 15 min in the same buffer containing 240 mM iodoacetamide. (alkylation).

7. Rinse strips in 1× SDS electrophoresis buffer and place onto the top of second-dimension gels in melted 0.5% agarose overlay, with the basic end of the strip towards the left hand side when the bonded plate is facing forward (Fig. 10.1B).

8. Run second-dimension gels until the dye front has completely run off to avoid fluorescence signals from bromphenol blue and free dye. For the Ettan DALT *twelve* system, this can be achieved by running 12% gels for 16 h at 2.2 W per gel.

9. Images are best acquired directly after the 2D run by scanning gels between their glass plates using a Typhoon 9400 imager or similar device. Ensure that both outer plate surfaces are clean and dry before scanning and that the bonded plate is the lower plate on the scanner bed. If the strip is placed correctly (Fig. 10.1B), the resulting image will not need to be rotated and give a consensus image with the acidic end to

the left. Alternatively, gels can be scanned after fixing with the gel facing up from the bonded plate in the scanner, giving the same orientation of the image.

10. Perform an initial low-resolution scan (1000 μm) for one gel on the Cy2, Cy3, and Cy5 channels with the photomultiplier tube (PMT) voltages set low (e.g., 500 V). The optimal excitation/emission wavelengths for fluorescence detection using the Typhoon 9400 are 488/520 nm for Cy2, 532/580 nm for Cy3, and 633/680 nm for Cy5, although other instrumentation may vary slightly. An image is then built up by the scanner for each channel and is converted to grey-scale pixel values.

11. Using ImageQuant software for the Typhoon 9400, establish maximum pixel values in various user-defined, spot-rich regions of each image and adjust the PMT voltages for a second low-resolution scan to give similar maximum pixel values (within 10%) on each channel and without saturating the signal from the most intense peaks (i.e., <90,000 pixels). As a guide, increasing the PMT voltage by 30 V roughly doubles the pixel value. Repeat scans may be required until values are within 10% for the three channels. PMT voltages can be increased further to enhance the detection of low-intensity features, whilst producing tolerable saturation of only a few of the most abundant protein features.

12. Once set for the first gel, use the same PMT voltages for the whole set of gels scanning at 100-μm resolution. A 24 × 20-cm gel image takes ~10 min to acquire per channel and two gels are scanned simultaneously. Images are generated as .gel files, the same format as .tif files.

13. Crop overlayed images in ImageQuant and import into DeCyder Batch Analysis software for subsequent BVA analysis according to the DeCyder software user

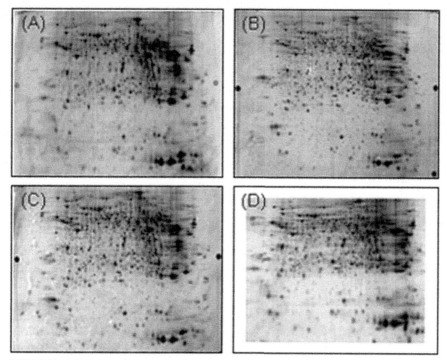

FIGURE 10.2 (A) Cy5 fluorescence image of 100 mg of mouse liver homogenate separated on a 24-cm pH 3–10 NL IPG strip and 12% PAGE gel. (B) SyproRuby poststained gel of 300 mg of mixed CyDye-labeled liver homogenate. (C) Colloidal CBB poststained gel of 300 mg of mixed CyDye-labeled liver homogenate. This is the same gel as shown in A. (D) Adobe Photoshop-generated Cy3/Cy5 coloured overlay of WT (red) and mutant (blue) mouse liver lysates showing differential expression.

manual. Differential expression can also be detected visually using Adobe Photoshop by overlaying coloured images (made in the Channel Mixer) and merging using the "Multiply" option in "Layers" (Fig. 10.2D).

Comments

1. Low-fluorescence glass plates should be used to reduce background.
2. Bind Saline is extremely resistant to removal, and cleaned plates previously treated are still likely to bind acrylamide with subsequent use. For this reason, dedicated treated plates are marked with a diamond pen and reused in the same orientation for subsequent experiments. Bind and Repel saline should be reapplied for subsequent runs.
3. Plates with bonded gels are best cleaned by scraping with a sturdy straight-edged decorator's scraper in warm water with detergent. It is important to remove all gel material, as this produces a fluorescent signal in the Cy3 channel when dried.
4. CyDye-labeled, gel-separated proteins can also be visualized on membranes following electroblotting. The blotted PVDF membrane is scanned using the Typhoon Imager immediately after transfer, wet and face down under a low-fluorescence glass plate. These membranes are subsequently used for

immunoblotting with specific antibodies, and the immunoblot signal is aligned directly with the CyDye signal. This alignment can be used for spot identification, validation of MS, or to identify post translationally modified proteins such as phosphoproteins. Note that gels must not be bonded and the plates used must never have been treated with bind saline.

D. Poststaining and Spot Excision

Bonded gels must be poststained to allow accurate picking (see earlier discussion). We have found that both Sypro Ruby and colloidal Coomassie brilliant blue (CBB) protein stains can be used in conjunction with CyDye labeling (Fig. 10.2). These general protein staining methods are sensitive down to the low nanogram level and are reported to be compatible with downstream mass spectrometric identification of proteins (Scheler *et al.*, 1998; Berggren *et al.*, 2000; Lopez *et al.*, 2000; Gharbi *et al.*, 2002). MS-compatible silver staining (Shevchenko *et al.*, 1996) is not recommended for bonded gels due to its insensitivity and variability from gel to gel.

Solutions

1. *Fixing solution*: 35% (v/v) methanol, 7.5% (v/v) acetic acid in distilled H_2O.
2. *Colloidal CBB fixing solution*: For colloidal Coomassie brilliant blue staining, fix gels in 35% (v/v) ethanol, 2% (v/v) phosphoric acid in distilled H_2O.
3. *Colloidal CBB staining solution*: 34% (v/v) methanol, 17% (w/v) ammonium sulphate, 3% (v/v) phosphoric acid in distilled H_2O.

Steps

1. All gel staining steps should be performed in a dedicated clean room to avoid contamination.

2. After CyDye fluorescence scanning, remove spacer plate and immerse gels in fixing solution and incubate overnight with gentle shaking. Fixed and bonded gels can now be stored for many months at 4°C by sealing in plastic bags with 50 ml of 1% (v/v) acetic acid. The CyDye fluorescence signal is also detectable after several months of storage.

3. For poststaining with Sypro Ruby protein stain (Berggren *et al.*, 2000), wash fixed gels for 10 min in distilled H_2O and incubate in Sypro Ruby stain for at least 3 h on a shaking platform in the dark. Pour off the stain and wash the gel in distilled H_2O or destain [10% (v/v) methanol, 6% (v/v) acetic acid] for three times 10 min. Drain and dry the outer surface of the glass plate and scan gel-side up in a Typhoon 9400 imager at the appropriate excitation/emission wavelength for the Sypro Ruby protein stain.

4. The colloidal CCB G-250 staining method has been modified from that of Neuhoff *et al.* (1988). Fix gels in colloidal CBB fixing solution for at least 3 h on a shaking platform. Wash three times for 30 min with distilled H_2O and incubate in CCB staining solution for 1 h. Add one crushed CCB tablet (25 mg) per 50 ml of staining solution (0.5 g/liter) and leave to stain for 2–3 days. No destaining step is required to visualize proteins. Stained gels can be imaged on a densitometer or on the Typhoon scanner using the red laser and no emission filters.

5. Align poststained and CyDye gel images to identify spots of interest for cutting. Alignment and spot identification can be carried out by comparing images by eye or using Adobe Photoshop to overlay images. A shift in molecular weight between poststained and CyDye images should be apparent due to the increased mass of the dye-labeled fraction of proteins (Gharbi *et al.*, 2002).

6. For automated spot picking, input and process poststained images in DeCyder software and create a pick list for the spots

of interest by comparing with the results of the BVA analysis. To facilitate sample tracking and later data matching with MS results, the poststained image can be imported and matched within the current experimental BVA work space. This means that any spot picked according to the poststained image will have the same master spot number as in the BVA quantitative analysis. Define the positions of the two reference markers in DeCyder (left then right) and export the pick list coordinate file (.txt) to the spot picker controller.

7. Excise chosen spots from the poststained gel. In the case of visible colloidal CBB-stained gels, this can be done manually with a glass Pasteur pipette or gel-plug cutting pipette or on a robotic picker incorporating a "click-n-pick" format, such as the Ettan Spot Picker. The gel is best submerged under 1–2 mm of distilled water, and picking performed in a dedicated clean room.

8. For automated picking using an Ettan Spot Picker, open the imported pick list and align the instrument with the reference markers according to the manufacturer's instructions. Pick and collect spots in 96-well plates in 200 µl of water, drain, and store at −20°C prior to MS analysis. Sample preparation and protein identification by mass spectrometry are detailed elsewhere.

Comment

Harsh fixatives (e.g., >35% methanol) should not be used on bonded gels as they cause over-shrinkage and cracking of gels.

References

Berggren, K., Chernokalskaya, E., Steinberg, T. H., Kemper, C., Lopez, M. F., Diwu, Z., Haugland, R. P., and Patton, W. F. (2000). Background-free, high sensitivity staining of proteins in one- and two-dimensional sodium dodecyl sulfate-polyacrylamide gels using a luminescent ruthenium complex. *Electrophoresis* **21**(12), 2509–2521.

Gharbi, S., Gaffney, P., Yang, A., Zvelebil, M. J., Cramer, R., Waterfield, M. D., and Timms, J. F. (2002). Evaluation of two-dimensional differential gel electrophoresis for proteomic expression analysis of a model breast cancer cell system. *Mol. Cell Proteomics* **1**(2), 91–98.

Lopez, M. F., Berggren, K., Chernokalskaya, E., Lazarev, A., Robinson, M., and Patton, W. F. (2000). A comparison of silver stain and SYPRO ruby protein gel stain with respect to protein detection in two-dimensional gels and identification by peptide mass profiling. *Electrophoresis* **21**(17), 3673–3683.

Neuhoff, V., Arold, N., Taube, D., and Ehrhardt, W. (1988). Improved staining of proteins in polyacrylamide gels including isoelectric focusing gels with clear background at nanogram sensitivity using Coomassie brilliant blue G-250 and R-250. *Electrophoresis* **9**(6), 255–262.

Patton, W. F. (2000). A thousand points of light: The application of fluorescence detection technologies to two-dimensional gel electrophoresis and proteomics. *Electrophoresis* **21**(6), 1123–1144.

Scheler, C., Lamer, S., Pan, Z., Li, X. P., Salnikow, J., and Jungblut, P. (1998). Peptide mass fingerprint sequence coverage from differently stained proteins on two-dimensional electrophoresis patterns by matrix assisted laser desorption/ionization-mass spectrometry (MALDI-MS). *Electrophoresis* **19**(6), 918–927.

Shevchenko, A., Wilm, M., Vorm, O., and Mann, M. (1996). Mass spectrometric sequencing of proteins silver-stained polyacrylamide gels. *Anal. Chem.* **68**(5), 850–858.

Tonge, R., Shaw, J., Middleton, B., Rowlinson, R., Rayner, S., Young, J., Pognan, F., Hawkins, E., Currie, I., and Davison, M. (2001). Validation and development of fluorescence two-dimensional differential gel electrophoresis proteomics technology. *Proteomics* **1**(3), 377–396.

Unlu, M., Morgan, M. E., and Minden, J. S. (1997). Difference gel electrophoresis: A single gel method for detecting changes in protein extracts. *Electrophoresis* **18**(11), 2071–2077.

Zhou, G., Li, H., DeCamp, D., Chen, S., Shu, H., Gong, Y., Flaig, M., Gillespie, J. W., Hu, N., Taylor, P. R., *et al.* (2002). 2D differential in-gel electrophoresis for the identification of esophageal scans cell cancer-specific protein markers. *Mol. Cell Proteomics* **1**(2), 117–124.

CHAPTER

11

Image Analysis and Quantitation

*Patricia M. Palagi, Daniel Walther, Gérard Bouchet,
Sonja Voordijk, and Ron D. Appel*

OUTLINE

I. INTRODUCTION

Many proteomics studies involve comparisons of two-dimensional electrophoresis (**2-DE**) **gels** to identify protein expression changes between different samples. They need efficient **image analysis** software to automatically analyse gel images and extract pertinent biological data. The major steps in such analyses include detection of protein spots in the gels, finding corresponding spots among gels, computation of **protein expression** modifications, and statistical interpretation.

Currently, at least 10 different commercial software packages for the analysis of 2-DE gel images are available commercially. Some of them are descendents of the first generation of tools and software to analyse 2-DE gels, such as Gellab (Lemkin and Lipkin, 1981), THYCO (Anderson *et al.*, 1981), and Melanie (Appel *et al.*, 1997). Although each of these new systems has its own philosophy and approaches, most of them provide the basic operations and functionalities necessary to carry out a complete gel study. The objective of this article is to describe the main steps in a 2-DE gel analysis necessary to find out differently expressed proteins as performed with Melanie software version 4.

II. MATERIALS AND INSTRUMENTATION

A. Software

The Melanie software (version 4) is developed at the Swiss Institute of Bioinformatics. It is commercialised under the name ImageMaster 2D Platinum by Amersham Biosciences in collaboration with GeneBio.[1] A demonstration version of the program and support documentation are freely available from GeneBio's Web site (www.genebio.com). Melanie Viewer, a reduced version of this software, can be freely downloaded from the ExPASy server (http://www.expasy.org).

To use the on-line manual and to access remote databases over the Internet with Melanie, Internet Explorer (Microsoft Corporation), Netscape Navigator (Netscape Communications Corporation), or Mozilla (The Mozilla Foundation) have to be installed on the computer.

B. Image Capture

Gel images may be produced with a large variety of image capture devices, ranging from flatbed document scanners, camera systems, densitometers, phosphor imagers, or fluorescence scanners. The default output format for most imaging equipment, and definitely the most appropriate one for further analysis by 2D software, is Tag Image File Format (TIFF, Aldus Corporation). This is the recommended format for use with Melanie, although the software can read some other file types.

The scanning resolution of the gel image is very important, as it influences the amount of visible details in the image. A low resolution corresponds to a large pixel size or a small number of pixels (or dots) per inch. When the image resolution is too low, the automatic spot detection becomes more difficult. However, when the scan resolution gets too high, the image file becomes very large, and this can slow down the gel analysis. A resolution of 100–200 μM (or about 250–150 dpi) is a good compromise.

The range of possible grey levels (intensity values) in a picture varies according to the image depth (number of bites used to represent a pixel). Images scanned with a higher image depth contain more information. In the case of an 8-bit image, for instance, one pixel has 256 possible grey values (0 to 255). A 16-bit image (65536 grey levels) may reveal subtler but often significant changes; however, it requires more memory. An image depth of at least 12 bits is judicious.

C. Computer Requirements

The Melanie software runs on any of the current Windows operating systems, i.e., 98/ME/NT/2000/XP.

The minimum recommended virtual memory is 256 MB, which is enough to open and process a large number of gels.

Melanie functions properly with a colour resolution of 8 bits (256 colours). However, to use its 3D View module, the colour resolution should be set to 24 bits (16.7 million grey values). It is also recommended to use a screen resolution of at least 1024 × 768 pixels.

III. PROCEDURES

A. Opening Gels and Setting up a Workspace

The investigation of six gels is explained step by step hereafter to describe the usual reasoning when carrying on a whole analysis of 2-DE gels.

---

[1] Melanie 4 is currently also available from Bruker Daltonics integrated with their PROTEINEER spII spot picking robot.

The examples and images shown in this article were generated with gels from a study of aortic smooth muscle cells from newborn (4-day-old) and aged (18-month-old) rats (Cremona *et al.*, 1995). Each population has three gels and are called henceforth newborn and aged.

To start a new work session, import the gel images with the import function and, if necessary, choose a reduction factor. It is highly recommended to setup a workspace as soon as gels have been imported. It facilitates to organize the gel experiments and to work in a personalized environment. The workspace holds information on the relationships between gels such as their organization into populations (classes) and their reference gel (the gel chosen to make the connection with the other gels). The workspace allows organizing gels into projects that reflect the structure and design of the experimental studies facilitating the subsequent work.

i. Click on the Melanie *Workspace* icon in the toolbar to display the Melanie *Workspace* window and create a new workspace.

ii. Inside the workspace, create a new project with the name *"Aortic smooth muscle cells."* Select the newborn gels and add a new class to the project. The selected gels will immediately be allocated to the new class.

iii. Repeat the same procedure for aged gels.

iv. Choose the best representative gel among the six gels and set it as the reference gel.

v. Position the mouse cursor on the class names and right click to open the classes and make these settings active.

B. Viewing and Manipulating Gels

The following usual operation is to adjust the image contrast to improve its visualization, i.e., to visually better differentiate the spots from the background. This kind of operation is often indicated because of differences between the images and the screen display depths. To adjust the gels contrast perform the following steps.

i. Select one or several gels and draw a region in these gels to get a preview of the contrast mapping modifications.

ii. Choose *View → Adjust Contrast → Current* from the menu.

iii. Select the gel for which you would like to display the histogram function by choosing it from the *Image* list.

iv. In the *Gel Display Settings* window, change the minimum and maximum grey levels by displacing the slider borders.

v. Click on OK to apply the visual changes to all selected gels.

In the case where many gels are opened simultaneously, their visible parts may become too small. Stacking gels, by displaying one gel on top of the others, thus creating a pile of gels, then becomes a good alternative to display and compare gels. To stack two or more gels, select them and drag them onto one of the display cells. The concept of stacking gels is very helpful to visually discover differences among gels and to add annotations. An example of stacked gels is given in Fig. 11.1.

A *Transparent* mode can also be used to visualise any similarities or differences between two gels by using a colour overlay of red and cyan. When the pixel colours of the two superimposed gels are added: overlapping spots appear as shades of grey, red and blue spots are present only in one of the gels, halos of red or blue around dark spots indicate that the protein is over- or underexpressed. This means that the fewer colours seen in the transparent mode, the more similar the gels are. To compare gels in the transparent mode:

i. Stack two or more gels.

ii. Select the stack reference.

iii. Choose *View → Stack → Show Transparent* from the menu.

Other useful functions are usually available at gel image analysis software. In case gels were scanned at the wrong orientation, they can be corrected with a rotate function. Cropping gels allows creating new gels with only defined regions. Scaling gels is particularly useful in the

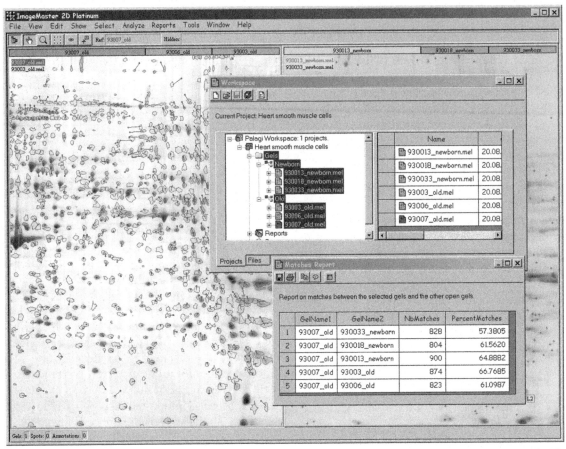

FIGURE 11.1 Spots are delimited by outlined shapes. Cell display at the left shows three gels in stacked mode. The paired spots in matched gels are linked by blue pair vectors. The Workspace window is displayed over the gels as well as a *Matches Report*.

case of very large images where a reduction in size may significantly decrease time and memory required for the analysis. A calibration procedure can be very useful to compensate for image differences caused by variations in experimental conditions (e.g., protein loading, staining) and scanning properties (e.g., image depth).

C. Detecting and Quantifying Spots

The elementary component of a gel is the spot, a shape that can be detected automatically by a spot detection algorithm or adjusted manually by the user. It delimits a more or less tiny region in the gel where a protein or a mixture of proteins is present. Each spot in a gel has an associated spot ID (a unique sequential number) automatically assigned to it when it is created. Moreover, a spot is automatically quantified, i.e., its optical density, area, and volume are computed. The spot detection algorithm of the Melanie software is optimised to identify a maximum number of proteins, while minimizing the number of artifacts detected. Three parameters have to be set to locate the spots automatically.

i. Smooth: it fixes the number of times the image is filtered before detecting spots, using a smooth-by-diffusion algorithm. The smooth parameter has to be optimized to detect all real spots and split as much as possible any overlapping spots.

ii. Min area: it eliminates spots that have an area smaller than a given threshold, eliminating dust particles that consist of a few very dark pixels (artifacts or noise).

iii. Saliency: it is based on the spot curvature and indicates how prominent a spot is. Real spots generally have high saliency values, whereas artifacts and background noise have small saliencies. Although the saliency is a very efficient quantity for filtering spots, it is also very dependent on the gels (e.g., gel resolution and image depth).

To detect spots automatically:

i. Select the gels from which spots will be detected.

ii. Draw a gel region with the *Region* tool on one or more selected gels in zones with representative spots.

iii. Choose *Edit → Spots → Auto-Detect* from the menu.

iv. The *Detect Spots* window appears on the screen, and spots in the active regions in selected gels are detected with the default parameters (Fig. 11.2).

v. Adjust the detection parameters if necessary. The default parameters are optimised to typical SDS–PAGE gels; however, refining the saliency parameter may be indicated. Each change in one of the spot detection parameters is immediately reflected in the selected region helping to choose the parameters. Using the cursor information window is very helpful to find optimal values for the spot filtering.

vi. Click on OK to detect all spots in the selected gels using the parameter values having been set. The spot shapes will be displayed over the gels.

Although it should be avoided to edit spots, they can be created, modified, merged, or deleted manually.

Once spots have been detected, the amount of protein present in each spot is computed automatically. Among the measured quantitative protein values, the most often used in analyses is the relative volume (%Vol). It is a normalized value that remains relatively independent of uninteresting variations between gels, particularly caused by varying experimental conditions. This measure takes into account variations due to protein loading and staining by considering the total volume over all the spots in the gel.

D. Annotating Spots and Pixels

Individual pixels and spots in a gel image may be indicated by annotations. In Melanie, annotations are active elements in the gel analyses and they can have several different purposes, e.g., be used to calibrate, align, and match gels or to be employed to mark spots with their proper information such as protein name, accession number, and so on. Annotations may also be used to mark spots with common characteristics, thus creating subsets. Annotations also offer the possibility to link and associate gel objects to external query engines or data sources of any format (text, html, spreadsheet, multimedia, 2-DE database entry) located locally or on the Internet.

An annotation is defined by its position on the gel (X and Y coordinates) and its set of labels. Each label belongs to a predefined or user-defined category. Among the available predefined annotation categories, some will be important for the explanations given in this chapter.

i. Ac: This category is provided to hold the accession number (AC) of the protein taken from a user-selected database, e.g., Swiss 2D-PAGE (Hoogland *et al.*, 2000) or Swiss-Prot (Boeckmann *et al.*, 2003), and can be the

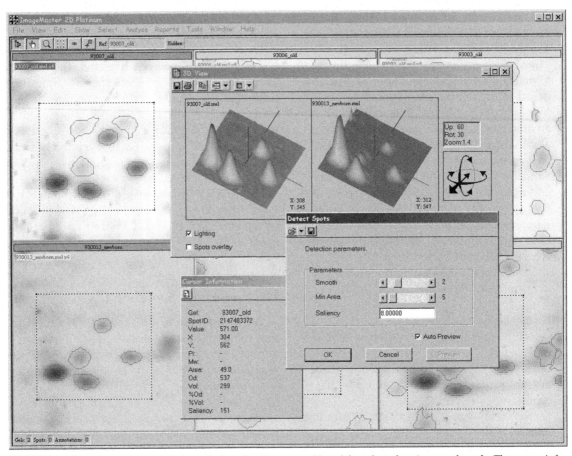

FIGURE 11.2   Three-dimensional view window details spot profiles of the selected regions on the gels. The cursor information window gives spot value information. Selected regions on the gels are updated in real time when adjusting spot detection parameters in the detect spots window.

link to the remote database query engine of Melanie. When such a link is defined, a double click on a label of this category displays the corresponding protein entry in the chosen database with the selected browser software.

ii. Landmark: This is used to mark pixels or spots in the gels as reference points, for the operations of gel alignment or matching, and for the calculation of corresponding locations between gels. Two annotations are considered to refer to the same point in different gels when they hold identical labels.

To create an annotation:

i. Activate the *Annotation* tool.
ii. Double click on the pixel or spot in the gel where the annotation should be located.
iii. In the pop-up window, enter the name of a new category or choose one of the existing categories by clicking on its name.
iv. When a new category is created, the *Create Category* window will appear.

v. Type the desired label in the next dialog box and click OK.

vi. The annotation is created and its label is displayed on the gel.

The predefined category *Set:* is used to mark spots with common properties by indicating that they belong to a set. The labels in such a category do not contain specific information. They only display the name of the set to which they belong.

To create a set:

i. Select one or many spots.

ii. Choose *Edit → Annotations → Add Labels* from the menu.

iii. In the pop-up window, click on the category called *Set:* and add a key word, which will be the name of the set. For instance, to mark spots that were found to be differently expressed and should be exported to a spot excision robot, the final category name might indicate *Set:ToPick*.

iv. A label containing the name of the set (e.g., *ToPick*) will be attached to the selected spots.

Figure 11.3 shows annotations of categories *Landmark*, *Set:Old*, *Set:Verified*, *Set:ToPick*, and *Comments*.

E. Matching Gels

After spots have been detected, and annotations of category *Landmark* have been possibly added to the gels, the next essential step is to match gel images, i.e., find the corresponding protein spots in different gels. A gel-matching algorithm compares two gel images to find *Pairs* of related spots, i.e., spots describing the same protein in both gels.

Matching two gels in Melanie means finding all the pairs between spots of the two gels. Matching several gels means picking out a *Reference gel* and then successively matching each gel with the reference gel. In this way, spots in all gels may be compared with spots in the reference gel.

All spots in selected gels that are paired with a given spot in the reference gel form a *Group*. A spot group is the basic element for analysing spot variations across gels and for producing reports and histograms, as well as for performing statistical and clustering analysis. Moreover, when several gels have been matched to a given reference gel, this reference gel provides a unique numbering scheme for spots across all gels. Indeed, each paired spot in a gel image may be associated to the corresponding Spot ID in the reference gel. The Spot ID in the reference gel is then called the *Group ID*.

To match two or more gels automatically:

i. Select the gels to be matched (including the match reference).

ii. Choose *Edit → Pairs → Auto-Match Gels* from the menu.

iii. Set the reference gel for matching in the pop-up dialog box, i.e., specify to what image the other gels should be matched and click *OK*.

iv. All selected gels are matched with the chosen reference gel.

When matching is completed, Melanie gives the total number of pairs found. In the example given in Fig. 11.1, gel *93007_old* is the reference gel and about 4200 pairs were found among this gel and the other 5. There are 840 pairs in average per gel as it can be seen on the *Matches Report* window.

F. Analysing and Reporting Data

There are numerous ways of finding variations in protein expression among gels with Melanie but only one of them is detailed in this article.

When populations of gels are known, e.g., when comparing gels of newborn-rat tissues against aged-rat tissue samples, the analysis is based on this classification. Consider that the classes are defined as described previously in

Section IIIA. The next step is to find out which are the characteristic spots of each class, i.e., proteins that are expressed differently. The class spot values may be summarized by statistical and overlapping descriptors, such as mean, standard deviation, gap, ratio, and normalization. To investigate groups of spots according to these descriptors one possible way is to

i. Select the gels and then select the groups to be studied with *Select → Groups → All*.
ii. Choose *Analyze → Classes Report*.
iii. In the pop-up list, select the %Vol value type to be displayed.
iv. Accept the default statistics (mean 100% and mean-squared deviation 100%) in the subsequent dialog box.
v. Change the *Displayed value* at the top of the *Classes Report* window for *Ratio*, rank the report in descending order (by clicking on the column headers), and select the rows showing a ratio from the highest value until 2.
vi. Create a new report by choosing the *Report from selection* option in the *Classes Report* window.
vii. In the new window report, change the *Displayed value* for *Gap*. Rank the gap values in descending order. Select all rows from the Gap report and create a *Classes + Groups Histograms*.
viii. Use the created reports and the *Select on Gels* function on the reports to verify the pertinence of the given results. Use the green arrows on the report menu to select rows one by one.
ix. In the Gap report, create an annotation of category "Set" with name *Verified* and type Boolean. When results are reliable, select the field *Verified* in the Gap report (Fig. 11.3).
x. When finished, select all rows in the Gap report, refine the selection using the column "Verified" with value 1, and reselect the spots on the gels and on the displayed reports with the function "Select on gels + reports."

Melanie also proposes statistical tests to help investigate the significance of the resulting spot groups: two-sample *t* test, Mann–Whitney *U* test, and Kolmogorov–Smirnov test. The principle of those tests is to calculate the probability of observing data sampled from populations with different means by chance or by fact. In order to get the statistical test results:

i. Choose *Analyze → Statistical Tests*.
ii. In the pop-up list, select the %Vol value type.
iii. Choose one or more of the statistical tests among the two-sample *t* test, Mann–Whitney *U* test, and Kolmogorov–Smirnov test to be displayed.
iv. Sort, for example, the *t* test values in the report in descending order by clicking on the column header on the top of the *Statistical Tests Report* window.
v. Reselect the first 30 groups that have the highest two-sample *t* test value to concentrate the analysis on the most significant spot differences between classes.
vi. Click the *Classes + Groups Histograms* button at the top of the *Statistical Tests Report* window and then on each histogram to check up the obtained results.
vii. To obtain another view of the results, click the *Classes + Groups Histograms* button at the top of the *Classes Histograms* window and then on each histogram to check up the obtained results.
viii. Mark the resulting spots with labels from the "Set:" category and name *t* test.

Based on the explained procedures, groups composed of spots whose quantification values are unusual may be located. The detected variations can result from protein expression changes among gels or can be due to an inadequate detection or matching operation. Therefore, this analysis is not only useful for investigating extracted data, but also for controlling them.

Among the other Melanie functions to analyse gels, factor analysis and heuristic clustering

FIGURE 11.3 Annotations of numerous categories such as *Landmark, Set:ToPick, Set:Old,* and *Set:Verified* are displayed on the gels. Group 6506 has its histogram highlighted in the *Classes+Groups Histogram* window as well as in the *Classes Report* window.

are two options to check when gels correspond to experimentally known populations. These functions do not rely on any class attribution already set up; they blindly classify gels according to a global similarity and identify the characteristic spots of each population.

G. Integrating Data

Importing and exporting data to and from 2-DE analysis software are fundamental procedures. To make data produced available for processing by other applications or to import information coming from external sources, Melanie uses the common XML format. The main interest of the XML format, besides being used directly by this software, is that external applications can easily extract necessary data.

Melanie exports all gel-related data into a single XML format file, which may include all available information on a set of gels, together with spots (shape, quantification, aligned coordinates), annotations, and pair information.

Spot coordinates can also be exported in XML format to spot excision robots. On the other hand, once the spots have been identified, by mass spectrometry analysis or Edman degradation, for example, the accession number of the identified proteins, as well as other identification data, may be imported from an XML file to annotate the gels.

IV. COMMENTS

Working with many gels at a time may be a tiring task, especially when the images are of bad quality. Melanie 4.0 tools for controlling and automating gel analyses may make repetitive tasks less tedious.

The *History* guarantees a better control over the gel analysis study; the operations that have been performed during the current and preceding work sessions can be checked. The *History* operator consists of a list of actions that have been carried out on the open gels, the parameters, and the selection criteria used at the different steps.

To display a history window
  i. Select the gels for which you would like to display the History.
 ii. Choose *Edit → History → Show* from the menu.
iii. To insert a marker in the history, choose *Edit → History → Insert Marker* from the menu.
 iv. To clear the list of actions, choose *Edit → History → Clear*.

The *History* function is directly related to the Script function. The *Script* operation enables the automation of parts of the analysis process. A script is a sequence of instructions that is carried out automatically by Melanie when it is run. It is a kind of program that may be encoded by the user without any programming knowledge, just by cutting and pasting some desired actions from the *History* to the *Script*. The easiest way to create a script is to
  i. Carry out the desired sequence of operations on a set of gels.
 ii. Display the *History* of these gels.
iii. Copy the required actions to a new script by selecting the actions and pressing the *New Script* icon in the *History* toolbar.

A *Script* window will then be displayed on the screen. It contains a list of action descriptors in the *Script* navigator (actions list), as well as a toolbar at the top of the window. The toolbar contains icons that correspond to standard functionalities, such as copy, paste, save, and print, which create and run the scripts.

In addition to *History* and *Script* operators, the Undo operator corrects mistakes and helps to better control the gel analysis processing. Any earlier state of an analysis may be selected in the action descriptor list of the Undo/Redo operator, which is a particular sequence of actions that can be cancelled any time. Through this multiple undo function, whole parts of the gel analysis may be recovered, avoiding errors.

V. PITFALLS

1. Spot editing: Quantitative protein data, especially the spot volume, are highly dependent on an optimal and reproducible definition of the spot borders and a correct splitting of partially overlapping spots. To guarantee reproducibility of quantitative work, it is highly recommended to create spots by using the automatic spot detection algorithm and to avoid manual editing as much as possible. At most, spots should be separated where necessary.
2. Be critical when matching gels: When the gels are very distorted or different, automatic matching may fail. The choice of landmarks or pairs is very important to obtain good matching results. During

matching with Melanie version 4.0, landmarks essentially correct global deformations of gels. Therefore, it is recommended not to put landmarks on spots in locally distorted regions because this can worsen the matching results around such regions.

References

Anderson, N. L., Taylor, J., Scandora, A. E., Coulter, B. P., and Anderson, N. G. (1981). The TYCHO system for computer analysis of two-dimensional gel electrophoresis patterns. *Clin Chem.* **27**(11), 1807–1820.

Appel, R. D., Hochstrasser, D. F., Roch, C., Funk, M., Muller, A. F., and Pellegrini, C. (1988). Automatic classification of two-dimensional gel electrophoresis pictures by heuristic clustering analysis: A step toward machine learning. *Electrophoresis* **9**, 136–142.

Appel, R. D., Palagi, P. M., Walther, D., Vargas, J. R., Sanchez, J. C., Ravier, F., Pasquali, C., and Hochstrasser, D. F. (1997). MelanieII: A third generation software package for analysis of two-dimensional electrophoresis images. I. Features and user interface. *Electrophoresis* **18**, 2724–2734.

Boeckmann, B., Bairoch, A., Apweiler, R., Blatter, M.-C., Estreicher, A., Gasteiger, E., Martin, M. J., Michoud, K., O'Donovan, C., Phan, I., Pilbout, S., and Schneider, M. (2003). The SWISS-PROT protein knowledgebase and its supplement TrEMBL in 2003. *Nucleic Acids Res.* **31**, 365–370.

Cremona, O., Muda, M., Appel, R. D., Frutiger, S., Hughes, G. J., Hochstrasser, D. F., Geinoz, A., and Gabbiani, G. (1995). Differential protein expression in aortic smooth muscle cells cultured from newborn and aged rats. *Exp. Cell Res.* **217**(2), 280–287.

Hoogland, C., Sanchez, J.-C., Tonella, L., Binz, P.-A., Bairoch, A., Hochstrasser, D. F., and Appel, R. D. (2000). The 1999 SWISS-2DPAGE database update. *Nucleic Acids Res.* **28**, 286–288.

Lemkin, P. F., and Lipkin, L. E. (1981). GELLAB: A computer system for 2D gel electrophoresis analysis. I. Segmentation of spots and system preliminaries. *Comput. Biomed. Res.* **14**(3), 272–297.

Two-Dimensional Gel Profiling of Posttranslationally Modified Proteins by *in vivo* Isotope Labeling

Pavel Gromov and Julio E. Celis

I. INTRODUCTION

The repertoire of posttranslational modifications, which some cellular proteins may undergo after synthesis, falls into two main categories: chemical modifications and processing. Chemical modification involves the linkage of a chemical group to the terminal amino acid or carboxyl groups of the backbone or to reactive groups in the side chains of internal residues. Protein processing involves the proteolytic removal of the polypeptide segments from the premature protein chain. In some cases, both types of protein modifications are closely coupled. Such posttranslational modifications may alter the activity, life span, interactions, and/or cellular localization of proteins, depending on the nature of the modification(s). The most common chemical modifications include acetylation, methylation, phosphorylation, glycosylation, lipid-mediated modifications, and ADP-ribosylation, as well as several others (Mumby, 2001; Fu *et al.*, 2002; Spiro, 2002; Corda *et al.*, 2002; Sinensky, 2000; Cohen, 2000; Casey, 1995 and references therein). New multiplexing tools suitable for general protein detection, including posttranslational variants, have been employed in proteome analysis over the years and are reviewed by Patton (2002).

Each modification causes changes in the molecular weight and often in the charge of a protein, a fact that makes two-dimensional (2D) PAGE well suited for the detection of many posttranslational modifications in combination with mass spectrometry and/or Western immunoblotting if the epitope structure of the modified molecule is not altered by the modification. Several mass spectrometric approaches have been developed for the identification and analysis of various posttranslational modifications and have been reviewed by Abersold and Mann (2003) and by Sickmann *et al.* (2002).

In many cases, however, the identification and analysis of posttranslational modifications can be achieved by *in vivo* or *in vitro* labeling of the proteins with an appropriated isotope-labeled metabolite. Once the radiolabeled ligand is covalently attached to the protein, it can be readily detected on a gel or in a blot using autoradiography. This article describes protocols for 2D gel mapping of posttranslationally modified proteins as applied to the analysis of phosphorylated, glycosylated, and lipidated proteins (palmytoylated, myristoylated, farnesylated, and geranylgeranylated) from human keratinocytes and transformed human amnion cells (AMA).

II. MATERIALS AND INSTRUMENTATION

A. Transformed Human Amnion Cells

AMA cells are grown in complete Dulbecco's modified Eagle's medium (DMEM) containing 10% fetal calf serum (FCS). General procedures for culturing these cells are described in details elsewhere (Celis and Celis, 1997).

B. Noncultured Human Keratinocytes

Noncultured unfractionated human epidermal keratinocytes are prepared from normal skin epidermis as described by Celis *et al.* (1995).

C. Reagents

DMEM (Cat. No. 31966-021) is from GIBCO. Penicillin/streptomycin (Cat. No. A2213) is from Biocrom KG. Dulbecco's modified Eagle's phosphate-deficient medium (Cat. No. 16-423-49) is from ICN. FCS (Cat. No. 04-001-1A) is obtained from Biological Industries. Tissue culture plates (24-well) (Cat. No. 662 160) are from Greiner. All other reagents and general tissue culture facilities are as described elsewhere (Celis and Celis, 1997). 2,5-Diphenyloxazole (PPO, Cat. No. D-4630) is from Sigma, and dimethylether (Cat. No. 823277) is from Merck. [^{35}S]Methionine (Cat. No. SJ 204), [^{32}P]orthophosphate (Cat. No. PBS 13A), [^{3}H]mannose (Cat. No. TRK364), [^{3}H]palmitic acid (Cat. No. TRK909), [^{3}H]myristic acid (Cat. No. TRK907), [^{3}H]farnesyl pyrophosphate (Cat. No. TRK917), and [^{3}H]geranylgeranyl pyrophosphate (Cat. No. TRK918) are from Amersham. Lovastatin is from Merck.

III. PROCEDURES

The protocols for metabolic labeling of posttranslationally modified proteins are illustrated using noncultured unfractionated human epidermal keratinocytes and AMA cells, but can be applied to other cultured cell lines. The volumes given in the following procedure are for labeling in a single well of a 24-well plastic culture plate. For larger tissue culture dishes or multi-well labeling, adjust all amounts accordingly.

A. Phosphorylation

Eukaryotic protein phosphorylation is a reversible covalent addition of phosphate to a protein molecule by means of formation of an ester bond between the phosphoryl group and serine, threonine, or tyrosine residues. Protein phosphorylation is one of the most common posttranslational modifications and is of crucial importance for many regulatory processes. Net phosphorylation is regulated by a complex cascade of protein kinases and phosphatases that catalyze phosphorylation and dephosphorylation reactions, respectively (Cohen, 2000 and references therein). While mass spectrometric methods have taken a leading role in the identification of phosphoproteins, protein radiolabeling with ^{32}P inorganic phosphate is still an effective and inexpensive method for the detection of the ^{32}P subproteome using 2D gels (Mason et al., 1998). This article presents a protocol for the detection of phosphorylated proteins that is based on the metabolic incorporation of [^{32}P]orthophosphate into cultured AMA cells and noncultured human keratinocytes, followed by 2D PAGE and autoradiography.

Solutions

1. *Complete Dulbecco's modified Eagle's medium*: To make 500 ml, mix 445 ml of DMEM medium, 5 ml of 10× stock penicillin/streptomycin, and 50 ml of FCS.
2. *Complete DMEM phosphate-free medium*: Prepare as described in the article by Celis and Celis (1997) using commercial DMEM phosphate-deficient medium.
3. *Hank's buffered saline solution*: To make 1 liter, dissolve 0.4 g of KCl, 0.06 g of KH_2PO_4, 0.0621 g of $NaHPO_4 \cdot 2H_2O$, and 8 g of NaCl

in 800 ml of distilled water. After dissolving, complete to 1 liter with distilled water.
4. *Lysis solution*: 9.8 M urea, 2% (w/v) Nonidet P-40 (NP-40), 2% ampholytes, pH 7–9, and 100 mM dithiothreitol, (DTT).
5. *[^{32}P]orthophosphate-labeling solution*: To prepare 0.25 ml, add 50 μl of commercial aqueous [^{32}P]orthophosphate solution (10 mCi/ml; HCl free) to 200 μl of complete DMEM phosphate-free medium.

Steps

1. Seed AMA cells in a 24-well tissue culture plate (approximately 3×10^4 cells per well) in 0.3 ml of complete DMEM medium containing 10% calf serum, 2 mM glutamine, and antibiotics.
2. Incubate the cells at 37°C in a humidified, 5% CO_2 incubator for approximately 24 h, or until the cells reach about 50% confluence.
3. Prior to labeling, aspirate the medium and wash twice with phosphate-free medium. Add 0.3 ml of phosphate-free medium and incubate the cells for 1 h.
4. Remove the phosphate-free medium and overlay with 0.25 ml of [^{32}P]orthophosphate-labeling solution (2 mCi/ml).
5. Incubate the cells at 37°C in a humidified, 5% CO_2 incubator for 8 h, or a shorter period if necessary.
6. Carefully remove the medium containing [^{32}P]orthophosphate from the plate and gently wash the cells twice with 1 ml of Hank's buffered saline solution. Dispose of radioactive solutions according to the safety procedures enforced in your laboratory.
7. Repeat the washing twice more.
8. Carefully remove excess Hank's buffered saline solution from the plate using an elongated Pasteur pipette.
9. Resuspend the cells in 50 μl of lysis buffer and run 2D gels.

10. Dry the gels and subject to phosphorimaging or to X-ray film autoradiography at −70°C using an amplifying screen.

Comments

Representative autoradiographs (gel) of phosphoproteins from human keratinocytes and AMA cells labeled metabolically with [^{32}P]orthophosphate are shown in Fig. 12.1. To facilitate the identification of phosphorylated proteins, we recommend adding small amounts of [^{35}S]methionine-labeled proteins from the same cell type to the ^{32}P-labeled protein mixture. 2D gels can then be autoradiographed using two films placed on top of each other. The first film, which is placed in direct contact with the dried gel, visualizes both ^{35}S and ^{32}P isotopes, whereas the second one reveals only ^{32}P.

Identification of ^{32}P-labeled proteins can also be done in combination with 2D gel blot immunodetection. In this case, following 2D gel electrophoresis, the proteins are electroblotted to the nitrocellulose membrane and the blot is probed with the antibody of interest, e.g., against phosphotyrosine, prior or after autoradiography.

B. Glycosylation

Protein glycosylation is perhaps one of the most abundant and structurally diverse types of posttranslational modification. Formation of the amino acid–sugar bond is a critical event in the biosynthesis of glycoproteins and leads to diverse biological functions (Spiro, 2002 and references therein). Glycoproteins can be detected by autoradiography after metabolic incorporation of ^{3}H or ^{14}C sugars into cultured cells or tissues (Chandra *et al.*, 1998). The procedure for revealing glycosylated proteins described here is based on the metabolic incorporation of [^{3}H]mannose into cultured AMA cells, followed by 2D gel electrophoresis.

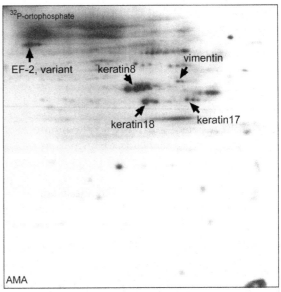

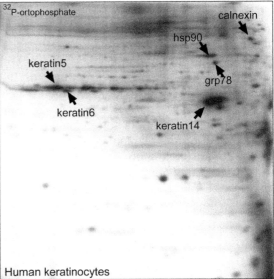

FIGURE 12.1 Two-dimensional (IEF) autoradiographs of whole protein extracts from AMA cells (top) and human keratinocytes (bottom) labeled with [^{32}P]orthophosphate. Several phosphoproteins are indicated as references.

Solutions

1. *Complete Dulbecco's modified Eagle's medium*: Prepare as described in Section III,A.
2. *Hank's buffered saline solution*: Prepare as described in Section III,A.

3. *[³H]Mannose–labeling solution*: 200 μCi/ ml. Evaporate 50 μCi of commercial [³H]mannose solution using a Speed–Vac centrifuge or by directing a gentle stream of nitrogen gas onto the surface of the solution (during this operation the temperature of the solution should not exceed 30°C). Resuspend in 0.25 ml of complete DMEM medium.

4. *Lysis solution*: 9.8 M urea, 2% (w/v) NP-40, 2% ampholytes, pH 7–9, and 100 mM DTT. Prepare as described in Section III,A.

5. *Amplifying solution*: 7% of 2,5-diphenoloxazole (PPO) in dimethylether. To make 100 ml, weigh 7 g of PPO and complete to 100 ml with dimethylether. Store at −20°C in a hermetic glass vessel.

Steps

1. Plate and grow AMA cells in complete DMEM medium as described in steps 1 and 2 in Section III,A.

2. Aspirate the medium using an elongated Pasteur pipette and replace it with 0.25 ml of the [³H]mannose-labeling solution.

3. Incubate the cells at 37°C in a humidified, 5% CO_2 incubator for 2 h.

4. Aspirate the labeling solution and wash the cells twice with Hank's buffered saline solution.

5. Carefully aspirate the excess of Hank's buffered saline solution using an elongated Pasteur pipette.

6. Resuspend the cells in 50 μl of lysis solution and run 2D gels.

7. Following 2D gel electrophoresis, transfer the proteins onto a nitrocellulose membrane by electroblotting.

8. Dry the nitrocellulose blot overnight at room temperature.

9. Pour 100 ml of amplifying solution in a rectangular glass container.

10. Immerse the nitrocellulose blot into the amplifying solution for 1s.

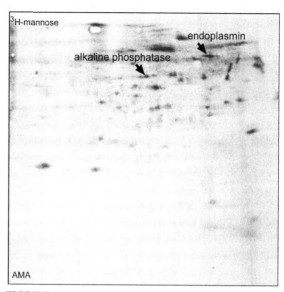

FIGURE 12.2   Two-dimensional (IEF) fluorograph of whole protein extracts from AMA cells labeled with [³H]mannose. Several glycoproteins are indicated as references.

11. Place the nitrocellulose blot on the filter paper with the protein-bearing side facing upward and dry for 30 min.

12. Expose the dried nitrocellulose blot to an X-ray film for 1–7 days at −80°C.

Comments

A representative 2D gel (IEF) fluorograph of AMA cell proteins labeled with [³H]mannose is shown in Fig. 12.2. Following exposure to an X-ray film, the membrane can be stained with amido black to aid in the identification of polypeptide spots. ³H-labeled polypeptides can also be revealed by fluorography of dried gels stained with silver and saturated with the amplifying solution. However, ³H fluorography from dried gels requires longer exposure times as compared to fluorography using nitrocellulose blots. To facilitate protein identification, we recommend adding a concentrated, unlabeled AMA protein extract to the ³H-labeled protein sample prior to 2D gel electrophoresis.

C. Palmitoylation and Myristoylation

Protein lipidation involves co- or posttranslational modification by specific lipids. For most lipid-modified proteins, attached lipids appear to direct or enhance the interaction with both membrane lipids and other proteins, resulting in their specific membrane localization (Casey, 1995). Lipid modified proteins are classified based on the identity of the attached lipid. Palmitoylation and myristoylation are the result of the cotranslational addition of the saturated 16-carbon fatty acid palmitate or 14-carbon fatty acid myristate, respectively, to a glycine residue at the N or C terminus of the protein. The procedure for revealing lipidated proteins described here is based on the metabolic incorporation of either [³H]palmitate or [³H]myristate into cultured AMA cells, followed by 2D gel electrophoresis.

Solutions

1. *Complete Dulbecco's modified Eagle's medium*: Prepare as described in Section III,A.
2. *Hank's buffered saline solution*: Prepare as described in Section III,A.
3. *[³H]Palmitic acid–labeling solution*: 200 μCi/ml. Evaporate 50 μCi of commercial [³H]palmitic acid solution (supplied in ethanol) using a Speed–Vac centrifuge or by directing a gentle stream of nitrogen gas onto the surface of the solution (during this operation the temperature of the solution should not exceed 30°C). Resuspend the label in 0.25 ml of complete DMEM medium.
4. *[³H]Myristic acid–labeling solution*: Prepare as described earlier for [³H]palmitic acid solution
5. *Lysis solution*: 9.8 M urea, 2% (w/v) NP-40, 2% ampholytes, pH 7–9, and 100 mM DTT. Prepare as described in Section III,A.
6. *Amplifying solution*: 7% of PPO in dimethylether. Prepare as described in Section III,B.

Steps

Grow, label, and handle AMA cells as described in Section III,B.

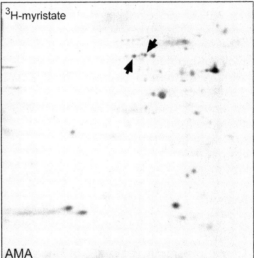

FIGURE 12.3 Two-dimensional (IEF) fluorographs of whole protein extracts from AMA cells labeled with [³H]palmitic acid and [³H]myristic acid. Proteins indicated with closed arrowheads incorporated both lipids.

Comments

Representative 2D gel blot fluorographs, with IEF in the first dimension, of [³H]palmitoylated (Fig. 12.3, top) and [³H]myristoylated (Fig. 12.3, bottom) proteins from cultured AMA cells are shown in Fig 12.3. Spots indicated with arrowheads are labeled with both fatty acids.

D. Isoprenylation: Farnesylation and Geranylgeranylation

Posttranslational modifications of protein with isoprenoids play important roles in targeting a number of proteins to the plasma membrane, as well as in protein–protein interactions, and membrane-associated protein traffic (Sinensky, 2000 and references therein). Protein isoprenylation consists in the covalent attachment of 15-carbon isoprenoid farnesyl or 20-carbon isoprenoid geranylgeranyl via a stable thioether bond to a cysteine residue located in the C-terminal "CAAX," "CC," or "CXC" boxes (Clarke, 1992; Cox and Der, 1992). The method for detecting isoprenylated proteins is based on the specific inhibition of endogenous mevalonate (prenoid precursor) synthesis by lovastatin, followed by metabolic labeling of isoprenylated proteins *in vivo* with either [³H]farnesyl- or [³H]geranylgeranyl pyrophosphate. The protocol described here follows closely those of Danesi *et al.* (1995) and Gromov *et al.* (1996).

Solutions

1. *Complete Dulbecco's modified Eagle's medium*: Prepare as described in Section III,A.
2. *Hank's buffered saline solution*: Prepare as described in Section III,A.
3. *10 mM lovastatin*
4. *[³H]Farnesyl-PP labeling solution*: 50 µCi/ml. Evaporate 12.5 µCi of commercial [³H]farnesyl-PP solution using a Speed–Vac centrifuge or by directing a gentle stream of nitrogen gas onto the surface of the solution (during this operation the temperature of the solution should not exceed 30°C). Resuspend in 0.25 ml of complete DMEM medium and add 0.5 ml of 10 mM lovastatin.
5. *[³H]Geranylgeranyl-PP labeling solution*: Prepare as described previously for [³H]farnesyl-PP
6. *Lysis solution*: 9.8 M urea, 2% (w/v) NP-40, 2% ampholytes, pH 7–9, and 100 mM DTT. Prepare as described in Section III,A.

7. *Amplifying solution*: 7% of PPO in dimethylether. Prepare as described in Section III,B.

Steps

1. Plate and grow AMA cells in complete DMEM medium until they reach about 80% confluence as desribed in steps 1 and 2 of Section III,A.
2. Aspirate the medium from the plate and replace it with 0.25 ml of fresh, complete DMEM medium. Add 1.25 µl of 10 mM lovastatin (final concentration is 50 µM).
3. Incubate the cells at 37°C in a humidified, 5% CO_2 incubator for 6 h.
4. Following incubation, remove the medium and replace it with 50 µCi/ml of either [³H]farnesyl- or [³H]geranylgeranyl-PP labeling solutions.
5. Incubate the cells at 37°C in humidified, 5% CO_2 incubator for 16 h.
6. Remove the labeling medium and rinse the cells twice with 0.5 ml of Hank's buffered saline solution.
7. Proceed as described in Section III,B, steps 5–12.

Comments

Lovastatin induces growth inhibition and apoptosis, especially at high concentration and after prolonged treatment (Perez-Sala and Mollinedo, 1994; Patterson *et al.*, 1994). Therefore, when labeling other cell types, it may be necessary to lower its concentration. Also, the efficiency of labeling may differ between cell types due to possible differences in the uptake of the isoprenoids.

Some isoprenylated proteins can be modified by both farnesyl and geranylgeranyl (Gromov *et al.*, 1996). Representative ³H fluorographs of AMA proteins labeled with [³H]farnesyl- or [³H]geranylgeranyl-PP are shown in Figs. 12.4, top and 12.4, bottom respectively.

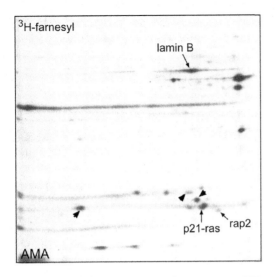

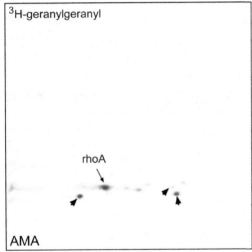

FIGURE 12.4 Two-dimensional (IEF) fluorographs of whole protein extracts proteins from AMA cells labeled with [³H]farnesyl-PP (top) and [³H]geranylgeranyl-PP (bottom). Several prenylated proteins are indicated with arrows. Proteins labeled with both isoprenoids are indicated with closed arrowheads.

IV. PITFALLS

1. Use as short a labeling time as possible in order to reduce the effect of secondary labeling.
2. Proteins that carry the same posttranslational modification may require different labeling times due to differences in protein metabolism.
3. Do not immerse the nitrocellulose filter into the amplifying solution longer than 1s as the dimethylether destroys the membrane.

References

Aebersold, R., and Mann, M. (2003). Mass spectrometry-based proteomics. *Nature* **422**, 198–207.

Casey, P. J. (1995). Protein lipidation in cell signaling. *Science* **268**, 221–225.

Celis, A., and Celis, J. E. (1997). General procedures for tissue culture. In *"Cell Biology: A Laboratory Handbook"* (J. E. Celis, N. Carter, T. Hunter, D. Shotton, K. Simons, J. V. Small, eds.), Vol. 1, pp 5–16. Academic Press, San Diego.

Celis, J. E., Rasmussen, H. H., Gromov, P., Olsen, E., Madsen, P., Leffers, H., Honoré, B., Dejgaard, K., Vorum, H., Kristenesen, D. B., Øsregaard, M., Haunsø, A., Jensen, N. A., Celis, A., Basse, B., Lauridsen, J. B., Ratz, G. P., Andersen, A. H., Walbum, E., Kjærgaard, I., Andersen, I., Puype, M., Damme, J. V., and Vandekerckhove, J. (1995). The human keratinocyte two-dimensional gel protein database (update 1995): Mapping components of signal transdauction pathways. *Electrophoresis* **16**, 2177–2240.

Chandra, N., Spiro, M., and Spiro, J. (1998). Identification of a glycoprotein from rat liver mitochondrial inner membrane and demonstration of its origin in the endoplasmic reticulum. *J. Biol. Chem.* **273**, 19715–19721.

Clarke, S. (1992). Protein isoprenylation and methylation at carboxy-terminal cystein residues. *Annu. Rev. Biochem.* **61**, 355–386.

Cohen, P. (2000). The regulation of protein function by multisite phosphorylation: A 25 year update. *Trends Biochem. Sci.* **25**, 596–601.

Cox, A. D., and Der, C. J. (1992). Protein prenylation: More than just glue? *Curr. Opin. Cell Biol.* **4**, 1008–1016.

Danesi, R., Mc Lellan, C. A., and Myers, C. E. (1995). Specific labeling of isoprenylated proteins: Application to study inhibitors of the post-translational farnesylation and geranylgeranylation. *Biochem. Biophys. Res. Commun.* **206**, 637–643.

Fu, M., Wang, C., Wang, J., Zafonte, B. T., Lisanti, M. P., and Pestell, R. G. (2002). Acetylation in hormone signaling and the cell cycle. *Cytokine Growth Factor Rev.* **13**, 259–276.

Gromov, P., and Celis, J. E. (1996). Identification of isoprenyl modified proteins metabolically labeled with [³H]farnesyl- and [³H]geranylgeranyl-pyrophosphate. *Electrophoresis* **17**, 1728–1733.

Mason, G. G., Murray, R. Z., Pappin, D., and Rivett, A. J. (1998). Phosphorylation of the ATPase subunits of the 26S proteasome. *FEBS Lett.* **430**, 269–274.

Mumby, M. (2002). A new role for protein methylation: Switching partners the phosphatase ball. *Sci. STKE* **79**, PE1.

Patterson, S. D., Grossman, J. S., D'Andrea, P., and Latter, G. I. (1994). Reduced numatrin/B23/nucleosphosmin labeling in apoptotic jurcat T lymphoblasts. *J. Biol. Chem.* **270**, 9429–9436.

Patton, W. F. (2002). Detection technologies in proteome analysis. *J. Chromatogr. B* **771**, 3–31.

Perez-Sala, D., and Mollinedo, F. (1994). Inhibition of isoprenoid biosynthesis induces apoptosis in human promyelocytic HL-60 cells. *Biochem. Biophys. Res. Commun.* **199**, 1209–1215.

Sickmann, A., Mreyen, M., and Meyer, H. E. (2002). Identification of modified proteins by mass spectrometry. *IUBMB Life* **54**, 51–57.

Sinensky, M. (2000). Recent advances in the study of prenylated proteins. *Biochim. Biophys. Acta* **1484**, 93–106.

Spiro, R. G. (2002). Protein glycosylation: Nature, distribution, enzymatic formation, and disease implications of glycopeptide bonds. *Glycobiology* **12**, 43R–56R.

PROTEIN–PROTEIN AND PROTEIN–SMALL MOLECULE INTERACTIONS

Immunoprecipitation of Proteins under Nondenaturing Conditions

Jiri Lukas, Jiri Bartek, and Klaus Hansen

I. INTRODUCTION

Immunoprecipitation of native proteins has proven to be a powerful and widely used approach in addressing questions related to the nature of a single protein or protein complexes existing under different biological conditions. A number of characteristics can be revealed using this method: (1) what is the relative molecular weight of the protein under study, (2) does it contain any posttranslational modifications, such as phosphorylation, acetylation, and glycosylation, (3) is the protein part of a larger multiprotein complex, (4) does it interact with nucleic acids or other ligands, and (5) does the level of the protein change upon growth factor stimulation, during progression through the different phases of the cell cycle, or during the transition between active proliferation and differentiation?

Combined with the recent improvements of protein microsequencing techniques and mass spectrometry, immunoprecipitation also gives the researcher an option to obtain sequence information from unknown proteins identified through coimmnunoprecipitation and thereby collect data on multiprotein complexes. Several papers have described high-throughput analysis of multiprotein complexes in yeast using the mass spectrometry approach in a technique called high-throughput mass spectrometric protein complex identification (HMS-PCI) (Ho *et al.*, 2002) and tandem-affinity purification (TAP) and mass spectrometry (Puig *et al.*, 2001; Gavin *et al.*, 2002). The future will surely bring this kind of screening technique into focus, as it can, compared to conventional two-hybrid screens (which detect only those binary interactions not influenced by posttranslational modifications), detect interactions that require protein modifications.

A critical prerequisite for successful analysis of immunoprecipitated native proteins is the quality of the primary antigen-specific antibodies. For the most straightforward interpretation of results, such a reagent should form specific immunocomplexes with the antigen in its native form without dissociating other associated proteins. In this context, one obvious possibility is to express the protein under investigation as an epitope-tagged protein from an exogenous promoter or even from an endogenous promoter (when possible), which gives the chance to immunoprecipitate the protein using specific antibodies against the tag epitope, such as Myc, Flag, and His. These tags can even be combined. Various strategies in how to obtain such reagents are described elsewhere (Harlow and Lane, 1988; Erica Golemis, 2002). This article shares experiences gained with immunoprecipitation under native conditions that have been obtained during studies of various proteins involved in cell cycle regulation (Lukas *et al.*, 1995; Hansen *et al.*, 2001). We provide a detailed description of an optimized immunoprecipitation protocol that could serve as a basis for isolating native proteins and protein complexes from mammalian cells.

II. MATERIALS AND INSTRUMENTATION

The following chemicals are from Sigma-Aldrich (see catalogue numbers in parentheses).

HEPES (H-7523), NaCl (S-3014), EDTA (E-5134), EGTA (E-4378), glycerol (G-5516), Tween 20 (P-1379), Triton X-100 (X-100), IGEPAL CA-630 (I7771), sodium dodecyl sulfate, SDS (L-4509), dithiothreitol, DTT (D-9779), β-glycerophosphate (G-6251), sodium fluoride, NaF (S-1504), sodium orthovanadate, Na_3-VO_4 (S-6508), leupeptine (hemisulfate salt, L-2884), aprotinin (A-1153), phenylmethylsulfonyl fluoride, PMSF (P-7626), bromophenol blue (B-7021), and myelin basic protein, MBP (M1891). Histone H1 is from Roche (223549).

Protein G–Sepharose 4 fast flow (17-0618-01) and protein A–Sepharose 4 fast flow (17-0974-01) are from Pharmacia Biotech.

Safe-Lock 1.5-ml polypropylene tubes are from Eppendorf (0030 123.328), and 15-ml conical polypropylene tubes are from Corning (430791). Cell scrapers (179693), 92-mm tissue culture dishes (150350), and 80-cm^2 tissue culture flasks (153732) are from Nunc. Protein concentration is measured by the Bio-Rad protein assay kit from Bio-Rad (500-0006).

III. PROCEDURES

Stock Solutions

1. *1M HEPES, pH 7.5*: To make 1 liter, dissolve 238.3 g of HEPES in 800 ml of distilled water, adjust the pH to 7.5 with 10N NaOH, and complete the volume to 1 liter. Sterilize by autoclaving; store at room temperature.
2. *1M Tris, pH 7.5*: To make 1 liter, dissolve 121.1 g of Trizma base in 800 ml of distilled water (room temperature). Adjust the pH to 7.5 by concentrated HCl and fill up to 1 liter with distilled water. Autoclave and store at room temperature.
3. *5M NaCl*: Disolve 292.2 g of NaCl in 800 ml of distilled water. Adjust the volume to 1 liter, autoclave, and store at room temperature.
4. *0.5M EDTA*: To make 100 ml, add 18.6 g EDTA to 80 ml of distilled water. Stir

vigorously on a magnetic stirrer. Adjust the pH to 8.0 with concentrated HCl and let the powder dissolve completely (the disodium salt of EDTA will not go into solution until the pH of the solution reaches approximately 8.0 by the addition of HCl). Adjust the final volume to 100 ml, autoclave, and store at room temperature.

5. *0.5M EGTA*: To make 100 ml, add 19.0 g of EGTA to 80 ml distilled water. Stir and adjust the pH to 8.0 with HCl (see EDTA preparation). Adjust the volume to 100 ml, autoclave, and store at room temperature.

6. *1M DTT (100× stock)*: Dissolve 5 g of DTT in 32 ml of distilled water. Sterilize by filtration (do not autoclave!), dispense into 1-ml aliquots, and store at −20°C.

7. *1M β-glycerophosphate (100× stock)*: To make 100 ml, dissolve 21.6 g of β-glycerophosphate in 80 ml of distilled water. Adjust the volume 100 ml, autoclave, and store at room temperature.

8. *0.5M NaF (500× stock)*: To make 100 ml, dissolve 2.1 g of NaF in a total amount of 100 ml of distilled water. Autoclave and store a room temperature.

9. *0.1M Na_3VO_4 (100× stock)*: Dissolve 200 mg of Na_3VO_4 in 10.8 ml of distilled water, sterilize by filtration, dispense in multiple (100 μl) aliquots, and freeze at −20°C. Once recovered from the freezer, use the batch instantly and only once. Do not refreeze repeatedly.

10. *10 mg/ml leupeptin (1000× stock)*: Dissolve 25 mg of leupeptin in 2.5 ml of precooled distilled water, dispense into 0.5-ml aliquots, and freeze at −20°C.

11. *2 mg/ml aprotinin (1000× stock)*: Dissolve 10 mg of aprotinin in 5 ml of distilled water, dispense into 0.5-ml aliquots, and freeze at −20°C. Do not refreeze repeatedly; once recovered, store the batch for a maximum of 1–2 weeks at 4°C.

12. *0.1M PMSF (1000× stock)*: To make 10 ml, dissolve 174 mg of PMSF in pure isopropanol (store at 4°C) (PMSF is highly unstable in aqueous solutions).

13. *2× Laemmli SDS sample buffer*: 100 mM Tris–HCl, pH 6.8, 200 mM DTT, 4% SDS, 20% glycerol, and approximately 0.2% bromphenol blue. To make 10 ml, mix 1 ml of Tris–HCl, pH 6.8, 2 ml of 1M DTT, 4 ml of 10% SDS, 2 ml glycerol, and 1 ml of distilled water. Add traces of bromophenol blue powder to obtain the desired blue color and mix well. Divide into 0.5- to 1-ml aliquots and freeze at −20°C.

Buffers

A number of different lysis buffers have been described in the literature to effectively extract native proteins from mammalian cells. The following sections offer three different protein extraction buffers that have been used repeatedly and successfully in cell cycle studies for the evaluation of protein–protein interactions and for functional assays such as measuring *in vitro* kinase activity of the immunoprecipitated protein complexes.

1. *Lysis buffer 1 (Matsushime et al., 1994)*: 50 mM HEPES, pH 7.5, 150 mM NaCl, 1 mM EDTA, 2.5 mM EGTA, 10% (v/v) glycerol, and 0.1% Tween 20. To make 1 liter of 1× basic stock solution, add 50 ml of 1M HEPES, pH 7.5, 30 ml of 5M NaCl, 2 ml 0.5M EDTA, 5 ml of 0.5M EGTA, 100 ml glycerol, and 1 ml Tween 20 into 812 ml of distilled water. Stir well on a magnetic stirrer and store at 4°C. Immediately prior to use, add DTT (1:1000 from a 1M stock); phosphatase inhibitors: NaF (1:500 from a 0.5M stock), β-glycerophosphate (1:100 from a 1M stock), and Na_3VO_4 (1:1000 from 0.1M stock); and protease inhibitors: leupeptin (1:1000 from a 10-mg/ml stock), aprotinin (1:1000 from a 2-mg/ml stock), and PMSF (1:1000 from a 0.1M stock). Keep on ice throughout the whole procedure.

2. *Lysis buffer 2 (Jenkins and Xiong, 1996)*: 50 mM Tris–HCl, pH 7.5, 150 mM NaCl, and 0.5% (v/v) IGEPAL CA-630. To make 1 liter of 1× basic stock, add 50 ml of 1M Tris–HCl, pH 7.5, 30 ml of 5M NaCl, and 5 ml IGEPAL CA-630 into 915 ml of distilled water. Store at 4°C. Immediately prior to use, add DTT, phosphatase inhibitors, and protease inhibitors as described for buffer 1.

3. *Lysis buffer 3 (Pagano et al., 1993)*: 50 mM Tris, pH 7.5, 250 mM NaCl, 5 mM EDTA, and 0.1% (v/v) Triton X-100. To make 1 liter of 1× basic stock, add 50 ml of 1M Tris–HCl, pH 7.5, 50 ml of 5M NaCl, 10 ml of 0.5M EDTA, and 1 ml Triton X-100 into 889 ml of distilled water. Store at 4°C. Immediately prior to use, add DTT, phophastase inhibitors, and protease inhibitors as described for buffer 1.

4. *Kinase assay buffer*: 20 mM HEPES, pH 7.2, 1 mM DTT, 10 mM MgCl$_2$, 10 mM MnCl$_2$, 2.5 mM EGTA, 1 mM NaF, 0.2 mM sodium orthovanadate, 2.5 μg/ml leupeptin, and 2 μg/ml aprotinin; should be prepared fresh from stock solutions upon use.

A. Cell Lysis

For all three lysis buffers, highly effective extraction of native proteins can be achieved by the following protocol.

Steps

1. Wash the cell monolayer twice with ice-cold phosphate-buffered saline (PBS) using 10 ml of PBS per washing step for a surface corresponding to a 92-mm diameter tissue culture dish or to a 80-cm^2 tissue culture flask.

2. Add 2.5 ml of ice-cold PBS and dislodge the cells with a cell scraper. Transfer the cell suspension into a prechilled 15-ml polypropylene test tube and repeat the same procedure with another 2.5 ml in order to recover the cells quantitatively. Spin for 5 min in a precooled (4°C) centrifuge (1000 g), discard the supernatant, wash the pellet briefly with 5 ml of cold PBS, and spin again. The cell pellet is now ready for lysis or, for many assays, can be frozen quickly by dipping the tube into liquid nitrogen and stored at −80°C until use.

3. Lyse the cells by adding 3–5 pellet volumes of ice-cold lysis buffer, vortex vigorously (4°C) for 10s, and keep on ice for an additional 30 min. Throughout this period, resuspend the cells by brief vortexing every 5–10 min (4°C).

In the case of buffer 1, efficient lysis has been reported (Matsushime *et al.*, 1994) that involves resuspension of the cell pellet in lysis buffer and subsequent brief sonication on ice. We have successfully reproduced this procedure in our laboratory using a Branson sonifier 250 (2 × 10-s pulses at output level position 6).

In several cases it can be advantageous to avoid the scraping of cells into PBS and to perform a more instant lysis procedure based on adding lysis buffer directly to the cell monolayer (200 μl per surface corresponding to a 92-mm diameter tissue culture dish; see previous discussion for further specifications) that has been washed previously three times with ice-cold PBS. Distribute the lysis buffer on the cell monolayer (after draining off PBS) and collect the lysate with a cell scraper. Transfer the cell lysate into a prechilled 1.5-ml Eppendorf tube. Incubate the lysate on ice for an additional 30 min with occasional brief vortexing (4°C) in order to obtain an efficient protein extraction.

4. Centrifuge the protein extract in a microfuge cooled down to 4°C for 15 min at 20,000 g to pellet cell debris. Transfer the cleared extract to a clean test tube prechilled on ice and measure the total amount of extracted protein (in our laboratory, we use the Bio-Rad protein assay system and exactly follow

the manufacturer's protocol with the protein standards supplied with the kit).

B. Preclearing

Pre-equilibrate the protein A(G)–Sepharose beads supplied by the manufacturer by three successive rounds of resuspension in $10\times$ bead volume of lysis buffer and gentle pelleting by brief (10 s) spinning in a microfuge. To eliminate nonspecific contaminants that can potentially associate with the beads, mix the cell lysate [up to 2 mg of total protein in total volume of 1 ml per tube with 50 μl of pre-eqiulibrated protein A(G)–Sepharose (in a 50% slurry)] and rotate in the cold room for 30 min. Pellet the beads by a 5-min centrifugation in a precooled microfuge ($20,000 g$) and carefully transfer the supernatant into a clean prechilled Eppendorf tube, leaving the beads behind. The protein extract is now ready for immunoprecipitation with specific antibodies. To improve the preclearing step, one can use general nonspecific control IgG (pre-coupled to proteinA/G-Sepharose) from the same species as the specific antibody used for the final immunoprecipitation. It is an advantage to use chemically cross-linked control IgG (Harlow and Lane, 1988).

C. Immunoprecipitation

Steps

1. To presaturate protein A(G)–Sepharose beads with antibodies, aliquot 10–20 μl of beads (50% slurry), pre-equlibrated with the chosen lysis buffer, into Eppendorf tubes containing 0.5 ml of lysis buffer. Add the desired antibodies in saturating amounts. The amount of antibody varies significantly depending on the titer and source of a particular batch and should be determined beforehand. As a rough guide, we recommend starting with 1 μg of purified immunoglobulin, 1–2 μl of mouse ascites,
100–200 μl of hybridoma supernatant, or 2–4 μl of crude rabbit antiserum per 10 μl of beads (note that protein A–Sepharose is particularly suitable for all antibodies of rabbit origin and for mouse IgG2 subclasses. For other subclasses of mouse immunoglobulins, use protein G–Sepharose in order to achieve high-affinity binding). Rotate slowly for 1 h in the cold room and then wash the beads three times in 2 ml of lysis buffer [to pellet the beads between the washing steps, centrifuge briefly in a cooled microfuge ($5000 g$)].

2. Add the protein extract to a 10- to 20-μl aliquot (50% slurry) of beads precoated with the desired antibody and adjust the volume with the lysis buffer to 0.5–1 ml. The total amount of protein input in each sample depends very much on the type of assay and the relative abundance of the protein under study. Thus for sensitive functional assays, such as measuring *in vitro* kinase activity, as little as 50 μg of total protein could be sufficient, but the usual input ranges between 200 μg and 2 mg of total extracted proteins. Close the tubes and rotate end over end in the cold room for 90 min up to 2–3 h (it is not recommended to immunoprecipitate overnight, as this will increase the risk of proteolysis and dephosphorylation of proteins, even in the presence of diverse inhibitors).

3. Pellet the beads for 10 s in a cooled microfuge ($5000 g$) and wash four times by resuspending the beads in 1 ml of of lysis buffer. Gently invert the tubes several times between each washing step.

4. After the last wash, aspirate the lysis buffer above the beads carefully (we recommend using a blunt-ended 25-gauge needle connected to a vacuum pump). For kinase reactions, continue to step 5, for other immunoprecipitations, continue to step 6.

5. For kinase reactions, wash the beads twice in 1 ml of kinase assay buffer and remove

excess liquid above the beads. Start the kinase reaction by adding kinase assay buffer including the appropriate protein substrate (from 1–2 µg per reaction) and ATP. The amount of nonlabeled ATP to be added to the reaction depends on the kinase under study but should normally be in the range of 15–200 µM (final concentration). Furthermore, it is convenient to include [h^{32}P]-ATP (1–10 µCi per reaction; >3000 Ci/mmol) in order to be able to quantitate the incorporation of phosphate in the target substrate by exposure on X-ray film or a PhosphorImager screen. The kinase reaction should take place at 30°C for 10–30 min (should be optimized for the kinase under investigation). Reactions are terminated by the addition of 2× LSB containing 5 mM EDTA and heating for 5 min at 95°C. In cases where a phosphospecific antibody has been developed, it is possible to omit the isotope and instead perform immunoblotting after electrophoresis.

6. For one-dimensional SDS-PAGE analysis, resuspend the beads in 30 µl of 2× Laemmli SDS sample buffer and heat the samples on a heating block (95°C) for 4–5 min. Finally, centrifuge the tubes in a microfuge for 1 min at room temperature and load the samples directly on the gel by use of thin gel-loading tips.

In case the size of the protein of interest is close to 50 or 25 kDa, which also corresponds to the size of the heavy and light chains of immunoglobulins used for the immunoprecipitation, it can be an advantage to use chemically cross-linked antibodies as described by Harlow and Lane (1988). Because the cross-linking of antibodies preferentially takes place between the heavy chain and protein A/G, it is recommended to elute the immunoprecipitated proteins using Laemmli sample buffer without DTT or other reducing agent and, after heating at 95°C for 4–5 min, transfer the eluted material to a clean tube, avoiding beads, and thereafter

add a similar volume of Laemmli sample buffer containing the appropiate amount of DTT or other reducing agent and then heat to 95°C again in order to reduce disulfide bonds in the immunoprecipitated proteins. The sample is now ready to load on the gel as described earlier. By using this two-step elution procedure, you can almost completely avoid disturbing signals from heavy and light chains upon Western blotting or staining of the gel.

Figure 13.1 shows an example of an immunoprecipitation of the cyclin-dependent kinase 4 (Cdk4) from primary mouse fibroblasts in complex with its associated subunits: D-type cyclins and Cdk inhibitors. Figure 13.2A shows an example of in vitro kinase assays using specific

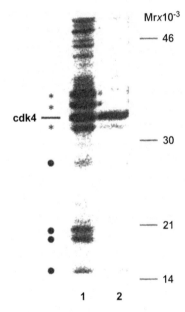

FIGURE 13.1 Primary mouse embryonic fibroblasts were labeled with [^{35}S]methionine, lysed in buffer 1, and immunoprecipitated with rabbit polyclonal antibody against the Cdk2 C-terminal peptide (kindly provided by Dr. M. Pagano). Lane 1 shows a labeled Cdk4 protein coprecipitated along with the associated subunits: D-type cyclins (marked by asterisks) and Cdk inhibitors (marked by closed circles). The parallel reaction in lane 2 was subsequently heated and incubated with SDS in order to dissociate specific protein–protein interactions and was reimmunoprecipitated with the same antibody.

antibodies against three different kinase complexes in immunoprecipitation: Erk1 and Erk2 combined, cyclin D/Cdk4(6), and cyclin E/ Cdk2. Three different substrates were used in the *in vitro* kinase reaction, which also included [h^{32}P]-ATP. The right-hand side of Fig. 13.2A shows the effect of the Cdk2-specific inhibitor

roscovitine when included in the final kinase reaction. Figure 13.2B shows a time course of activation of cyclin D/Cdk4(6) after the release of serum-starved T98G cells (a human glioblastoma cell line). It is obvious that the activity of the complex increases with time after release as cells progress through G1 and approach the G1/S phase transition. The bottom part of Fig. 13.2B indicates that the loading on the gel was equal for all samples by staining for the substrate GST-Rb.

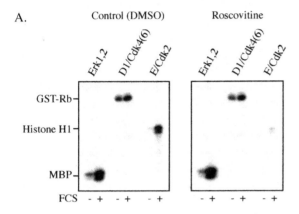

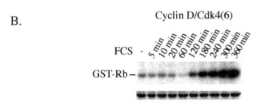

FIGURE 13.2 Human glioblastoma cells (T98G) were starved (0.1% FCS) for 48h before readdition of serum (10%). After 5 min a lysate was prepared for measuring Erk1 and Erk2 activation; at 6h a lysate was made to measure cyclin D/Cdk4(6) activation; and at 12h a lysate was made in order to measure cyclin E/Cdk2 activation. The following antibodies were used: Santa Cruz antibodies against Erk1 (Santa-Cruz SC-093) and Erk2 (Santa-Cruz SC-154) were mixed; DCS 6 mouse monoclonal antibody to cyclin D1 was used to immunoprecipitate cyclin D1/Cdk4(6) complexes; and the mouse monoclonal antibody to cyclin E, HE172, was used to immunoprecipitate cyclin E/Cdk2 complexes. In the *in vitro* kinase reactions, myelin basic protein (MBP) was used as a substrate for Erks, whereas the GST-Rb fusion protein was used as a substrate for Cdk4(6) and histone H1 as a substrate for Cdk2. The kinase reactions in A were performed in the absence (DMSO control) or presence of the Cdk2 inhibitor roscovitine (5 μM final concentration).

References

Gavin, A. C., *et al.* (2002). Functional organization of the yeast proteome by systematic analysis of protein complexes. *Nature* **415**, 141–147.

Golemis, E. (2002). "Protein–Protein Interactions." Cold Spring Harbor Laboratory Press, Cold Spring Harbor, NY.

Hansen, K., Farkas, T., Lukas, J., Holm, K., Rönnstrand, L., and Bartek, J. (2001). Phosphorylation-dependent and -independent functions of p130 cooperate to evoke a sustained G1 block. *EMBO J.* **20**, 422–432.

Harlow, E., and Lane, D. (1988). "Antibodies: A Laboratory Manual." Cold Spring Harbor Laboratory Press, Cold Spring Harbor, NY.

Ho, Y., *et al.* (2002). Systematic identification of protein complexes in *Saccharomyces cerevisiae* by mass spectrometry. *Nature* **415**, 180–183.

Jenkins, C. W., and Xiong, Y. (1996). Immunoprecipitation and immunoblotting in cell cycle studies. In "Cell Cycle Materials and Methods" (M. Pagano, ed.), pp. 250–264. Springer Lab Manual.

Lukas, J., Bartkova, J., Rohde, M., Strauss, M., and Bartek, J. (1995). Cyclin D1 is dispensable for G1 control in retinoblastoma gene-deficient cell independently of cdk4 activity. *Mol. Cell. Biol.* **15**, 2600–2611.

Matsushime, H., Quelle, D. E., Shurtleff, S. A., Shibuya, M., and Sherr, C. J. (1994). D-type cyclin-dependent kinase activity in mammalian cells. *Mol. Cell. Biol.* **14**, 2066–2076.

Pagano, M., Pepperkok, R., Lukas, J., Baldin, V., Ansorge, W., Bartek, J., and Draetta, G. (1993). Regulation of the cell cycle by the cdk2 protein kinase in cultured human fibroblasts. *J. Cell Biol.* **121**, 101–111.

Puig, O., Caspary, F., Rigaut, G., Rutz, B., Bouveret, E., Bragado-Nilsson, E., Wilm, M., and Seraphin, B. (2001). The tandem affinity purification (TAP) method: A general procedure of protein complex purification. *Methods* **24**(3), 218–229.

Xiong, Y., Hannon, G., Zhang, H., Casso, D., Kobayashi, R., and Beach, D. (1993). p21 is a universal inhibitor of the cyclin kinases. *Nature* **366**, 701–704.

Nondenaturing Polyacrylamide Gel Electrophoresis as a Method for Studying Protein Interactions: Applications in the Analysis of Mitochondrial Oxidative Phosphorylation Complexes

*Joél Smet, Bart Devreese, Jozef Van Beeumen,
and Rudy N. A. Van Coster*

I. INTRODUCTION

Under native PAGE conditions, polypeptides retain their higher order structure, enzymatic activity, and interaction with other polypeptides. The migration of proteins depends on many factors, including size, shape, and native charge. The resolution of nondenaturing electrophoresis is generally not as high as that of

SDS–PAGE, but the technique is useful when the native structure or enzymatic activity of a protein must be preserved following electrophoresis. One straightforward approach to native electrophoresis is to omit the sodium-dodecylsulfate (SDS) and the reducing agent dithiothreitol (DTT) from the standard Laemmli SDS protocol (Amersham Bioscience Web site: http://www4.amershambioscience.com). The separation of water-soluble proteins using the Laemmli system can simply be done by replacing SDS with Triton X. The published methods can be devided into two classes: (i) relying on the own charge of the protein to determine the anodic or cathodic migration at a given pH and (ii) using charged, mild detergents to induce a charge shift so that all proteins binding the detergent migrate in the same direction. The methods for native electrophoresis of membrane proteins, however, suffer from many drawbacks. Any new method designed to separate them must include a charge shift method. Two components were introduced for that purpose: Coomassie blue G for the separation of multiprotein complexes ("blue native PAGE") and taurodeoxycholate for the separation of lower molecular mass proteins ("native PAGE") (Schägger and von Jagow, 1991).

The blue native PAGE technique is particularly useful for the characterization of mitochondrial oxidative phosphorylation (OXPHOS) enzymes. Combining this separation technique with histochemical staining makes it possible to quantify mitochondrial enzymes *in situ* (Zerbetto *et al.*, 1997). It allows to evaluate the enzymatic activities of the complexes I, II, IV, and V in heart and skeletal muscle, liver and cultured skin fibroblasts and to detect deficiencies. Also, the amount of protein in the complexes I, II, III, IV, and V can be evaluated using silver or Coomassie staining. Often the complexes are even visible in the gel without additional staining. When the background is high, immunoblotting has to be used to visualize the amount of complex proteins (Van Coster *et al.*, 2001).

Combining blue native PAGE with SDS–PAGE reveals a two-dimensional pattern showing the individual subunits of the five OXPHOS multienzyme complexes. The implementation of mass spectrometric techniques, e.g., mass fingerprinting and mass spectrometric sequence analysis, allows the unambiguous identification of the individual subunits (Devreese *et al.*, 2002).

II. MATERIALS AND INSTRUMENTATION

Tris–HCl (Cat. No. T1378), sucrose (Cat. No. S0389), glycerol (Cat. No. G6279), phenylmethylsulfonyl fluoride (PMSF, Cat. No. S0389), dimethyl sulfoxide (DMSO, Cat. No. D5879), Tricine (Cat. No. T0377), aminocaproic acid (Cat. No. A7824), bis–Tris (Cat. No. 9754), dodecyl β-D-maltoside (Cat. No. D4641), nitro blue tetrazolium (NBT, Cat. No. N6876), phenazinemethosulfate (PMS, Cat. No. P9625), succinic acid (Cat. No. S2378), 3,3'-diaminobenzidine (DAB, Cat. No. D8001), catalase (Cat. No. C9322), cytochrome c (Cat. No. C7752), magnesium sulfate (Cat. No. M7506), lead(II) nitrate (Cat. No. L7281), adenosinetriphosphate (ATP, Cat. No. A5394), Tween 20 (Cat. No. P1379), and the Gelbond film (Cat. No. E0389) are from Sigma. Serva Blue G (Cat. No. 35050) is from Serva. Acrylamide (Cat. No. 161-0101), *N,N'*-methylenebisacrylamide (bisacrylamide, Cat. No. 161-0201), ammonium persulfate (APS, Cat. No. 161-0700), *N,N,N',N*-tetramethylenediamine (TEMED, Cat. No. 161-0800), the gel-staining Coomassie Bio-Safe solution (Cat. No. 161-0787), the blotting grade blocker (Cat. No. 170-6404), the gel-drying solution (Cat. No.161-0752), and the transparent membranes (Cat. No. 165-0963) are from Bio-Rad. SDS (Cat. No. 44215) is from BDH Laboratories. Nicotinamide adenine dinucleotide, reduced form (NADH, Cat. No. 107735), is from Roche. The silver stain kit GelCode SilverSNAP (Cat. No. 24602) and the

Bradford reagent for protein determination (Cat. No. ZZ23238) are from Pierce. Whatman paper (Cat. No. 3030672) is from VWR. PVDF immunoblot membranes (Cat. RPN2020F), ECL films (Cat. No. RPN 3103K), the secondary IgG HRP-linked whole AB to mouse (Cat. No. NA931), and the ECL plus kit (Cat. No. RPN2132) are from Amersham Biosciences. The different primary antibodies against OXPHOS subunits are from Molecular Probes (http://www.probes.com). The homogeniser Model L42 is from Schwaben Präsizion Nordlingen. The glass/glass dual tube tissue homogenizers of different volumes are from Kontes, distributed by Helma Benelux. The air-driven ultracentrifuge airfuge (Cat. No. 340401) and the microcentrifuge tubes (Cat. No. 344718) are from Analis. Minigels (gel size 8.3×7.3 cm) are run on a Mini-Protean three electrophoresis cell (Cat. No. 165-3302). All parts, such as glass plates, combs, casting stand, spacers, electrode assembly, tank lid with power cables, and power supply (Model 200/2.0), are from Bio-Rad. The Mini Trans-Blot cell (Cat. No. 170-3930) and the Hydrotech gel-drying system (Model 583, Cat. No. 165-1745) equipped with a Hydrotech vacuum pump (Cat. No. 165-1782) are from Bio-Rad. The hot plate (Präzitherm, type PZ 28-2 T), the Inolab pH level 1 pH meter, and the orbital shaker (a GFL model 3006) are from VWR. Pipettes (P1000, P100, and P25) are from Hamilton. The Sharp JX-330 scanner is from Amersham Biosciences.

III. PROCEDURES

A. Isolation of Mitochondria from Tissues and Cultured Fibroblasts

This procedure is performed according to Scholte *et al.* (1992). In order to isolate a sufficient amount of mitochondria, at least 100 mg (wet weight) of tissue (muscle, heart, liver) is needed. Starting from cultured fibroblasts, after harvesting, a pellet volume of 100 μl (approximately three T75 Falcon flasks) is needed.

B. Solubilization of Mitochondria

Solutions

1. *Protease inhibitor solution*: 1 mM PMSF in DMSO. Prepare a 25-ml solution by dissolving 4.3 mg PMSF into 25 ml DMSO, aliquot in 1-ml portions, and store at −20°C.
2. *Solution A*: 750 mM aminocaproic acid, 50 mM Bis–Tris, and 20 μM PMSF. To make 25 ml, dissolve 2.46 g aminocaproic acid and 0.26 g Bis–Tris and add 20 μl of 1 mM PMSF in DMSO solution; adjust pH to 7.4 and adjust volume to 25 ml. Divide into single-use aliquots of 1 ml and store at −20°C. Samples can be stored under these conditions up to 3 months.
3. *Solution B*: Laurylmaltoside 10%. Dissolve 0.5 g in 5 ml distilled water, divide into single-use aliquots of 150 μl, and store at −20°C. Samples can be stored under these conditions maximally for 3 months.

Steps

1. Place the mitochondrial pellets (stored at −80°C) on ice.
2. Add a mixture of 96 μl solution A and 12 μl solution B equivalent to a mitochondrial pellet resulting from approximately. 100 mg of tissue or a fibroblast pellet of 100 μl.
3. Dissolve the pellet by pipetting up and down and by vortexing the tube strongly.
4. Pipette the solubilized material into a microcentrifuge tube (max 200 μl/tube).
5. Centrifugate in the airfuge at 100,000 g for 15 min.
6. Pipette the clear supernatant containing the oxidative phosphorylation enzyme complexes into an Eppendorf tube; avoid pipetting material from the pellet, which is discarded.

7. Use 5 μl of the solution to determine the protein content according to the Bradford (1970) method.

8. Store at −80°C if the supernatant cannot be submitted to electrophoresis within 2 h after preparation.

C. First-Dimension BN-Polyacrylamide Gel Electrophoresis

This is a procedure according to Schägger (1995).

Solutions

1. *Cathodal buffer (colorless)*: 50 mM tricine, 15 mM bis–Tris. To make 1 liter, dissolve 9.0 g tricine and 3.1 g bis–Tris in distilled water, adjust pH to 7.0 and complete to 1 liter with distilled water. Store for a maximum of 2 months at 4°C.

2. *Cathodal buffer (colored)*: Colorless buffer + 0.02% Serva Blue G. Add 0.2 g Serva Blue G to 1 liter of colorless cathodal buffer.

3. *Anodal buffer (5× concentrated)*: 250 mM bis–Tris. To make 1 liter, dissolve 52 g bis–Tris in distilled water, adjust pH to 7.0, and complete to 1 liter. Dilute prior to use 1/5 with distilled water. Store for 2 months maximum at 4°C.

4. *Acrylamide/bisacrylamide mixture (49.5% T, 3% C)*: To make a 250-ml solution, add 3.13 g bisacrylamide to 250 ml of the 40% solution from the supplier. Divide into 10-ml aliquots, which are stored at −20°C for 3 months maximum.

5. *Gel buffer (3× concentrated)*: 1.5 M aminocaproic acid and 150 mM bis–Tris. To make 100 ml, dissolve 19.7 g aminocaproic acid and 3.1 g bis–Tris, adjust pH to 7.0, and complete to 100 ml with distilled water. Divide into 10-ml aliquots, which are stored at −20°C for 3 months maximum.

6. *10% ammonium persulfate solution*: To make 10 ml, dissolve 1 g APS in 10 ml distilled water and divide into 1-ml aliquots, which are stored at −20°C for 3 months maximum. Use within 24 h.

7. *Sample loading buffer*: 5% Serva Blue G and 750 mM aminocaproic acid. Prepare a solution of 1 ml by dissolving 50 mg Serva Blue G and 98 mg aminocaproic acid in a final volume of 1 ml distilled water. Divide into single-use 100 μl aliquots and store them for 3 months maximum.

Steps

1. Assemble the Mini-Protean gel apparatus according to the manual of the supplier. Gel dimensions are 8.0 × 7.3 × 0.1 cm.

2. Prepare a 5% and 13% T gel solution in order to make a 5–13% resolving gel gradient (see Table 14.I).

3. Finally add the APS and TEMED to the 5 and 13% solutions and vortex both solutions.

4. Pipette 2× 100 μl of the 5% solution into the 13% solution and vortex.

5. Pipette 2× 100 μl between the glass plates at the left and right side of both gels.

6. Repeat steps 4 and 5 until you reach a level of approximately 1 cm under the comb.

7. Overlay with water and leave to polymerize. After a minimum of 1 h, place at 4°C overnight.

8. Prepare the 4% stacking gel solution.

9. Remove the water using Whatman paper and pour the solution on top of the resolving gel. Insert the comb (e.g., 10-well comb) and let polymerize for 45 min.

10. Poor the cold (<10°C) colored cathodal buffer between the two glass plates at a level 0.5 cm below the bottom of the sample wells.

11. Add 5 μl of sample loading buffer to 100 μl of the solubilized mitochondrial sample, which is kept on ice, and vortex.

12. Load the calculated volume (depending of the amount of protein) of the sample into

TABLE 14.I

Reagents	5% T	13% T	Stacking gel 4% T
A–B mixture (49.5% T–3% C)	0.8 ml	1.7 ml	0.4 ml
Gel buffer (3×)	2.1 ml	1.8 ml	1.5 ml
Glycerol	—	0.9 ml	—
AD	3.4 ml	1.0 ml	2.6 ml
Total volume	**6.3 ml**	**5.4 ml**	**4.5 ml**
APS (10%)	35 μl	18 μl	36 μl
TEMED	3.5 μl	1.8 μl	3.6 μl

the sample wells (20–100 μg of protein/lane). Two lanes are needed if you want to evaluate the activities by activity staining (see Section III,F).

13. Completely fill the space between the glass plates with colored cathodal buffer.
14. Place the gel assembly into the anodal buffer tank and poor the anodal buffer into the tank up to a level above the bottom end of the glass plates.
15. Cool the apparatus to 4°C and connect the electrodes to the power supply.
16. Run the gels at 75 V for 1 h.
17. Replace the colored cathodal buffer by the colorless buffer without rinsing the cathodal buffer chamber and run the gels at 200 V until the Serva Blue tracking dye has run off the gel completely (approximately 3 h). The gels can be used immediately for protein staining (see Section III,D), immunoblotting (see Section III,E), activity staining (see Section III,F), or can be stored for several months at −80°C.

D. Protein Staining

1. Coomassie Staining

Steps

1. Fix the gels in a 50/40/10 mixture of distilled water/methanol/acetic acid by shaking on an orbital shaker for 30 min.
2. Stain the gels by transferring them into the Bio-Safe Coomassie staining solution and place them on the shaker for 60 min.
3. Destain the gels by transferring them in a tray filled with distilled water and put on the shaker for 60 min, refresh, and continue destaining on the shaker overnight.
4. Put the gels in a drying solution for 30 min.
5. The gels then can be dried on Whatman paper or between two transparent membranes according to the manufacturer's protocols. Normally a 2-h drying time at 80°C is sufficient (Fig. 14.1).

2. Silver Staining

Steps

1. Fix the gels in 50 ml of a 50/40/10 mixture of distilled water/methanol/acetic acid by shaking on an orbital shaker for 30 min.
2. Wash the gels twice for 5 min with 50 ml 10% ethanol solution and twice for 5 min with distilled water.
3. Stain the gels with 50 ml of a 50/1 mixture of the staining/enhancing solution for 30 min.
4. Rinse with distilled water for 1 min.
5. Transfer the gels into 50 ml of a 50/1 mixture of developer/enhancer solution for 2 to 5 min.

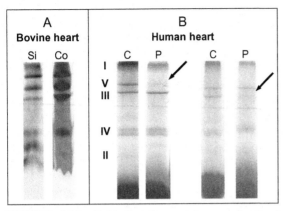

FIGURE 14.1 (Left) Complex proteins are visualized using protein staining (silver and Coomassie). (Right) Complex proteins are visible prior to any protein staining. Arrows mark complexes with low amounts of protein.

6. When the bands are clearly visible, stop the development by placing the gels in 50 ml of a 5% acetic acid solution.
7. Dry the gels on Whatman paper or between two transparent membranes according to the manufacturer's protocols. Normally a 2-h drying time at 80°C is sufficient (Fig. 14.1).

E. Immunoblotting

Solutions

1. *Towbin blotting buffer* (Towbin *et al.*, 1979): 10× concentrate (0.25 M Tris and 1.92 M glycine, pH 8.5). To make 1 liter (10×), dissolve 30.3 g Tris and 144 g glycine in distilled water, complete to 1 liter.
2. *Phosphate-saline buffer*: 10× concentrate (0.8 M Na$_2$HPO$_4$, 0.2 M NaH$_2$PO$_4$, 1 M NaCl). To make 250 ml (10×), dissolve 28.4 g Na$_2$HPO$_4$, 6.9 g NaH$_2$PO$_4$, and 14.5 g NaCl, complete to 250 ml with distilled water.
3. *Washing buffer (0.1% Tween 20)*: Dilute 100 ml 10× phosphate-saline buffer with 900 ml distilled water and add 1 ml Tween 20, adjust pH to 7.5.

4. *Blocking reagent*: 5% blocking reagent in washing buffer. To make 100 ml solution, dissolve 5 g blocking reagent in a final volume of 100 ml washing buffer.
5. *Primary antibody dilution*: Make a dilution of the primary OXPHOS antibodies into the blocking solution. Minimally, 20 ml is needed for incubation of the two membranes. When using the ECLplus system, a concentration of 1 µg/20 ml primary antibody is, in most cases, sufficient to detect even very low amounts of protein.
6. *Secondary antibody dilution*: Make a 1/2000 dilution of the secondary HRP antibody: 10 µl of antibody solution in 20 ml blocking solution.
7. *ECLplus detection reagent*: Place reagents A and B at room temperature for 1 h. Just prior to the detection procedure, mix reagents A and B in a 40:1 ratio. For two membranes, 4 ml of solution A and 100 µl of solution B are needed. Keep away from light.

Steps

1. Cut the PVDF membranes and Whatman papers (two per gel) in a size such that they completely cover the gel.
2. Dip the PVDF membranes briefly into 100% methanol and incubate them, together with the Whatman papers and the sponges, for 30 min in the Towbin blotting buffer.
3. Fill the blotting cassette with sponge–paper–gel–membrane–paper–sponge (in this order) and close the cassette. Make sure no air bubbles are trapped between the gel and the membrane.
4. Place the cassette into the holder, with the membrane facing the cathode. Insert the ice box and fill the blotting chamber with 600 ml Towbin blotting buffer.
5. Connect the electrodes and run for 1 h at 100 V.
6. Remove the membranes from the cassette, dip them briefly into methanol, and place them in 20 ml washing buffer for 30 min.

7. Incubate the membranes in 20 ml blocking solution for at least 1 h on the orbital shaker.
8. Remove the blocking solution, add 20 ml of the primary antibody solution, and incubate on the orbital shaker for a minimum of 1 h.
9. Remove the primary antibody solution and wash the membranes at least three times for 5 min with 30 ml of washing buffer on the shaker.
10. Add 20 ml of the 1/2000 dilution of the secondary HRP antibody and incubate using the orbital shaker for at least 1 h.
11. Remove the secondary antibody solution and wash the membranes at least three times for 5 min with 30 ml of washing buffer.
12. Place the membrane, protein side facing up, in a tray and add the detection reagent on the whole surface of the membrane using a pipette. Let react for 5 min. Take the membranes with a forceps, let excess reagent drip off, and fix the membranes between two sheets of transparent membrane in a film cassette.
13. Place an ECL hyperfilm on the membrane for several time intervals in order to obtain the best signal-to-noise result on the film (Fig. 14.2).
14. After detection, wash the membranes three times for 5 min with 30 ml of washing buffer on the shaker; these can then be stored in a sealed plastic bag for several months at −20°C.

F. In-Gel Activity Staining

This is the procedure according to Zerbetto *et al.* (1997), with certain modifications.

Staining Solutions

1. *Complex I (10×):* 2 mM Tris–HCl, pH 7.4, 0.1 mg/ml NADH, and 2.5 mg/ml NBT. To make 25 ml of a 10× solution, dissolve 6.0 mg Tris–HCl, 2.5 mg NADH, and 62.5 mg

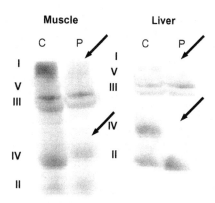

FIGURE 14.2 Immunoblotting of OXPHOS complexes using one specific antibody to one subunit of each complex and ECLplus detection. Arrows mark low CRM of the complex in patient lanes (P) as compared to controls (C).

NBT in distilled water. Adjust pH to 7.4, complete to 25 ml, and divide into 3-ml single-use aliquots, which are kept at −80°C.
2. *Complex II:* 4.5 mM EDTA, 10 mM KCN, 0.2 mM PMS, 84 mM succinic acid, and 10 mM NBT in a 1.5 mM phosphate buffer, pH 7.4. Prepare 0.1 M phosphate buffer, pH 7.4, by mixing a 100-ml solution of 1.36 g KH_2PO_4 and a 100-ml solution of 1.74 g K_2HPO_4. Prepare a 1.5 mM buffer solution by adding 98.5 ml distilled water to 1.5 ml 0.1M buffer. To make 25 ml solution, dissolve 33 mg EDTA, 16 mg KCN, 1.5 mg PMS, 0.56 g succinic acid, and 200 mg NBT in the 0.1 M phosphate buffer, adjust pH to 7.4, complete to 25 ml, and divide into 3-ml single-use aliquots, which are kept at −80°C.
3. *Complex IV:* 5 mg DAB dissolved in 9 ml phosphate buffer (0.05M, pH 7.4), 1 ml catalase (20 μg/ml), 10 mg cytochrome c, and 750 mg sucrose. Prepare a 0.05M phosphate buffer by adding 13.5 ml distilled water to 13.5 ml 0.1 M phosphate buffer (see complex II). Dissolve 2 mg catalase in 100 ml distilled water. To make 30 ml solution, dissolve 15 mg DAB, 30 mg cytochrome c, and 2.25 g sucrose, add 3 ml of the catalase solution,

complete to 27 ml with 0.05M phosphate buffer, adjust pH to 7.4, and divide into 5-ml single-use aliquots, which are kept at −80°C.

4. *Complex V*: 35 mM Tris–HCl, 270 mM glycine, 14 mM MgSO$_4$, 0.2% Pb(NO$_3$)$_2$, and 8 mM ATP, pH 7.8. To make 25 ml solution, dissolve 0.11 g Tris–HCl, 0.51 g glycine, 42 mg MgSO$_4$, 50 mg Pb(NO$_3$)$_2$, and 0.11 g ATP, adjust pH to 7.8, complete to 25 ml, adjust pH to 7.4, and divide into 3-ml single-use aliquots, which are kept at −80°C.

Steps

1. Thaw the staining solutions. Prepare a 1/10 dilution of the 10× complex I staining solution by adding 9 ml distilled water to 1 ml of the (10×) solution. Pour approximately 1 ml of each staining solution for one gel into the four staining trays (one tray/complex) (dimensions: e.g., 8 × 15 cm).
2. The gels were kept at −80°C, first thaw the gel quickly by immersing the plate in a water bath. Divide the gel as shown in Fig. 14.3. Using a sharp scalpel knife, divide the gels into four parts (dotted lines). The upper part of the first lane is used for complex I staining and the lower part for complex IV staining. The upper part of the second lane is used for complex V staining and the lower part for complex II staining.
3. Place the gels into the appropriate staining solutions. Cover with the rest of the staining solution and place on a hot plate at 37°C.
4. Keep the gels humid by adding 1 ml of an isotonic saline solution (0.87 g NaCl per 100 ml distilled water) at regular time intervals.
5. The bands of complexes I and II reach their maximum intensity after 3–4 h and the bands of complexes IV and V after 4–6 h.
6. Thereafter, fix the gels for 15 min in 50 ml of a mixture of 50% methanol and 10% acetic acid, except gels stained for complex V, which are rinsed with distilled water.

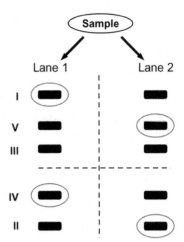

FIGURE 14.3 Schematic drawing of the five OXPHOS complexes and cutting of the gel.

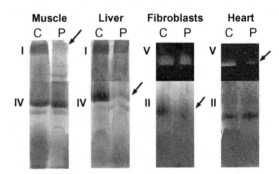

FIGURE 14.4 Activity staining of OXPHOS complexes in different tissues. Arrows mark deficient complexes in patients (P) in comparison with controls (C).

7. Place the gels between two Gelbond films and scan in transmission mode using a JX-330 scanner. To evaluate the white complex V band, scan the gel in the reflection mode using a black background. Staining for complex I results in a dark blue/purple band, and staining of Complex V gives a white band, both visible in the upper part of the gel. In the lower part of the gel, staining for complex IV results in a brown band and complex II staining results in a dark purple band. (Fig. 14.4).

8. The gels then can be dried on Whatman paper or between two transparent membranes according to the manufacturers protocols. Normally 2 h of drying time at 80°C is sufficient.

References

Bradford, M. M. (1970). *Anal. Biochem.* **74**, 248–254.

Devreese, B., Vanrobaeys, F., Smet, J., Van Beeumen, J., and Van Coster, R. (2002) Mass spectrometric identification of mitochondrial oxidative phosphorylation subunits separated by two-dimensional blue-native polyacrylamide gel electrophoresis. *Electrophoresis* **23**(15), 2525–2533.

Schägger, H. (1995). Quantification of oxidative phosphorylation enzymes after blue native electrophoresis and two-dimensional resolution: Normal complex I protein amounts in Parkinson's disease conflict with reduced catalytic activities. *Electrophoresis* **16**(5), 763–770.

Schägger, H., and von Jagow, G. (1991). Blue native electrophoresis for isolation of membrane protein complexes in enzymatically active form. *Anal. Biochem.* **199**(2), 223–231.

Scholte, H. R., Ross, J. D., Blow, W., Boonman, A. M., van Diggelen, O. P., Hall, C. L., Huijmans, J. G., Luyt-Houwen, I. E., Kleijer, W. J., de Klerk, J. B., *et al.* (1992). Assessment of deficiencies of fatty acyl-CoA dehydrogenases in fibroblasts, muscle and liver. *J. Inherit. Metab. Dis.* **15**(3), 347–352.

Towbin, H., *et al.* (1979). *Proc. Natl. Acad. Sci. USA* **76**, 4350.

Van Coster, R., Smet, J., George, E., De Meirleir, L., Seneca, S., Van Hove, J., Sebire, G., Verhelst, H., De Bleecker, J., Van Vlem, B., Verloo, P., and Leroy, J. (2001). Blue native polyacrylamide gel electrophoresis: A powerful tool in diagnosis of oxidative phosphorylation defects. *Pediatr. Res.* **50**(5), 658–665.

Zerbetto, E., Vergani, L., and Dabbeni-Sala, F. (1997). Quantification of muscle mitochondrial oxidative phosphorylation enzymes via histochemical staining of blue native polyacrylamide gels. *Electrophoresis* **18**(11), 2059–2064.

Analysis of Protein–Protein Interactions by Chemical Cross-Linking

*Andreas S. Reichert, Dejana Mokranjac,
Walter Neupert, and Kai Hell*

I. INTRODUCTION

Protein–protein interactions can be studied by a variety of approaches. Frequently employed methods include copurification and coimmunoprecipitation of protein complexes. For studying molecular interactions *in vivo*, yeast two-hybrid assays (Toby and Golemis, 2001) and fluorescence-based approaches using fluorescence resonance energy transfer and bioluminescence resonance energy transfer have

been developed (reviewed in Boute *et al.*, 2002; Lippincott-Schwartz *et al.*, 2001). A rather versatile method capable of detecting even weak and transient interactions is chemical cross-linking. The basis of this technique is to covalently link closely apposed proteins or protein domains by chemical cross-linkers. A large number of cross-linkers are available that differ in their selectivity of reactive groups, spacer arm length, bi- or trifunctionality, membrane permeability, solubility, use of iodination, and the presence of affinity tags such as biotin

(Pierce Biotechnology, Inc. provides a list of a wide choice of available reagents). In order to select a suitable cross-linker, these characteristics have to be considered. Chemical properties of the protein, such as the presence of functional groups and its chemical environment, have to be taken into account. Most importantly, cross-links between two proteins can only be established when suitable side chains are present in the right distance and are accessible. Therefore, the characterization of protein–protein interactions by chemical cross-linking often requires empirical testing of a variety of different cross-linkers and reaction conditions. Photoreactive derivates can be used independent of the presence of functional side chains in the proteins to be cross-linked; these can be incorporated co- or post-translationally (reviewed in Brunner, 1996).

Another important issue is to identify an unknown protein that was cross-linked to the protein of interest. When a known protein is suspected to be cross-linked, several approaches can be employed. One possibility is to check whether the cross-link can be immunoprecipitated by using an antibody against the partner protein in question. Another way is to perform the cross-linking experiment in a mutant background that either lacks the candidate protein or contains a variant that can be distinguished from the endogenous protein (e.g., by the presence of a protein tag or by a variation of its apparent size). For identification of unknown binding partners, it is advisable to enrich the cross-link by affinity chromatography and subsequently identify the protein, e.g., by mass spectrometry (reviewed in Farmer and Caprioli, 1998). In that respect, one may consider the use of proteins that contain an affinity tag (e.g., GST or His_{6-10}) or to use biotinylated and/or cleavable cross-linkers.

In the following example the interaction of a mitochondrially targeted precursor protein with components of the mitochondrial inner membrane translocase complex TIM23 is investigated by chemical cross-linking. To this end, the precursor protein $pb_2\Delta19(167)DHFR_{K5}$ (Schneider

et al., 1994) is radiolabeled upon translation *in vitro* and incubated with isolated energized mitochondria from *Neurospora crassa* (for a review about mitochondrial import, refer to Neupert, 1997). The precursor protein is arrested during the *in vitro* import reaction at a certain stage spanning both the outer and the inner membrane. Such a trapping step is often necessary to increase the specificity and yield of the cross-linking reaction. After addition of the homobifunctional, noncleavable, amine-specific cross-linker 1.5-difluoro-2,4-dinitrobenzere (DFDNB) cross-links can be detected by SDS–PAGE and autoradiography (Fig. 15.1). In order to identify to which of the known proteins of the TIM23 complex the precursor was cross-linked, the samples were immunoprecipitated using antibodies against Tim17 and Tim44, subunits of the TIM23 complex.

II. MATERIALS AND INSTRUMENTATION

NADPH (Cat. No. 1045), NADH (Cat. No. 1051), and HEPES (Cat. No. 1009) are from Gerbu, Germany. Methotrexate (MTX, Cat. No. A6770), sorbitol (Cat. No. S-1876), and glycerol (Cat. No. G-7757) are from Sigma-Aldrich. KCl (Cat. No. 104936), Mg/acetate · $4H_2O$ (Cat. No. 105819), KH_2PO_4 (Cat. No. 104873), $Na_2EDTA·2H_2O$ (Titriplex, Cat. No. 108418), glycine (Cat. No. 104201), $MnCl_2·4H_2O$ (Cat. No. 105927), KOH (Cat. No. 105033), Tris (Cat. No. 108382), NaCl (Cat. No. 106404), $NaH_2PO_4·H_2O$ (Cat. No. 106349), Triton X-100 (Cat. No. 108603), dimethyl sulfoxide (DMSO, Cat. No. 102952), ethanol (Cat. No. 100983), and NaOH (Cat. No. 106467) are all from Merck KG, Germany. Sodium dodecyl sulfate (SDS, Cat. No. 20760), bromphenol blue (15375), and phenylmethyl sulforyl fluoride (PMSF, Cat. No. 32395) are from Serva, Germany. ATP·$3H_2O$ (Cat. No. 635316) and protein A–Sepharose CL-4B (Cat. No. 170963) are from Roche,

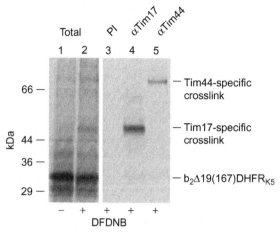

FIGURE 15.1 Cross-linking of arrested b2Δ19(167)DHFR$_{K5}$ to Tim17 and Tim44. Radiolabelled b2Δ19(167)DHFR$_{K5}$ was arrested during import into isolated mitochondria from *Neurospora crassa*. After import, a portion (1/12) of mitochondria were mock treated in the absence of the cross-linker DFDNB (lane 1). The rest was subjected to cross-linking with 200 μM DFDNB for 30 min on ice and another portion (1/12) of the sample was withdrawn (lane 2). Mitochondria were reisolated and used for SDS–PAGE directly (lanes 1 and 2). For immunoprecipitation of cross-linked species, the remaining portion (10/12) of the sample was lysed, split in three aliquots, and incubated with preimmune serum (lane 3) and antibodies raised against Tim17 (lane 4) or Tim44 (lane 5). Immunoprecipitates were harvested and subjected to SDS–PAGE and autoradiography. Exposure times for lanes 1 and 2 were approximately 10 times shorter than for lanes 3 to 5.

Switzerland, and Amersham Biosciences, Sweden, respectively. DFDNB (Cat. No. 21525) is from Pierce Biotechnology, Inc.

III. PROCEDURES

A. Arrest of a Radiolabeled Precursor Protein as a Translocation Intermediate

Solutions

1. *0.1 mM methotrexate (MTX)*: First make a 10 mM stock solution by dissolving 4.54 mg in 1 ml DMSO. Store at −20°C. Dilute the stock solution 100-fold with distilled water to 0.1 mM MTX prior to use.
2. *100 mM NADPH*: Dissolve 8.87 mg in 100 μl distilled water. Make fresh each time.
3. *200 mM ATP*: Dissolve 60.5 mg ATP in 500 μl distilled water and adjust with 10 M KOH to pH 7. Make aliquots and store at −20°C.
4. *200 mM NADH*: Dissolve 14.2 mg NADH in 100 μl distilled water. Make fresh each time.
5. *2× import buffer*: Dissolve 2.38 g HEPES, 18.2 g sorbitol, 1.19 g KCl, 0.43 g Mg acetate·4H$_2$O, 54.4 mg KH$_2$PO$_4$, 186.12 mg Na$_2$EDTA·2H$_2$O, and 99 mg MnCl$_2$·4H$_2$O in 90 ml distilled water. Adjust pH to 7.2 with KOH and fill up with distilled water to 100 ml. Store at −20°C and thaw each time before use.

Steps

1. The precursor protein was synthesized in the presence of [^{35}S]methionine using reticulocyte lysate from Promega, USA, according to the manufacturer's instructions. Preincubate 50 μl of the obtained lysate containing the radiolabelled precursor protein with 0.5 μl of methotrexate (0.1 mM) and 1.25 μl of 100 mM NADPH for 10 min at 25°C.
2. Mix 600 μl 2× import buffer with 24 μl methotrexate (MTX, 0.1 mM), 60 μl NADPH (100 mM), 24 μl ATP (200 mM), 30 μl NADH (200 mM), and 390 μl distilled H$_2$O thoroughly. Add 60 μl mitochondria (10 mg/ml) freshly prepared from *N. crassa* (Sebald *et al.*, 1979), 12 μl of pretreated lysate, and mix gently.
3. Incubate for 15 min at 25°C. Stop the import reaction by placing the sample on ice.

B. Cross-Linking

Solutions

1. *20 mM DFDNB*: Dissolve 2.04 mg in 500 μl DMSO. Make fresh immediately before use.

2. *1M glycine, pH 8.0*: Dissolve 7.51 g in 90 ml distilled water, adjust with KOH to pH 8.0, and fill up to a total volume of 100 ml with distilled water. Store at $-20°C$.
3. *SHKCl*: Dissolve 10.93 g sorbitol, 476 mg HEPES, and 596 mg KCl in 90 ml distilled water, adjust with KOH to pH 7.2, and fill up to 100 ml with distilled water. Store at 4°C.
4. *Laemmli buffer (without β-mercaptoethanol)*: Dissolve 1 g SDS, 5 ml glycerol, and 0.36 g Tris in 40 ml of distilled water. Adjust to pH 6.8 with HCl, add 4 mg bromphenol blue, and fill up to 50 ml with distilled water. Store at room temperature.

Steps

1. Split the import reaction into a 100-μl aliquot and an 1100-μl aliquot, representing 50 and 550 μg mitochondrial protein, respectively.
2. Add 1 μl DMSO to the 50-μg aliquot for mock treatment (total, no cross-link reagent) and 11 μl DFDNB (20 mM stock in DMSO) to the latter aliquot.
3. Incubate the samples for 30 min on ice.
4. Stop the cross-linking reaction by the addition of 1M glycine, pH 8.0, to a final concentration of 0.1M and incubate 10 min on ice.
5. Remove a 100-μl aliquot from the sample containing the DFDNB (total, plus cross-link reagent). The rest of the sample (1000-μl aliquot) can be used for identification of the cross-linked protein. As an example, see Section IIIC. Centrifuge the samples (totals, plus and minus cross-link reagent) for 10 min at 21,000 g at 4°C to reisolate mitochondria. Wash samples once carefully with 1 ml SHKCl and centrifuge again as described earlier.
6. Remove the supernatant and resuspend the pellet in 25 μl Laemmli buffer.
7. Resolve mitochondrial proteins and cross-linked products by SDS–PAGE. The

radiolabeled cross-linked products can be visualized by autoradiography.

C. Identification of the Cross-Linked Product by Immunoprecipitation

Solutions

1. *1M NaPi buffer, pH 8.0*: Dissolve 13.8 g NaH$_2$PO$_4$·H$_2$O in 90 ml distilled water. Adjust pH to 8.0 with NaOH and fill up with distilled water to 100 ml. Store at 4°C.
2. *1M NaCl*: Dissolve 58.44 g NaCl in 1 liter distilled water. Store at room temperature.
3. *10% (w/v) SDS*: Dissolve 10 g SDS in distilled water and bring to a final volume of 100 ml. Store at room temperature.
4. *20% (w/v) Triton X-100*: Dissolve 10 g Triton X-100 in distilled water and bring to a final volume of 50 ml. Store light protected at room temperature.
5. *TBS buffer*: Dissolve 1.21 g Tris and 9 g NaCl in 950 ml distilled water. Adjust pH to 7.4 with HCl and fill up with distilled water to 1 liter.
6. *SHKCl*: See solutions in Section IIIB.
7. *0.2M PMSF*: Dissolve 34.8 mg PMSF in 1 ml ethanol. Prepare fresh each time.
8. *SDS lysis buffer*: Mix 20 μl 1M NaPi buffer, pH 8.0, 100 μl 1M NaCl, 100 μl 10% (w/v) SDS, and 775 μl distilled water. Finally add 5 μl 0.2M PMSF.
9. *IP buffer*: Add 0.5 ml 20% (w/v) Triton X-100 to 50 ml TBS buffer.

Steps

1. For three immunoprecipitation reactions, take 75 μl protein A–Sepharose CL-4B (PAS) beads and wash three times with 1.5 ml TBS buffer.
2. Take 25 μl of PAS beads in 500 μl TBS buffer per reaction and add affinity-purified antibodies (approximately 10 μg of IgGs)

against Tim17 or Tim44, or preimmune serum, respectively.

3. Incubate for at least 30 min at 4°C under gentle shaking.

4. Before immunoprecipitation, wash the PAS beads with the bound antibodies twice with 1 ml TBS and once with 500 μl IP buffer.

5. Take the 1000-μl aliquot from the cross-link reaction (Section IIIB, step 5) for immunoprecipitation and centrifuge for 10 min at 21,000 g at 4°C.

6. Wash isolated mitochondria with 1 ml SHKCl and centrifuge again as in step 5.

7. Lyse mitochondria by resuspension in 50 μl SDS lysis buffer and shake gently for 15 min at room temperature.

8. Dilute with Triton X-100-containing IP buffer to 1 ml.

9. Centrifuge the sample for 30 min at 125,000 g in a TLA45 rotor at 2°C to remove nonsolubilized and aggregated material.

10. Take the supernatant and add equal amounts to PAS beads coupled to antibodies against Tim17, Tim44, or preimmune serum, respectively (see earlier). Fill up to 1 ml with IP buffer.

11. Rotate supernatant for 2 h at 4°C.

12. Wash PAS beads twice with 1 ml IP buffer and once with 1 ml TBS.

13. Elute bound material by adding 40 μl Laemmli buffer. Keep sample for 3 min at 95°C and remove eluate from the PAS beads.

14. Analyze the precipitated material by SDS–PAGE and autoradiography.

IV. COMMENTS

One limitation of this technique is that the amounts of cross-linked as related to noncross-linked protein species are normally less than 1% in case of transient interactions but can exceed 10% with stable interactions. Therefore, it is common to halt a biological process at a defined step to increase the cross-linking yield. In the example given earlier, the precursor protein is arrested at a stage, where it spans the outer and the inner mitochondrial membrane via the translocation machineries. Complete import into mitochondria is inhibited by the C-terminal DHFR domain of the precursor protein, which is stably folded due to the presence of the substrate analogue methotrexate and the cosubstrate NADPH. Translocation intermediates can also be generated by depletion of matrix ATP or depletion of membrane potential. Import at low temperature prolongs the time of interaction. Another strategy could be to use mutants, which still have the potential to transport the precursor to a certain intermediate stage but not further.

DFDNB is a homobifunctional aryl halide-containing cross-linker, which contains two reactive fluorine atoms and reacts with amines. It should be noted that DFDNB is not completely specific for amine groups, but can also react with amino acids containing thiol, phenolate, and imidazolyl groups. Commonly used compounds to cross-link amines are the family of N-hydroxysuccinimide esters (NHS esters). NHS esters react with deprotonated primary amines present at the N terminus of proteins or in the side chain of lysine residues within proteins. It is important that the cross-linking reaction is performed under alkaline conditions within the pH range of 7.5 to 9 to reduce the protonation of the amine groups.

Cross-link efficiencies between interacting proteins are not predictable. Therefore, cross-link conditions have to be optimized for each protein–protein interaction. In addition to the test of various cross-linkers (e.g., BMH, DSG, DSG, EDC, MBS, and SPDP) with different reactivities and spacer lengths, the concentration of the cross-linker can be titrated and the temperature and reaction time can be modified. We normally test cross-linkers at 12°C, room temperature, or on ice in a concentration range of 0.05 to 1 mM.

It is also possible to introduce additional functional groups within the protein to facilitate the cross-link reaction. In the example, additional lysines were added in a position within the precursor protein, where they were likely to be in close contact to the translocation components.

V. PITFALLS

1. As stated earlier, DFDNB and NHS esters react with primary amines. Buffers containing primary amines such as Tris cannot be used as reaction buffers. They would react with the cross-linking reagent and quench the cross-linking reaction. Indeed, Tris at pH 7.5–8 is used as a quenching reagent similar to glycine. In addition, the reaction buffers should not contain high amounts of unspecific proteins, such as bovine serum albumine, whose lysines would compete for the cross-linker. When using thiol-specific reagents, reducing agents such as dithiothreitol or β-mercaptoethanol have to be omitted. For example, some [^{35}S]methionine preparations contain β-mercaptoethanol for stabilization. These should not be used for labeling of proteins *in vitro* if the obtained lysate is directly used for a cross-linking reaction with a thiol-specific cross-linker.

2. Hydrolysis of most cross-linking reagents occurs quite rapidly. Therefore the cross-linkers have to be kept dry during storage to prevent hydrolysis. It is recommended to store the cross-linker under nitrogen once the vial is opened. As this is not always practicable, we store them in an exsiccator. To protect the cross-linker against condensing water, make sure that the vial is equilibrated to room temperature before opening.

3. In an immunoprecipition experiment the detection of cross-linked products can be complicated by the coelution of IgG chains. Under reducing conditions the light and heavy chains of IgGs are separated and have apparent molecular masses of approximately 25 and 50–60 kDa, respectively. In contrast, under nonreducing conditions, when light and heavy chains are not separated, the IgGs have a molecular mass of larger than 150 kDa. This has to be considered when a cross-linked protein is in the respective size range.

References

Boute, N., Jockers, R., and Issad, T. (2002). The use of resonance energy transfer in high-throughput screening: BRET versus FRET. *Trends Pharmacol. Sci.* **23**, 351–354.

Brunner, J. (1996). Use of photocrosslinkers in cell biology. *Trends Cell Biol.* **6**, 154–157.

Farmer, T. B., and Caprioli, R. M. (1998). Determination of protein-protein interactions by matrix-assisted laser desorption/ionization mass spectrometry. *J. Mass Spectrom.* **33**, 697–704.

Lippincott-Schwartz, J., Snapp, E., and Kenworthy, A. (2001). Studying protein dynamics in living cells. *Nature Rev. Mol. Cell Biol.* **2**, 444–456.

Neupert, W. (1997). Protein import into mitochondria. *Annu. Rev. Biochem.* **66**, 863–917.

Schneider, H. C., Berthold, J., Bauer, M. F., Dietmeier, K., Guiard, B., Brunner, M., and Neupert, W. (1994). Mitochondrial Hsp70/MIM44 complex facilitates protein import. *Nature* **371**, 768–774.

Sebald, W., Neupert, W., and Weiss, H. (1979). Preparation of *Neurospora crassa* mitochondria. *Methods Enzymol.* **55**, 144–148.

Toby, G. G., and Golemis, E. A. (2001). Using the yeast interaction trap and other two-hybrid-based approaches to study protein-protein interactions. *Methods* **24**, 201–217.

16

Peroxisomal Targeting as a Tool to Assess Protein–Protein Interactions

Trine Nilsen, Camilla Skiple Skjerpen, and Sjur Olsnes

I. INTRODUCTION

Assessing protein–protein interactions is essential in order to elucidate the molecular mechanisms of the cell. Several methods have been developed to identify, monitor, or confirm such interactions, but they all suffer from limitations, such as being purely *in vitro* methods, not adaptable for mammalian systems, or laborious to implement. By exploiting the peroxisomal targeting machinery it is possible to assay for protein–protein interactions in living mammalian cells in a simple and affordable way. Two peroxisomal targeting signals (PTSs) have been described where PTS-1 consists of a C-terminal tripeptide (typically Ser-Lys-Leu) and PTS-2 of an N-terminal nonapeptide (Gould *et al.*, 1989; Swinkels *et al.*, 1991). Protein oligomers can be imported into the lumen of peroxisomes and, as a consequence, proteins lacking a PTS can be imported in a "piggyback fashion" (McNew and Goodman, 1994; Titorenko *et al.*, 2002).

This specific feature is exploited in the protein–protein interaction assay described here. Initially the protein of interest is targeted to the peroxisomes by adding PTS-1 to its extreme C terminus. This targeting is confirmed by colocalising the expressed PTS-tagged protein with the peroxisomal marker catalase (Fig. 16.1). After confirming peroxisomal targeting, cells are cotransfected with the PTS-tagged protein and a potential interacting partner lacking a PTS. Following coexpression and binding of the two overexpressed proteins in the cytosol, interacting partners can be colocalised in the peroxisomal lumen. Such colocalisation indicates that the two proteins in question bind to each other *in vivo* (Skjerpen *et al.*, 2002).

II. MATERIALS AND INSTRUMENTATION

NH_4Cl (Cat. No. 11145-1), $NaH_2PO_4 \cdot H_2O$ (Cat. No. 17157-1), $Na_2HPO_4 \cdot 12H_2O$ (Cat. No 16579-1), NaCl (Cat. No. 16404-1), and glycerol (Cat. No. 14094-1) are from VWR International.

Tris–HCl (Cat. No. T-1503), digitonin (Cat. No. D-1407), and Triton X-100 (Cat. No. T-9284) are from Sigma. The FuGENE 6 transfection reagent (Cat. No. 1 815 091) is from Roche Molecular Biochemicals, and paraformaldehyde (Cat. No. 762 40) is from Fluka Chemika. Mowiol (Cat. No. 475904) is from Calbiochem. The confocal microscope used for collecting images is a Leica TCS NT from Wezlar, Germany, and the software used for image processing is Adobe Photoshop 5.0 (Mountain View, CA).

III. PROCEDURES

A. Plasmid Preparation

Steps

1. Clone the cDNAs encoding the proteins to be analysed into plasmid vectors suitable for transient transfection.
2. Purify the plasmid DNA. Anion-exchange chromatography, using disposable columns such as Qiagen, is recommended in order to produce high-quality DNA.

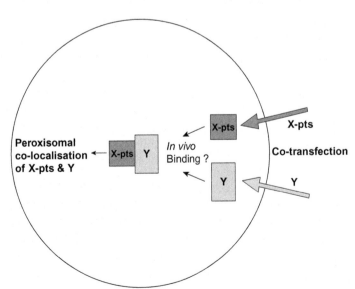

FIGURE 16.1 Peroxisomal colocalisation of two interacting proteins, X and Y. Protein X is targeted to the peroxisomes by the addition of a peroxisomal-targeting sequence (PTS). X-pts and its potential interacting partner, Y, are then cotransfected into cells. If there is an *in vivo* interaction between X-pts and Y in the cytosol, Y can be cotransported into the peroxisomes after the binding takes place in a "piggyback" fashion. In this case the two proteins will colocalise in the peroxisomes as seen when merging the immunofluorescent images of X-pts and Y, which will yield yellow-coloured peroxisomes.

B. Cotransfection and Confocal Microscopy Analysis

Solutions

1. *Phosphate-buffered saline (PBS)*: To make 1 litre, dissolve 0.16 g $NaH_2PO_4 \cdot H_2O$, 1.98 g $Na_2HPO_4 \cdot 12H_2O$, and 8.1 g NaCl in water
2. *40 μg/ml digitonin*: To make a stock solution of 400 μg/ml, dissolve 16 mg digitonin in 40 ml PBS. Sterile filter and store in aliquots at −20°C. Dilute 1:10 to get a working concentration of 40 μg/ml.
3. *3% paraformaldehyde*: To make 150 ml of 3% paraformaldehyde, heat 90 ml H_2O to 60°C and add 4.50 g paraformaldehyde. Stir for 3 h in a sterile hood and make sure not to overheat the solution. Add 2N NaOH drop wise until the solution is clear. Add 50 ml 3× PBS (to make 250 ml 3×PBS, dissolve 0.12 g $Na_2HPO_4 \cdot 12H_2O$, 1.49 g $Na_2HPO_4 \cdot 12H_2O$, and 6.08 g NaCl in water, no pH adjustment) and adjust the pH to 7.2 with HCl. Add water until the total volume is 150 ml. Filter the solution and store aliquots at −20°C.
4. *50 mM NH_4Cl*: Make a 0.5M stock solution by dissolving 4.01 g NH_4Cl in 150 ml PBS. Store at room temperature and dilute 1:10 to obtain a working solution of 50 mM.
5. *0.1% Triton X-100*: To make a stock solution of 10%, add 1 ml Triton X-100 to 9 ml H_2O. Sterile filter and store the aliquots at −20°C. Dilute 1:100 to get a working solution of 0.1%.
6. *Mowiol*: Add 6.7 g 87% glycerol and 2.4 g Mowiol to 6 ml H_2O and 12 ml 0.2M Tris–HCl, pH 8.5. Mix for 10 min at 50°C. Centrifuge the solution for 15 min at 5000 rpm in a Sorvall RC5C centrifuge. Sterile filter and store the aliquots at −20°C.

Steps

1. Plate cells on glass coverslips in a plastic dish (typically ~2 × 10^5 cells/3.5-cm^2

well, depending on cell type) and incubate overnight at 37°C.
2. Cotransfect the cells with approximately 0.5–1.0 μg DNA of each construct by conventional methods for transient transfection, such as the FuGENE 6 transfection reagent. Allow expression for 20–24 h at 37°C.
3. All the following steps are performed at room temperature. Wash the cells in PBS for 5 min on the bench.
4. Incubate with digitonin dissolved in PBS (20–40 μg/ml, depending on cell density and cells type) for 10 min and wash the cells carefully in PBS for 5 min.
5. Fix the cells in 3% paraformaldehyde in PBS for 20–50 min at room temperature or at 4°C overnight.
6. Quench the autofluorescence with 50 mM NH_4Cl in PBS for 10 min.
7. Incubate the cells in 0.1% Triton X-100 in PBS for 5 min.
8. Block unspecific antibody binding sites with 5% FCS in PBS for 20 min.
9. Incubate the cells with primary antibodies. Place the coverslips with cells facing down for 20 min on a piece of Parafilm where 15 μl of the primary antibody solution has been placed. Ensure that catalase detection is included to be able to confirm peroxisomal targeting of the complex.
10. Transfer the coverslips back to the plastic dish and wash three times in PBS for 5 min.
11. Incubate with secondary antibodies as described in step 9.
12. Wash twice for 5 min in PBS.
13. Rinse the coverslips briefly in water and mount on a clean glass slide with Mowiol.
14. Collect separate immunofluorescence images of the two proteins and catalase in cotransfected cells using a confocal microscope.
15. Process the images using suitable software such as Adobe Photoshop 5.0. Merge images representing the PTS-tagged protein

and catalase to confirm peroxisomal targeting in the relevant cell. Then merge the image of the protein lacking a PTS with either catalase or its potential interacting partner to analyse for colocalisation.

IV. COMMENTS

The immunofluorescent protocol described here is designed specifically to visualise the peroxisomes against a low background of cytosolic material. Digitonin is a cholesterol-specific detergent and will therefore leave noncholesterol-containing compartments such as the peroxisomes intact. Because the plasma membrane contains a substantial amount of cholesterol, the cytosolic material can be depleted from the cell by digitonin treatment prior to fixation. The antibodies used in immunoflourescence are given access to the organelle lumen by Triton X-100 treatment.

V. PITFALLS

1. Before initiating cloning, ensure that detection of both expressed proteins by immunofluorescence will be feasible. If no antibodies suitable for immunofluorescence microscopy are available, tags such as Myc or green fluorescence protein should be added to the construct. Adding an autofluorescent marker to at least one of the proteins will facilitate the immunofluorescent procedure as it requires triple staining.
2. It is recommended to use cell lines that are tranfected easily, as this assay is dependent on cotransfection. In principle, any transfectable cell line can be used depending on the purpose of the study.
3. Some cells lines, such as COS-1 cells, are more susceptible to digitonin treatment than others and tend to loosen from the coverslip. In such cases it is recommended to decrease the digitonin concentration.
4. If one or both of the proteins in question are associated with intracellular membranes, it may be difficult, if not impossible, to redirect these proteins to the peroxisomes. A solution to this problem can be to delete the protein domain responsible for the attachment and use the truncated soluble form in this assay. Such an approach has been successful in the detection of protein–protein interaction studies (Nilsen *et al.*, 2004).
5. It may be difficult to determine whether the colocalisation between two proteins is specific, especially if the confocal images have a high background. In order to ensure specificity, merge the two images askew by the approximate diameter of one peroxisome in the relevant image. Using Adobe Photoshop 5.0, select and copy an image approximately one peroxisomal diameter smaller on two sides of the rectangle than the full size image. Then merge it onto its corresponding image of the interacting protein. The two images will now be askew, as merging is centered automatically. When merging a red (X-pts) and a green (Y) image askew, yellow-stained peroxisomes as observed in conventional merged images will appear as red and green (Fig. 16.2).

Acknowledgment

T.N. and C.S.S. are fellows of the Norwegian Cancer Society.

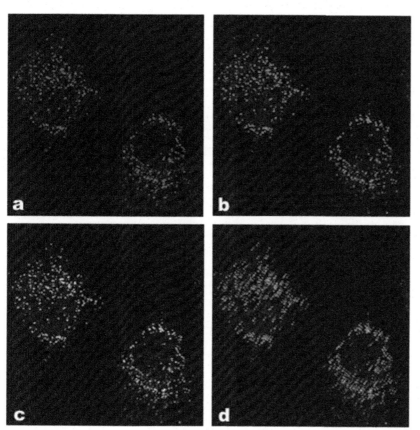

FIGURE 16.2 Askew imaging ensures detection of protein colocalisation specificity in merged images. (a) Confocal image representing protein X-pts. (b) Confocal image representing protein Y. (c) Merged confocal image of a and b. (d) Confocal image merged askew.

References

Gould, S. J., Keller, G. A., Hosken, N., Wilkinson, J., and Subramani, S. (1989). A conserved tripeptide sorts proteins to peroxisomes. *J. Cell Biol.* **108**, 1657–1664.

McNew, J. A., and Goodman, J. M. (1994). An oligomeric protein is imported into peroxisomes in vivo. *J. Cell Biol.* **127**, 1245–1257.

Nilsen, T., Skjerpen, C. S., and Olsnes, S. (2004). Peroxisomal targeting as a tool to assay protein-protein interactions into living cell. *J. Biol. Chem.* **279**, 4794–4801.

Skjerpen, C. S., Nilsen, T., Wesche, J., and Olsnes, S. (2002). Binding of FGF-1 variants to protein kinase CK2 correlates with mitogenicity. *EMBO J.* **21**, 4058–4069.

Swinkels, B. W., Gould, S. J., Bodnar, A. G., Rachubinski, R. A., and Subramani, S. (1991). A novel, cleavable peroxisomal targeting signal at the amino-terminus of the rat 3-ketoacyl-CoA thiolase. *EMBO J.* **10**, 3255–3262.

Titorenko, V. I., Nicaud, J. M., Wang, H., Chan, H., and Rachubinski, R. A. (2002). Acyl-CoA oxidase is imported as a heteropentameric, cofactor-containing complex into peroxisomes of *Yarrowia lipolytica. J. Cell Biol.* **156**, 481–494.

Biomolecular Interaction Analysis Mass Spectrometry

Dobrin Nedelkov and Randall W. Nelson

I. INTRODUCTION

Biomolecular interaction analysis mass spectrometry (BIA/MS) is multidimensional methodology for functional and structural protein analysis (Krone *et al.*, 1997; Nelson *et al.*, 1997a,b, 2000a,b). In essence, BIA/MS represents a synergy of two individual technologies: surface plasmon resonance (SPR) sensing and matrix-assisted laser desorption/ionization time-of-flight (MALDI-TOF) mass spectrometry. SPR is employed for quantification, whereas MS is utilized to delineate the structural features of the analyzed proteins. Proteins are affinity captured and quantified from solution via ligands covalently attached on the SPR sensor surface. Because the SPR detection is nondestructive, proteins retrieved on the SPR sensing surface can be further analyzed via mass spectrometry, either directly from the sensor/chip surface (as described in most of the publications from our laboratory) or separately, following elution and microrecovery (Gilligan *et al.*, 2002; Nelson *et al.*, 1999; Sonksen *et al.*, 1998, 2001). The combination of SPR with MS overcomes the limitation of nondiscriminatory SPR detection and allows for elucidation of non-specific binding (NSB) of other biomolecules to the surface-immobilized biomolecules (or to the underivatized sensor surface itself), binding of protein fragments, protein variants (existing due to posttranslational modifications and point mutations), and complexed (with other molecules) proteins. The BIA/MS approach has been utilized for isolation, detection, and identification of epitope-tagged proteins (Nelson *et al.*, 1999), detection of food pathogens (Nedelkov *et al.*, 2000), analysis of human urine protein biomarkers (Nedelkov and Nelson, 2001a), delineation of *in vivo* assembled multiprotein complexes (Nedelkov and Nelson, 2001c; Nedelkov *et al.*, 2003), and screening for protein functionalities (Nedelkov and Nelson, 2003). Detection of attomole amounts of proteins from complex biological mixtures is possible via the combined SPR-MS approach (Nedelkov and Nelson, 2000a). The modus operandi of the BIA/MS approach is illustrated via the analysis of apolipoproteins A-I and A-II (apoA-I and apoA-II) from human plasma sample.

II. MATERIALS AND INSTRUMENTATION

Biacore X Biosensor (Biacore AB, Uppsala, Sweden) is utilized for the protein affinity retrieval and SPR analysis. Ligands (e.g., antibodies) are immobilized on the carboxymethyl-dextran surface of a CM5 research grade sensor chip (Biacore AB, Cat. No. BR. 1000-14) using amine-coupling kit chemicals (Biacore AB, Cat. No. BR-1000-50).

A chip cutter with a circular heated cutter head (made in our laboratory) is used for excising a chip/plastic mount of a defined circular shape that fits into an appropriately configured MALDI mass spectrometer target (Nedelkov and Nelson, 2000b).

The MALDI matrix is applied to the chip via an aerosol-spraying device (also developed in our laboratory). The device consists of an aspirating/sheath gas needle ($\sim$30-μm orifice), backed by $\sim$30 psi of compressed air (Nedelkov and Nelson, 2000b). The air/matrix solution ratio can be adjusted to produce a fine mist of matrix solution that is aimed at the entire surface of the cutout chip. The matrix of choice is α-cyano-4-hydroxycinnamic acid (ACCA, Aldrich, Milwaukee, WI, Cat. No. 47,687-0), which is further processed by powder-flash recrystallization from a low-heat-saturated acetone solution of the original stock.

MALDI-TOF mass spectrometry analysis from the chip surface is performed on a custom-made MALDI-TOF mass spectrometer (Intrinsic Bioprobes, Inc.). The instrument consists of a linear translation stage/ion source capable of precise targeting of each of the flow cells under a focused laser spot. Ions generated during a 4-ns laser pulse (357 nm, nitrogen) are accelerated to a potential of 30 kV over a single-stage ion extraction source distance of $\sim$2 cm before entering a 1.5-m field-free drift region. The ion signals are detected using a two-stage hybrid (channel plate/discrete dynode) electron multiplier. Time-of-flight spectra are produced by signal averaging of individual spectra from 50 to 100 laser pulses (using a 500-MHz; 500-Ms/s digital transient recorder). Custom software is used in acquisition and analysis of mass spectra. All spectra are obtained in the positive ion mode.

III. PROCEDURES

A. SPR Analysis and Protein Affinity Retrieval

Ligands, Solutions, and Buffers

1. *Ligands*: When using antibodies (or other proteins) as ligands, dilute the original stock solutions with 10 mM acetate buffer, pH 5.0,

to a final ligand concentration of 0.01–0.1 mg/mL.

2. *Buffers and solutions*: Ultrapure, molecular biology grade, sterile water (American Bioanalytical, Natick, MA, Cat. No. 7732-18-5) is used for solution making. HBS-EP [$0.01M$ HEPES, pH 7.4, $0.15M$ NaCl, 0.005% (v/v) polysorbate 20, $3 mM$ EDTA] is typically used as a running buffer in the SPR biosensor. From the amine-coupling kit, make $400 mM$ solution of EDC [N-ethyl-N'-(dimethylamino propyl)carbodiimide] in the ultrapure water, $100 mM$ solution of N-hydroxysuccinimide (NHS), and use the $1M$ ethanolamine (pH 8.5) solution as supplied. Make $60 mM$ HCl for ligand surface regeneration.

Steps

1. Remove a new CM5 sensor chip from its packaging, take out the chip from the plastic housing cassette, wash it with five 200-μl aliquots of water, dry, and put the chip back in the cassette.

2. Insert the chip into the Biosensor (via the dock command), prime, set the flow rate at 5μl/min, and let it equilibrate with the HBS-EP running buffer at 5μl/min for 10–20 min.

3. Set the flow to a single flow cell and start the amine-coupling surface preparation procedure from the Biosensor software. Mix 35μl of the EDC solution and 35μl of the NHS solution in a small vial, and inject 35μl of the mixture over the flow cell surface to activate the carboxyl groups of the dextran matrix.

4. After the surface activation injection has ended, inject 70μl of the antibody solution. Monitor the SPR response for a sharp increase that will indicate binding of the antibody to the chip surface.

5. Inject 35μl of the ethanolamine solution to block free (unreacted) esters.

6. Inject 20μl of the HCl solution to release any noncovalently attached antibody.

7. Measure the SPR response (in resonance units, RU) at the end of the EDC/NHS injection and subtract it from the final SPR response measured after the HCl injection to yield an accurate estimate on the total amount of antibody immobilized on the surface of the flow cell. Because 1 RU equates to 1 pg of proteinacious material per $1 mm^2$ of the flow cell surface (the FC dimensions are $0.5 \times 2 mm$), a response of $\sim 15,000$ RUs at the end of the injection, indicating the immobilization of ~ 100 fmol antibody ($MW_{antibody} \sim 150,000$, 1 RU = 1 pg protein), is generally satisfactory.

8. Repeat steps 2–6 for immobilization of the other antibody in the second flow cell. Alternatively, leave the second flow cell underivatized and use as control.

9. Switch the flow to both flow cells, and let the antibody surface equilibrate with the running buffer for 10–20 min.

10. Inject a 50-μl aliquot of sample (e.g., plasma sample diluted appropriately) and record the SPR response at the end of the injection.

11. Stop the flow of the buffer (set the flow rate at 0μl/min) and quickly remove (undock) the chip from the biosensor.

12. Wash the chip with three 200-μl aliquots of ultrapure water and dry it under a stream of nitrogen.

B. MALDI-TOF Mass Spectrometry Analysis of Chips

Solution

Make a fresh solution of the MALDI matrix [aqueous solution of ACCA in 33% (v/v) acetonitrile and 0.4% (v/v) trifluoroacetic acid]

Steps

1. Place a piece of transparent tape on the glass side of the chip so that it covers both the

glass chip itself and the surrounding plastic support area.

2. Warm up the chip cutter and place the chip in the positioning holder, with the active side facing down (Fig. 17.1A). Rotate the holder to position the chip underneath the heater head (Fig. 17.1B) and gently push down the circular heated head onto the chip. After few seconds, release back the cutter head, rotate the holder, and remove the chip.

3. To separate the round chip/support piece from the rest of the plastic support, gently press on one side of the circle cutout.

4. Remove the tape from the back of the chip cutout and position the chip on a flat surface in a chemical hood, with the active side facing up.

5. Fill in the solution holder on the sprayer device with matrix solution and adjust appropriately the air-to-matrix solution ratio on a test surface. Then, position the device ~5 cm from the chip surface and, in one swift motion, spray the matrix evenly over the entire chip surface (Fig. 17.2A). The matrix mist should moisten, but not completely wet the chip surface (i.e., the tiny matrix droplets should stay as individual drops on the surface and not be connected into one large liquid drop). The matrix droplets will desorb the proteins from their respective capturing affinity ligand and, upon rapid drying, the matrix/protein mixture will be redeposited on the same area from where the proteins were captured originally in the SPR analysis (Figs. 17.2B and 17.2C).

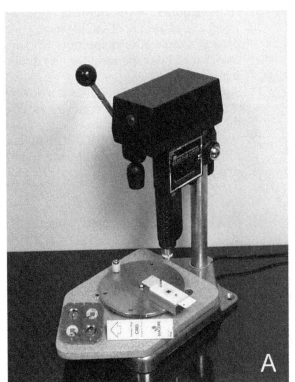

FIGURE 17.1   (A and B) IBI's chip cutter device. (C) The cutout chip fits into appropriately configured MALDI targets.

To prevent fast initial evaporation, cover briefly (for 10–20 s) the chip with a small cap and then let it go to dryness.

6. Place a double-stick tape on the back of the chip and place the chip in the MALDI target so that it is firmly positioned and attached in the probe holder (Fig. 17.1C). Carefully insert the probe into the mass spectrometer. Target each flow cell individually with the laser and acquire mass spectra.

7. Following the MS analysis, chips can be stored shortly at room temperature for further analysis and reevaluation. After a week or so, the quality of mass spectra obtained from the stored chips deteriorates significantly.

D. Example

As an example of concerted BIA/MS analysis, we show the investigation of two human apolipoproteins from human plasma. Antibodies to apolipoprotein A-I (Cat. No. 11A-G2B, 1 mg/ml) and apolipoprotein A-II (Cat. No. 12A-G1B, 1 mg/ml) were purchased from Academy Biomedical (Houston, TX). Human blood was obtained from a single subject recruited within Intrinsic Bioprobes Inc. (IBI), following a procedure approved by the IBI's Institutional Review Board (IRB), and after signing of an informed consent form. Human blood (45 μl) was drawn under sterile conditions from a lancet-punctured finger with a heparinized microcolumn (Drummond Scientific Co., Broomall, PA), mixed with 200 μl of HEPES-buffered saline (HBS-EP) buffer [0.01M HEPES, pH 7.4, 0.15M NaCl, 0.005% (v/v) polysorbate 20, 3 mM EDTA], and centrifuged for 30 s (at 7000 rpm, 2500 g) to pellet red blood cells. The supernatant (plasma) was further diluted 10-fold with HBS-EP buffer to yield plasma sample diluted 100-fold.

Figure 17.3a shows the immobilization of anti-apoA-I and anti-apoA-II in flow cell 1 (FC1) and flow cell 2 (FC2), respectively. The SPR responses indicate immobilization of ~100 fmol of antibody in each flow cell. Figure 17.3b shows a sensorgram resulting from the injection of a 50-μl aliquot of the 100-fold diluted plasma over both flow cells in series. Responses of ~500 RUs are indicated in both flow cells. Figures 17.3c and 17.3d show mass spectra obtained from the surfaces of the two flow cells following plasma sample injection. The mass spectrum obtained from FC1 contains multiply charged ions from apoA-I, in line with the immobilized antibody specificity. Two minor signals due to apoC-I and apoC-I' are also seen and can be attributed to nonspecific binding to the carboxymethyldextran

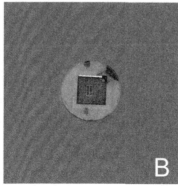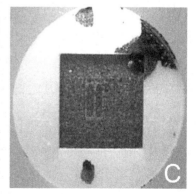

FIGURE 17.2 (A) Post-SPR spraying of the chip with MALDI matrix. (B and C) Dried-out protein/matrix mix chip surface showing the preserved spatial resolution between the two flow cells.

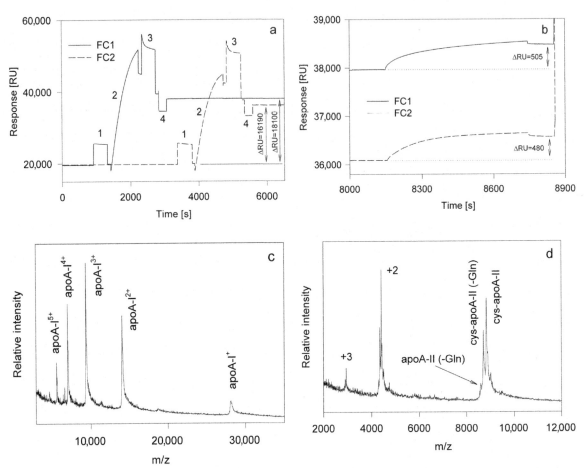

FIGURE 17.3 **(a)** SPR sensorgrams showing the immobilization of anti-apoA-I and apoA-II antibodies in FC1 and FC2, respectively, via (1) EDC/NHS surface activation, (2) antibody injection, (3) ethanolamine blocking, and (4) HCl for non-covalently attached antibody stripping. **(b)** SPR sensorgram showing the injection of 50 μl of 100-fold diluted human plasma over both flow cells. **(c)** Mass spectrum taken from surface FC1 showing the presence of multiply charged apoA-I ions. **(d)** Mass spectrum taken from the surface of FC2 showing the presence of multiple forms of apoA-II.

surface (Nedelkov and Nelson, 2001b). The spectrum obtained from the surface of FC2 contains multiply charged signals from apoA-II. Three major peaks are observed for each charge state: (1) cysteinylated form of apoA-II (cys-apoA-II, MW 8827) (apoA-II contains a single cysteine residue, which is cysteinylated readily *in vivo*), (2) cys-apoA-II missing one terminal glutamine residue [cys-apoA-II (-Gln), MW 8699] (both the C and the N apoA-II termini residues are glutamines), and (3) an apoA-II missing one terminal Gln residue [apoA-II (-Gln), MW 8580]. The native apoA-II (MW 8708) signal is most likely concealed in the mass spectrum by the strong cys-apoA-II (-Gln) peak. The apoA-II homodimer was barely observed in the spectrum (region not shown). Similar results were obtained in another study using a different approach to apolipoprotein extraction and affinity retrieval (Niederkofler *et al.*, 2003).

IV. COMMENTS

Protein modifications (at the native MW level) can be assessed rapidly via BIA/MS, as shown in the apoA-II example. Similarly, protein complexes can be delineated by the observance of signals from the constituent complex components in the MS analysis, and SPR detection can be utilized to monitor specific protein–protein interactions.

V. PITFALLS

1. Contrary to typical kinetic SPR analysis, it is recommended that high ligand densities are utilized in the initial stage of BIA/MS so that ample amounts of analyte are captured for subsequent MALDI-TOF MS analysis.
2. Lower flow rates (promoting mass transfer effects) should also be utilized to increase the binding of the analyte to the immobilized ligands. At high flow rates (60–100 μl/min) low-concentration analytes will not be captured in amounts permissible to downstream MS analysis.
3. Following sample analysis, the chip should be removed promptly from the biosensor to avoid losses of the captured proteins (especially when the interaction system under study exhibits fast dissociation phase). When possible, the undocking should be executed from within the root control software of the biosensor (OS9). For the same reasons, lengthy postcapture washes of the chip should be avoided.
4. Generally, higher quality mass spectra are obtained with ACCA for proteins smaller than ~25 kDa, whereas sinapic acid yields better results for higher molecular mass (>25 kDa) proteins. For the BIA/MS, however, ACCA is superior in that it is a better energy-absorbing matrix and,

consequently, requires less laser power. Lower laser power means that more spectra can be obtained from a single spot and the fast burning through the matrix/sample layer is avoided. Generally, the appearance and the intensity of the multicharged ion analyte signals obtained with ACCA are better indicators of the presence and the mass of the on-chip retained analyte during the BIA/MS analysis.

5. Reapplication of more matrix (following initial application and MS analysis) does not yield better signals and generally results in a decreased signal-to-noise ratio.

References

Gilligan, J. J., Schuck, P., and Yergey, A. L. (2002). Mass spectrometry after capture and small-volume elution of analyte from a surface plasmon resonance biosensor. *Anal. Chem.* **74**, 2041–2047.

Krone, J. R., Nelson, R. W., Dogruel, D., Williams, P., and Granzow, R. (1997). BIA/MS: Interfacing biomolecular interaction analysis with mass spectrometry. *Anal. Biochem.* **244**, 124–132.

Nedelkov, D., and Nelson, R. W. (2000a). Exploring the limit of detection in biomolecular interaction analysis mass spectrometry (BIA/MS): Detection of attomole amounts of native proteins present in complex biological mixtures. *Anal. Chim. Acta* **423**, 1–7.

Nedelkov, D., and Nelson, R. W. (2000b). Practical considerations in BIA/MS: Optimizing the biosensor-mass spectrometry interface. *J. Mol. Recogn.* **13**, 140–145.

Nedelkov, D., and Nelson, R. W. (2001a). Analysis of human urine protein biomarkers via biomolecular interaction analysis mass spectrometry. *Am. J. Kidney Dis.* **38**, 481–487.

Nedelkov, D., and Nelson, R. W. (2001b). Analysis of native proteins from biological fluids by biomolecular interaction analysis mass spectrometry (BIA/MS): Exploring the limit of detection, identification of non-specific binding and detection of multi-protein complexes. *Biosens. Bioelectron.* **16**, 1071–1078.

Nedelkov, D., and Nelson, R. W. (2001c). Delineation of *in vivo* assembled multiprotein complexes via biomolecular interaction analysis mass spectrometry. *Proteomics* **1**, 1441–1446.

Nedelkov, D., and Nelson, R. W. (2003). Delineating protein-protein interactions via biomolecular interaction analysis-mass spectrometry. *J. Mol. Recogn.* **16**, 9–14.

Nedelkov, D., Nelson, R. W., Kiernan, U. A., Niederkofler, E. E., and Tubbs, K. A. (2003). Detection of bound and free IGF-1 and IGF-2 in human plasma via biomolecular interaction analysis mass spectrometry. *FEBS Lett.* **536**, 130–134.

Nedelkov, D., Rasooly, A., and Nelson, R. W. (2000). Multitoxin biosensor-mass spectrometry analysis: A new approach for rapid, real-time, sensitive analysis of Staphylococcal toxins in food. *Int. J. Food Microbiol.* **60**, 1–13.

Nelson, R. W., Jarvik, J. W., Taillon, B. E., and Tubbs, K. A. (1999). BIA/MS of epitope-tagged peptides directly from *E. coli* lysate: Multiplex detection and protein identification at low-femtomole to subfemtomole levels. *Anal. Chem.* **71**, 2858–2865.

Nelson, R. W., Krone, J. R., and Jansson, O. (1997a). Surface plasmon resonance biomolecular interaction analysis mass spectrometry. 1. Chip-based analysis. *Anal. Chem.* **69**, 4363–4368.

Nelson, R. W., Krone, J. R., and Jansson, O. (1997b). Surface plasmon resonance biomolecular interaction analysis mass spectrometry. 2. Fiber optic-based analysis. *Anal. Chem.* **69**, 4369–4374.

Nelson, R. W., Nedelkov, D., and Tubbs, K. A. (2000a). Biomolecular interaction analysis mass spectrometry: BIA/MS can detect and characterize proteins in complex biological fluids at the low- to subfemtomole level. *Anal. Chem.* **72**, 404A–411A.

Nelson, R. W., Nedelkov, D., and Tubbs, K. A. (2000b). Biosensor chip mass spectrometry: A chip-based proteomics approach. *Electrophoresis* **21**, 1155–1163.

Niederkofler, E. E., Tubbs, K. A., Kiernan, U. A., Nedelkov, D., and Nelson, R. W. (2003). Novel mass spectrometric immunoassays for the rapid structural characterization of plasma apolipoproteins. *J. Lipid Res.* **44**, 630–639.

Sonksen, C. P., Nordhoff, E., Jansson, O., Malmqvist, M., and Roepstorff, P. (1998). Combining MALDI mass spectrometry and biomolecular interaction analysis using a biomolecular interaction analysis instrument. *Anal. Chem.* **70**, 2731–2736.

Sonksen, C. P., Roepstorff, P., Markgren, P. O., Danielson, U. H., Hamalainen, M. D., and Jansson, O. (2001). Capture and analysis of low molecular weight ligands by surface plasmon resonance combined with mass spectrometry. *Eur. J. Mass Spectrom.* **7**, 385–391.

Ligand Blot Overlay Assay: Detection of Ca^{+2}- and Small GTP-Binding Proteins

Pavel Gromov and Julio E. Celis

I. INTRODUCTION

Protein-targeting interactions play a central role in most biological processes. Their detection and analysis *in vitro* can provide important information on specificity, affinity, and structure/function relationships that are realized *via* these interactions. Protein blot overlay assays, also known as "Western–Western," "Far–Western," "ligand," or "affinity" blotting, are powerful techniques for detecting and analyzing proteins or protein motifs involved in cellular-targeting processes (Clegg *et al.*, 1998 and references therein).

These methods are based on the principle that proteins, or protein fragments resolved by electrophoresis and transferred to an immobilizing matrix such as nitrocellulose or a nylon membrane, can react with putative binding ligands. This article describes protocols for identifying Ca^{+2}- and GTP-binding proteins using whole cellular protein extracts from noncultured human psoriatic keratinocytes and COS-1 cells.

II. MATERIALS AND INSTRUMENTATION

A. ^{45}Ca Overlay Assay

Imidasole (Cat. No. I-0250) is from Sigma, MgCl$_2$ · 6H$_2$O (Cat. No. 105832) is from Merk. ^{45}CaCl$_2$ (Cat. No. CES3) is from Amersham. Nitrocellulose membrane sheets (Hybond C, Cat. No. RPN. 203C) are from Amersham. X-ray films (X-Omat UV, 18 × 24 cm, Cat. No. 524 9792) are from Kodak.

B. α-^{32}GTP Overlay Assay

Tween 20 (Cat. No. 822 184) and MgCl$_2$ · 6H$_2$O(Cat. No. 105832) are from Merk. Dithiothreitol (DTT, Cat. No. D-0632) and ATP (Cat. No. A-2383) are from Sigma. Tris base (Cat. No. 648311) is from Calbiochem. [α-^{32}P]GTP (10 mCi/ml, Cat. No. PB 10201) and nitrocellulose membranes (Hybond C, Cat. No. RPN. 203C) are from Amersham. X-ray films (X-Omat UV, 18 × 24 cm, Cat. No. 524 9792) are from Kodak. All other reagents and materials are as described elsewhere (Celis and Celis, 1997).

III. PROCEDURES

The protocol for blot overlay detection of Ca^{+2}- and GTP-binding proteins is exemplified using whole cellular protein extracts from noncultured human psoriatic keratinocytes and COS-1 cells, but can be applied to a variety of other cultured cells, tissue samples, and biological fluids.

A. ^{45}Ca Overlay Assay

Calcium ion is a universal intracellular signal that acts as a important second messenger for many cellular processes and whose effect is modulated by specific calcium-binding proteins (Berridge *et al.*, 1998). According to well-conserved structural elements, these proteins can be grouped into different families, including annexins, C2 domain proteins, and EF-hand proteins (Celio *et al.*, 1996; Maki *et al.*, 2002; Heizmann *et al.*, 2002). The calcium overlay assay (Maruyama *et al.*, 1984) as described here is essentially a specific application of a general metal ion-binding assay (Aoki *et al.*, 1986) and is widely used for studying various calcium-binding proteins (Son *et al.*, 1993), including those containing EF hands (Hoffmann *et al.*, 1993). Proteins are separated by means of one (1D)- or two-dimensional (2D) gel electrophoresis transferred to a nitrocellulose membrane, and overlaid with radioactive ^{45}Ca. Calcium-binding proteins are detected by autoradiography or phosphorimaging.

Solutions

The volumes given in the protocol are for 2D gel nitrocellulose blots (14 × 16 cm). The volumes for 1D gel blots (0.6 × 16 cm) are given in parentheses.

1. *Washing buffer*: 60 mM KCl, 5 mM MgCl$_2$, and 10 mM imidazole–HCl, pH 6.8. Prepare stock solutions in glass-distilled water. To make 1 liter of 3 M KCl, dissolve 223.6 g in 700 ml of water and bring to 1 liter. To make 1 liter of 1 M MgCl$_2$, dissolve 203.1 g in 700 ml of water and bring to 1 liter. Just before use, prepare 40 ml of the 1 M imidazole solution by dissolving 2.72 g of imidazole in 20 ml of distilled water and adjust to pH 6.8 with HCl. Complete to 40 ml with distilled water. To make 500 ml (20 ml) of washing buffer, enough for one 14 × 16-cm (0.6 × 14-cm) membrane, combine the following: 10 ml (0.4 ml) of 3 M KCl, 2.5 ml (0.1 ml) of 1 M MgCl$_2$, 5 ml (0.2 ml) of 1 M imidazole, and complete to 500 ml (20 ml) with distilled water.

2. *Probing buffer*: Add 15 μl (2 μl) of $^{45}CaCl_2$ to 15 ml (2 ml) of washing buffer (final concentration of 1 μCi/ml)
3. *Aqueous ethanol*: Prepare 150 ml (5 ml) of 67% aqueous ethanol per membrane. Add 50 ml (1.67 ml) of distilled water to 100 ml (3.33 ml) of 96% ethanol.

Steps

1. Work with radioactivity according to the safety procedures enforced in your laboratory.
2. Transfer proteins from the gels to the nitrocellulose membrane.
3. Place the nitrocellulose sheet in a rectangular glass container (19 × 24 cm) containing 100 ml of washing buffer. Wear gloves when handling the membrane. The nitrocellulose sheet should be placed with the protein-bearing side facing upward. Rinse the nitrocellulose sheet twice with washing buffer.
4. Remove the membrane from the blotting chambers and wash briefly with the blotting buffer. The membrane can be dried and stored in a plastic bag at room temperature until further use.
5. Soak the membrane in 150 ml (2 ml) of washing buffer and shake gently for 20 min.
6. Remove the washing buffer and add 150 ml (2 ml) of fresh washing buffer. Shake gently for 20 min.
7. Repeat step 6.
8. Place the membrane or strip in a plastic bag and add 15 ml (2 ml) of probing buffer. Seal the bag and incubate for 10 min with gentle agitation. Alternatively, incubate the membrane in an appropriate rectangular glass or plastic container containing 150 ml (2 ml) of probing buffer. The incubation can be extended up to 1 h without adverse effects. Handle radioactive material with care.

9. Transfer the membrane to an appropriate glass or plastic container and add 150 ml (2 ml) of aqueous ethanol. Shake for 5 min with gentle agitation. Dispose of the radioactive solutions according to the safety procedure in your laboratory.
10. Carefully remove the nitrocellulose sheet or strip from the container using plastic tweezers and air dry for at least 4 h at room temperature.
11. Expose the dried membrane to X-ray film for 1 h to 3 days at room temperature. The membrane can be stained with amido black after autoradiography to facilitate protein identification.

Figure 18.1 shows an isoelectric focusing two-dimensional gel of total protein extracts from psoriatic keratinocytes transferred to nitrocellulose membrane and probed with $^{45}CaCl_2$.

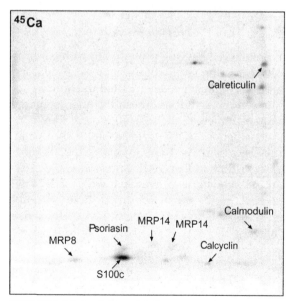

FIGURE 18.1 Calcium-binding proteins expressed by psoriatic keratinocytes. Autoradiograph of a $^{45}Ca^{2+}$ blot overlay of a 2D gel of whole protein extracts from noncultured, psoriatic keratinocytes.

Comments

No significant differences in calcium binding are observed when using membranes that had been dried before probing or membranes that had been used directly after blotting.

To decrease nonspecific binding, competing metal ions such as Mg^{+2} may be added in steps 5–8 to a final concentration of 10 mM.

The use of a rotating roller system for incubation of nitrocellulose membranes facilitates the handling of the membrane, reduces considerably the volume of reagents used, and provides even detection of the calcium-binding proteins.

Pitfalls

1. Use a fresh imidazole solution.
2. Do not dry the membrane between pieces of filter paper as it leads to severe background problems.

B. α-^{32}GTP Overlay Assay

Small GTP-binding proteins constitute a rapidly increasing family of monomeric regulatory switches that have been adopted to control various cellular activities. These include proliferation and differentiation (Ras), protein transport and secretion (Rab), cytoskeletal (Rho), and nuclear assembly (Ran) (Manser, 2002; Dasso 2002; Pruitt and Der, 2001; Jaffe and Hall 2002). Unlike oligomeric G proteins, which are composed of α, β, and γ subunits, these proteins are able to bind GTP specifically when separated by SDS–PAGE and blotted onto nitrocellulose or nylon membranes. The blot overlay nucleotide-binding assay allows detection of small GTP-binding proteins in polypeptide mixture separated by one or 2D PAGE. The protocol presented here follows closely those described by McGrath *et al.* (1984), Doucet and Tuana (1991), and Gromov and Celis (1994).

Solutions

The volumes given are for nitrocellulose blot sheets of 14 × 16 cm in size.

1. *10% Tween 20*: To make 50 ml, weigh 5 g of Tween 20 and complete to 50 ml with distilled H$_2$O. Mix carefully. Store at room temperature.
2. *100 mM ATP*: To make 1 ml, add 55.1 mg of ATP and complete to 1 ml with distilled water. Aliquot in 100-µl portions and store at −20°C.
3. *0.1 M MgCl$_2$*: To make 50 ml, weigh 0.102 g of MgCl$_2$ · 6H$_2$O and complete to 50 ml with distilled water. Store at 4°C.
4. *Washing buffer*: 50 mM Tris–HCl, pH 7.6, 10 µM MgCl$_2$, and 0.3% Tween 20. To make 1 liter, add 6.055 g of Tris base to 800 ml of distilled H$_2$O and titrate with HCl. Add 30 ml of a 10% stock solution of Tween 20 and 100 ml of a 0.1 M stock solution of MgCl$_2$. After dissolving, complete to 1 liter with distilled H$_2$O and store at 4°C.
5. *ATP overlay buffer*: 50 mM Tris–HCl, pH 7.6, 10 µM MgCl$_2$, 0.3% Tween 20, 100 mM DTT, and 100 µM ATP. To make 100 ml, dissolve 1.54 g DTT in 90 ml of washing buffer and add 100 µl of a 0.1 M stock solution of ATP. After dissolving, complete to 100 ml with washing buffer and store at 4°C.
6. *Binding buffer*: 50 mM Tris–HCl, pH 7.6, 10 µM MgCl$_2$, 0.3% Tween 20, 100 mM DTT, 100 µM ATP, and 1 nM [α-^{32}P]GTP [1 µCi [α-^{32}P]GTP/ml]. To make 100 ml, add 10 µl of a 10-mCi/ml solution of [α-^{32}P]GTP to the ATP overlay buffer. Prepare directly in the binding container prior to use (see step 4).

Steps

1. Transfer proteins from the gels to the nitrocellulose membrane.
2. Place the nitrocellulose sheet in a rectangular glass container (19 × 24 cm)

containing 100 ml of washing buffer. Wear gloves when handling the membrane. The nitrocellulose sheet should be placed with the protein-bearing side facing upward. Rinse the nitrocellulose sheet twice with washing buffer.

3. Remove the washing buffer from the container and fill it with 50 ml of ATP overlay buffer. Place the container on an orbital shaker and incubate for 10 min with gentle agitation at room temperature.

4. Remove the blotting container from the orbital shaker platform and add 10 μl of a 10-mCi/ml solution of [α-^{32}P]GTP. Incubate with gentle agitation for 60 min at room temperature. Handle radioactive material with care.

5. Remove the binding buffer from the container and fill it with 50 ml of washing buffer. Soak the nitrocellulose sheet for 10 min at room temperature. Dispose of radioactive solutions according to the safety regulations in your laboratory.

6. Repeat the washing procedure twice at room temperature (10 min per wash). Use as much washing buffer as possible.

7. Carefully remove the nitrocellulose sheet from the container using plastic tweezers and air dry for at least 4 h at room temperature.

8. Place the air-dried nitrocellulose sheet into a cassette for autoradiography (12–72 h at −80°C) using an X-ray film and an intensifying screen.

Comments

The use of a rotating roller system for incubating the membranes is recommended.

DTT (100 mM final concentration in steps 3 and 4) enhances the GTP-binding ability of small GTP-binding proteins. This improves considerably the signal-to-noise ratio as well as the sensitivity of the procedure (Gromov and Celis, 1994).

To reduce background, we recommend ATP as an effective competitor for nonspecific binding of GTP to the nitrocellulose. Alternatively, use bovine serum albumin (0.3% final concentration) in the overlay and binding buffer as recommended by McGrath et al. (1984) and Doucet and Tuana (1991).

Bound [α-^{32}P]GTP can be removed from the nitrocellulose blot without detectable loss of proteins by incubating the membrane in a solution containing 50 mM Tris–HCl, pH 7.4, and 1% SDS for 30 min at room temperature. The blot can be then reprobed with [α-^{32}P]GTP. In this case do not dry the membrane after probing and perform autoradiography, keeping the membrane in the plastic bag.

To facilitate the identification of [α-^{32}P]GTP-binding proteins on whole protein extracts separated by 2D gel electrophoresis, we recommend adding a small amount of [^{35}S]methionine-labeled proteins from an appropriate source (e.g., keratinocytes) to the protein mixture prior to electrophoresis. Following the [α-^{32}P]GTP-binding overlay assay, the nitrocellulose blot is subjected to autoradiography using two films placed on top each other. The first film, which is placed in direct contact with the blot, visualizes both ^{35}S and ^{32}P isotopes, while the second one reveals only ^{32}P. The positions of detected spots may be compared with those in the master keratinocyte database (http://proteomics.cancer.dk). Using the protocol described here, it is possible to detect many small GTP-binding proteins in various cell types and tissues. Representative 2D autoradiographs of [α-^{32}P]GTP-binding proteins from COS-1 cells that transiently express ADP-ribosylation factor, rab11a, and p21-ras are shown in Fig. 18.2.

Pitfalls

1. Do not dry the nitrocellulose membrane after protein transfer as it substantially reduces the efficiency of GTP binding.

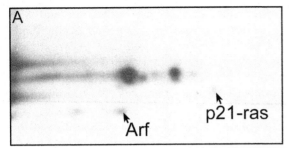

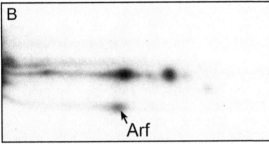

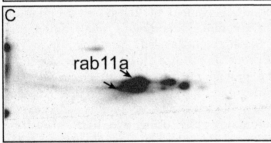

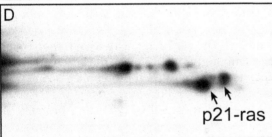

FIGURE 18.2 Two-dimensional blot autoradiographs of [α-^{32}P]GTP-binding proteins from COS-1 cells that transiently express several small GTP-binding proteins. (A) Control, nontransfected cells, (B) ADP-ribosylation factor (Arf), (C) rab11a, and (D) p21-ras.

2. Make sure that the solution covers the whole nitrocellulose sheet surface during agitation. Avoid scratching or tearing of the nitrocellulose membrane during manipulation.
3. Use high-grade ATP. Small traces of cold GTP, which may contaminate commercial ATP, decrease the efficiency of [α-^{32}P]GTP binding.

References

Aoki, Y., Kunimoto, M., Shibata, Y. Y., and Suzuki, K. T. (1986). Detection of metallothionein on nitrocellulose membrane using Western blotting technique and its application to identification of cadmium-binding proteins. *Anal. Biochem.* **157**, 117–122.

Berridge, M. J., Bootman, M. D., and Lipp, P. (1998). Calcium—a life and death signal. *Nature* **395**, 645–648

Celio, M., Pauls, T., and Schwaller, B. (1996). *"Guidebook to the Calcium-Binding Proteins"*, pp. 1–238. Oxford Univ. Press., Oxford.

Celis, A., and Celis, J. E. (1997). General procedures for tissue culture. *In* "Cell Biology: A Laboratory Handbook" (J. E. Celis, N. Carter, T. Hunter, D. Shotton, K. Simons, and J. V. Small eds.), Vol. 1, pp 5–16. Academic Press.

Clegg, R. A. (ed.) (1998). Protein targeting protocols.*In* "Methods in Molecular Biology," Vol. 88. Humana Press, Totowa, NJ.

Dasso, M. (2002). The Ran GTPase: Theme and variations. *Curr. Biol.* **12**, R502–R508.

Daucet, J.-P., and Tuana, B. S. (1991). Identification of low molecular weight GTP-binding proteins and their sites of interaction in subcellular fractions from skeletal muscle. *J. Biol. Chem.* **266**, 17613–17620.

Gromov, P. S., and Celis, J. E. (1994). Some small GTP-binding proteins are strongly downregulated in SV40 transformed human keratinocytes. *Electrophoreis* **15**, 474–481.

Hoffmann, H. J., Olsen, E., Etzerodt, M., Madsen, P., Thogersen, H.-G., Kruse, T., and Celis, J. E. (1994). Psoriasin binds calcium and is differentially regulated with respect to other members of the S100 protein family. *J. Invest. Dermatol.* **103**, 370–375.

Jaffe, A. B., and Hall, A. (2002). Rho GTPases in transformation and metastasis. *Adv. Cancer Res.* **84**, 57–80.

Heizmann, C. W., Fritz, G., and Schafer, B. W. (2002). S100 proteins: Structure, functions and pathology. *Front Biosci.* **7**, 1356–1368.

Maki, M., Kitaura, Y., Satoh, H., Ohkouchi, S., and Shibata, H. (2002). Structures, functions and molecular evolution of the penta-EF-hand Ca^{2+}-binding proteins. *Biochim. Biophys. Acta* **1600**, 51–60.

Manser, E. (2002). Small GTPases take the stage. *Dev. Cell* **3**, 323–328.

Maruyama, K., Mikawa, T., and Ebashi, S. (1984). Detection of calcium-binding proteins by ^{45}Ca autoradiography on nitrocellulose membrane after sodium dodecyl sulfate gel electrophoesis. *J. Biochem.* **95**, 511–519.

McGrath, J. P., Capon, D. J., Goeddel, D. V., and Levinson, A. D. (1984). Comparative biochemical properties of normal and activated human ras p21 protein. *Nature* **310**, 644–649.

Pruitt, K., and Der, C. J. (2001). Ras and Rho regulation of the cell cycle and oncogenesis. *Cancer Lett.* **171**, 1–10.

Son, M., Gunderson, R. E., and Nelson, D. L. (1993). A 2nd member of the novel Ca^{2+}-dependent protein-kinase family from paramecium-tetraurelia; purification and characterization. *J. Biol. Chem.* **268**, 5840–5948.

Modular Scale Yeast Two-Hybrid Screening

Christopher M. Armstrong, Siming Li, and Marc Vidal

I. INTRODUCTION

The observation that most transcription factors can be separated into a DNA-binding domain (DB) and a transcriptional activation domain (AD) led to the development of the yeast two-hybrid (Y2H) system as an *in vivo* screen or selection to identify and characterize protein interactions (Fields and Song, 1989). Using the Y2H, one can identify potentially interacting proteins (X–Y heterodimers or X–X homodimers) by generating two different hybrid proteins: one with protein X fused to DB and the other with protein Y fused to AD (see Fig. 19.1). If protein X and Y interact, the AD can be brought to the promoter by DB-X and thereby activate the gene driven by that promoter (usually a selectable or screenable marker). By fusing a library to the AD (Fields and Song, 1989) and, in some special cases, to DB (Du *et al.*, 1996), it is possible to screen for proteins and identify potential interactors.

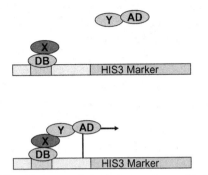

FIGURE 19.1 Two-hybrid interactions. (Top) Two fusion proteins are created, DB-X and AD-Y. If protein X fails to interact with protein Y, then the activation domain will not be brought to the promoter and the marker gene (in this case HIS3) will fail to activate. (Bottom) Protein X and protein Y interact successfully, bringing the activation domain to the promoter and activating the marker gene successfully.

Two-hybrid screens have been used quite successfully by scientists interested in identifying potential interacting partners to their protein of interest. While this one gene at a time approach has been fruitful, the ease of use of Y2H has allowed it to be scaled up to perform screens on more of a proteome scale (Ito *et al.*, 2000; Uetz *et al.*, 2000; Walhout *et al.*, 2000). While whole genome two-hybrids may be beyond the scope or interest of most laboratories, medium size screens (on the level of 20 to 50 genes) can be done easily by one or a few scientists (Davy *et al.*, 2001; Drees *et al.*, 2001; Boulton *et al.*, 2002). Screens of this size allow scientists to approach problems on a more modular scale, i.e., studying most of the genes involved in a process rather than a few at a time (Hartwell *et al.*, 1999), thereby addressing more global questions than traditional single gene approaches allow.

While the Y2H is quite powerful, it is important to remember that genes identified in a Y2H screen are only potential interactors and further experiments are necessary to validate the relevance of the interactions. Techniques such as GST-pulldowns or immunoprecipitations can help confirm that the interaction exists, while analysis of expression patterns and phenotypes can help establish the functional relevance of the predicted interactions. The power of the Y2H is its ability to act as a starting point for the identification of interacting proteins.

Many variations of the method exist, but the fundamentals are the same for all of them. Here we use the strains and vectors described in Vidal *et al.* (1996), but the protocols work just as well with other strains so long as you take into account any changes in selectable markers and reporter assays that may be an issue. The strain used is MaV203, which has three screening markers: *GAL1::lacZ*, *SPAL10::URA3*, and *GAL1::HIS3* (Vidal *et al.*, 1996). Initial screening uses the *HIS3* marker to identify potential positives. Secondary screening is then done with other markers to test the strength of the interactions. The techniques here are derived in part from the techniques described in Walhout and Vidal (2001), but have been streamlined to make it possible to work with many baits in parallel.

II. MATERIALS AND INSTRUMENTATION

Bacto agar is obtained from Labscientific (Cat. No. A466). Bactopeptone (Cat. No. DF0118170), yeast extract (Cat. No. DF0127-17-9), ammonium sulfate (Cat. No. A702-500), *N,N*-dimethylformamide (Cat. No. D119-1), 3-mm glass beads (Cat. No. 11-312A), and 125-mm Whatman filter papers (Cat. No. 09-868C) are from Fisher Scientific. 3-Amino-1,2,4-triazole (3-AT, Cat. No. A-8056), salmon testes DNA (Cat. No. D9156), glucose (Cat. No. G8270), polyethylene glycol MW 3350 (Cat. No. P-4338), lithium acetate (Cat. No. L-4158), 5-bromo-4-chloro-3-indolyl-β-d-galactopyranoside (X-Gal, Cat. No. B-4252), 2-mercaptoethanol (Cat. No. M-3148), yeast synthetic drop-out medium amino acid supplement (without histidine, leucine, tryptophan, and uracil) (Cat. No. Y-2001), tryptophan (Cat. No. T-0254), uracil (Cat. No. U-0750), and histidine (Cat. No. H-6034) are from Sigma. 5-Fluoroorotic acid (5FOA, Cat. No. 1555) is from BioVectra. Zymolyase-20T (Cat. No. 120491) is from Seikagaku. Airpore tape sheets (Cat. No. 19571) are from Qiagen. Nitrocellulose filters (Cat. No. WP4HY13750) are from Osmonics. Eight-and-a-half-inch replica velvets (#2008) and 6-in. replica blocks (#4006) can be obtained from Cora Styles Needles 'N Blocks (www.corastyles.com).

MaV203 yeast strain (MATα, leu2-3,112, trp-901, his3Δ200, ade2-1, gal4Δ, gal80Δ, SPAL10::URA3, GAL1::lacZ, GAL1::HIS3-@LYS2, can1^R, cyh2^R) can be obtained from Invitrogen as part of the ProQuest two-hybrid system with Gateway Technology (Cat. No. 10835–031) or directly from the Vidal laboratory (http://vidal.dfci.harvard.edu/).

III. PROCEDURES

A. Preparation of Yeast Culture Plates and Medium

1. *YPD liquid media*: Add 10 g of yeast extract and 20 g of Bactopeptone to 1 liter of dH$_2$O and autoclave. Add 50 ml of 40% glucose before use.
2. *20 mM uracil*: Dissolve 1.21 g of uracil in 500 ml of ddH$_2$O. Filter to sterilize. Note that it may be necessary to heat the water to get the uracil into solution.
3. *100 mM histidine*: Dissolve 7.76 g of histidine in 500 ml of ddH$_2$O. Filter to sterilize and wrap the bottle in foil.
4. *40 mM tryptophan*: Dissolve 4.08 g of tryptophan in 500 ml of ddH$_2$O. Filter to sterilize and wrap the bottle in foil and keep at 4°.
5. *Synthetic complete (SC) liquid medium*: To prepare a liter of SC medium, add 1.4 g of

yeast synthetic drop-out medium amino acid supplement, 1.7 g of yeast nitrogen base, and 5 g of ammonium sulfate into 925 ml of water. Adjust the pH with 10N NaOH to a final pH of 5.9. Autoclave media to sterilize. Add 50 ml 40% glucose before use. This media is -Leu, -Trp, -Ura, -His. To supplement with a missing amino acid/nucleotide, add 8 ml of the appropriate solution (e.g., to make SC-Leu-Trp, add 8 ml of 20 mM uracil and 8 ml of 100 mM histidine).

6. *Synthetic complete (SC) agar plates*: To prepare a liter of SC plates (ten to twelve 15-cm petri dishes), add 1.4 g of yeast synthetic drop-out medium amino acid supplement, 1.7 g of yeast nitrogen base, and 5 g of ammonium sulfate into 425 ml of water. Adjust pH to 5.9 with 10N NaOH. At the same time, in a separate flask, add 20 g of bacto agar into 500 ml of water. Autoclave both agar and SC solutions. After autoclaving, mix the flasks. Cool to 55°C in a water bath. Add 50 ml of 40% glucose and 8 ml of the appropriate amino acid solution before pouring.
7. *SC-Leu plates*: Prepare 1 liter of SC/agar and add 8 ml of 20 mM uracil, 8 ml of 100 mM histidine, and 8 ml of 40 mM tryptophan before pouring. Pour into ten to twelve 15-cm petri plates.
8. *SC-Leu-Trp plates*: Prepare 1 liter of SC/agar and add 8 ml of 20 mM uracil and 8 ml of 100 mM histidine before pouring. Pour into ten to twelve 15-cm petri plates.
9. *SC-Leu-Trp-His+3-AT plates (3-AT plates)*: Prepare 1 liter of SC/agar and add 8 ml of 20 mM uracil and 1.18 g of 3-AT powder before pouring (this makes 20 mM 3-AT plates). Pour into ten to twelve 15-cm petri plates.
10. *SC-Leu-Trp+5-FOA plates (5-FOA plates)*: Prepare 1 liter of SC/agar and add 8 ml of 20 mM uracil and 8 ml of 100 mM histidine before pouring. Add 2 g of 5-FOA powder

and stir to dissolve. It will take a while for 5-FOA to go into solution. Pour into ten to twelve 15-cm petri plates.

11. *YPD plates*: Add 10 g of yeast extract and 20 g of Bactopeptone to 500 ml of dH$_2$O. At the same time, in a separate flask, add 20 g of bacto agar to 500 ml of dH$_2$O. Autoclave both flasks and mix afterward. Cool to 55°C in a water bath. Add 50 ml of 40% glucose and pour into ten to twelve 15-cm petri plates.

B. Preparation of DB-ORF Bait Strains

Any cloning strategy can be used to generate DB-ORF fusion vectors. The Gateway recombination method (Walhout *et al.*, 2000) is useful for the cloning of large numbers of DB-ORF baits. For the strains and plasmids that we describe, the DB plasmid has a *LEU2* marker and the AD plasmid has a *TRP1* marker, but other variations of the two-hybrid system use other markers. It is important to make sure that the selective plates you use match the markers of the vectors you are using. After obtaining the constructs, the next step is to introduce them into the two-hybrid strain (MaV203 in this method). Several transformation protocols that are optimized for the scale of the transformation to be performed are given throughout the article, but any yeast transformation protocol should work. The following protocol is optimized for transforming large numbers of baits in a 96-well plate.

Solutions

1. *1 M lithium acetate stock solution*: Add 51 g of lithium acetate into 500 ml of ddH$_2$O. Autoclave to sterilize.
2. *10× TE stock solution*: 100 mM Tris–HCl (pH 7.5), 10 mM EDTA, autoclave
3. *50% PEG stock solution*: Dissolve 125 g of polyethylene glycol in warm ddH$_2$O and finalize to 250 ml. Sterilize by filtration (the

PEG solution is very viscous and takes a long time to filter).

4. *TE/LiAc*: To make 50 ml, add 5 ml of 10× TE and 5 ml of 1 M LiAc into 40 ml of sterile ddH$_2$O
5. *TE/LiAc/PEG*: To make 50 ml, add 5 ml of 10× TE and 5 ml of 1 M LiAc into 40 ml of 50% PEG
6. *Boiled ssDNA*: Boil the 10 mg/ml of salmon testes DNA for 5 to 10 min and chill on ice before transformation

Steps

1. Start an overnight culture of MaV203 yeast by scratching a small clump of cells from a patch into at least 0.5 ml of media for each bait you plan to transform (a minimum of 5 ml of media should be used).
2. The next day, take 0.5 ml of the overnight culture for each transformation.
3. Spin down the cells at 2000 rpm for 5 min.
4. Wash the cells by adding 0.25 ml of ddH$_2$O for each transformation.
5. Spin down the cells and wash in 100 μl of TE/LiAc for each transformation.
6. Spin down the cells and resuspend the cell pellet in 20 μl of TE/LiAc for each transformation.
7. Add 2 μl of boiled ssDNA for each transformation.
8. Aliquot 22 μl of yeast into the wells a 96-well plate.
9. Add 50–100 ng of the appropriate DB-ORF DNA to each well. A transformation without DB-ORF DNA serves as a negative control.
10. Add 100 μl of TE/LiAc/PEG to each well and mix by pippetting.
11. Incubate the plates at 30°C for 30 min.
12. Heat shock at 42°C for 15 min.
13. Spin down and remove the TE/LiAc/PEG solution with a multichannel pipette.
14. Add 120 μl of ddH$_2$O to each well, but be careful not to resuspend the cells. Remove

105 μl of ddH$_2$O from each well and resuspend the cells in the remaining 15 μl of liquid.

15. Using a multichannel pipette, spot 6–7 μl onto two 15-cm SC-Leu plates. You should be able to spot all 96 wells onto a 15-cm plate (see Fig. 19.2). The second plate is to have a copy in case spots run together or one plate is contaminated. You can spot onto additional plates if necessary.

16. Incubate for 2 to 3 days at 30°C.

17. Plates can be stored at 4°C for up to 2 months. Fifteen percent glycerol stocks can also be created and stored at −80°C indefinitely.

C. Introduction of cDNA Libraries

Once the bait strains have been created, an AD-cDNA library needs to be transformed into the yeast. Typically, for most cDNA libraries, at least 10^6 independent clones need to be screened to get good coverage of a library, thereby requiring a high efficiency transformation procedure. If a large number of baits are being screened or if you are using a normalized ORFeome library (Reboul et al., 2003), it may be preferable to screen as few as 2×10^5 colonies.

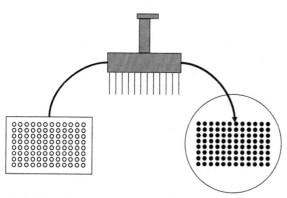

FIGURE 19.2 Spotting a 96-well plate onto a 15-cm agar plate. If a 12-tip multichannel pipette is used, all 8 rows of a 96-well plate can fit easily onto an agar plate as diagrammed.

While more potential interactions will be identified if you screen more colonies, screening a smaller number can still identify many interesting interactions and will allow you to screen a larger number of baits as well. We give protocols for both large and small numbers of colonies to be screened. Positives are screened for their ability to activate the *HIS3* marker. In MaV203, *HIS3* has a low level of activity, leading to moderate growth on SC-His media. To overcome this, the medium is supplemented with 3-AT, an inhibitor of *HIS3*, to reduce the background growth (Durfee et al., 1993).

Solutions and Media

See transformation solutions in Section IIIB. SC-Leu liquid media, SC-Leu-Trp plates, and SC-Leu-Trp-His + 3-AT plates (referred to as 3-AT plates from here on) are needed.

Protocol 1. Introduction of AD Library into Y2H Strain (30 Plate Scale)

Steps

1. Grow the DB-ORF baits in 3 ml of SC-Leu yeast liquid media at 30°C for approximately 24 h.

2. Resuspend the cells well by vortexing and inoculate 80–100 μl in 250 ml of YPD and incubate at 30°C for 15 to 18 h until the OD$_{600}$ reaches 0.3 to 0.6. *Note*: It may be necessary to vary the amount of cells added and the time of incubation for the cells to be ready at the right time.

3. Harvest the cells by centrifuging for 5 min at 1800 rpm.

4. Wash the cell pellet in 50 ml of ddH$_2$O.

5. Wash the cells in 10 ml of TE/LiAc.

6. Resuspend the cells in $5 \times$ OD$_{600}$ ml of TE/LiAc (e.g., if the OD$_{600}$ = 0.5, then add 2.5 ml of TE/LiAc).

7. Transfer 1.6 ml of resuspended cells into a 15-ml Falcon tube and add 160 μl of boiled ssDNA and 20–30 μg of cDNA library.

8. Add TE/LiAc/PEG to final volume of 9 ml. Mix by inverting several times and aliquot at 1 ml into 1.5-ml Eppendorfs.
9. Incubate at 30°C for 30 min to 1 h.
10. Heat shock at 42°C for 15 min.
11. Transfer the cells into a 15-ml Falcon tube and spin down at 1800 rpm for 5 min.
12. Remove the supernatant and resuspend the cells in 9 ml of sterile ddH$_2$O.
13. Take 10 μl and add to 10 ml of ddH$_2$O to create a 1:1000 dilution.
14. To about thirty 15-cm 3-AT plates, add approximately twenty-five 3-mm glass beads. Plate 300 μl of cells from step 12 on each 15-cm 3-AT plate. Spread the cells evenly by shaking the plates with glass beads. Remove beads when done. The beads can be washed, autoclaved, and reused.
15. To measure transformation efficiency, plate 300 μl of the 1:1000 diluted cells on a SC–Leu-Trp plates. Count the colonies 2 to 3 days after plating.
16. Incubate the 3-AT plates for 4 to 5 days at 30°C and take to Section III,D.

Protocol 2. Introduction of AD Library in Y2H Strain (Three Plate Scale)

Steps

1. Grow the DB-ORF baits in 3 ml of SC-Leu yeast medium at 30°C for approximately 24 h.
2. Resuspend the cells well by vortexing and inoculate 10–15 μl in 35 ml of YPD and incubate for 15 to 18 h at 30°C until the OD$_{600}$ reach 0.3 to 0.6. This can be done in 50-ml Falcon tubes. *Note*: It may be necessary to vary the amount of cells added and the time of incubation for the cells to be ready at the right time.
3. Harvest the cells by centrifuging for 5 min at 1800 rpm.
4. Wash the cell pellet in 1.5 ml of ddH$_2$O by vortexing and transfer to a 2.0-ml Eppendorf.

5. Spin at highest speed for 5 s in a microcentrifuge.
6. Wash the cells in 1 ml TE/LiAc and spin again.
7. Resuspend the cells in 275 μl of TE/LiAc.
8. Add 30 μl of boiled ssDNA and 3–5 μg of the normalized AD-library.
9. Add 1.5 ml TE/LiAc/PEG and mix by inverting several times.
10. Incubate at 30°C for 30 min to 1 h.
11. Heat shock at 42°C for 15 min.
12. Spin down the cells.
13. Remove the supernatant and resuspend the cells in 900 μl of sterile ddH$_2$O.
14. Take 10 μl and add to 10 ml of ddH$_2$O to create a 1:1000 dilution
15. To three 15-cm 3-AT plates, add approximately twenty-five 3-mm glass beads. Plate 300 μl of cells from step 12 on each 15-cm 3-AT plate. Spread the cells evenly by shaking the plates with glass beads. Remove beads when done. The beads can be washed, autoclaved, and reused.
16. Calculate the transformation efficiency by plating 300 μl of 1:1000 diluted cells on SC-Leu-Trp plates.

D. Isolation of Two-Hybrid Positives

After 4 to 5 days of growth, colonies with interacting proteins should have grown enough to isolate them. At times it is difficult to differentiate a true positive from a large background colony, but later in Section III,E you can use the phenotypic assays to help differentiate them. Colonies are initially grown on 3-AT plates after picking to allow true positives to outgrow any background cells that might have been picked up accidentally.

Solutions

1. SC-Leu-Trp-His + 3-AT plates
2. SC-Leu-Trp liquid media

Steps

1. Get the plates from Section II,C. See Fig. 19.3 for an example of what a screening plate might look like (with positive colonies growing in a field of background colonies).
2. Use a toothpick to pick colonies that grow above the background. Patch the colony in a small streak onto a 15-cm 3-AT plate. Typically we divide the plate into 96 sectors that correspond to a 96-well plate so that the positives can be stored and manipulated in a 96-well format in the future (see Fig. 19.4). If you only have a small number of positives you will not fill up a plate, but if you have many baits and a large number of positives then you could fill up many plates with potential positive clones.
3. Incubate the plates at 30°C for 2 to 3 days.

4. When the patches have grown up, use a multichannel pipette with tips to scrape a small clump of cells into 120 µl of SC–Leu-Trp medium in U-bottom 96-well plates (see Fig. 19.5). Seal the plates with airpore tape.
5. Incubate the plates at 30°C for 2 days before making a 15% glycerol stock.
6. Take the remaining culture to Section III,E for phenotypic assays.

E. Phenotypic Assays

Picking positive colonies is only the first step in identifying potential interacting proteins. It is frequently difficult to determine which of the colonies are true positives and which are merely large growing background colonies. It is important to retest the potential positives and see if they are able to activate expression of a variety of different two-hybrid reporters. We generally look for colonies that are able to activate at least two of the three two-hybrid reporters. It is also important to include at least one positive and one negative control on the plates. A good negative control would be the empty AD and

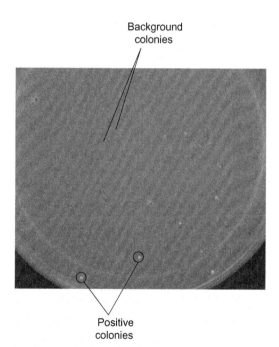

FIGURE 19.3 An example of a 3-AT screening plate before colonies have been picked. Two of the colonies that are likely two-hybrid positives have been circled.

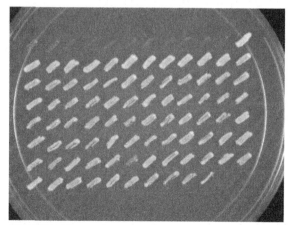

FIGURE 19.4 An example of positive colonies that have been picked and patched onto a 3-AT plate. By patching them in the same pattern and dimensions as a 96-well plate, it makes it easier to manipulate the samples at later stages.

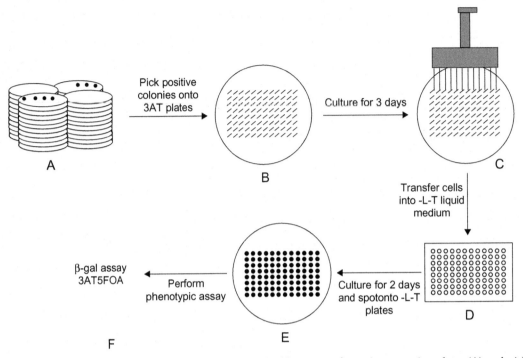

FIGURE 19.5 Diagram of picking positives. First gather the library transformation screening plates (A) and pick the positives onto a fresh 3-AT plate (B). After 3 days of growth, use a multichannel pipette with tips (C) to transfer the positives to liquid SC-Leu-Trp in a 96-well plate (D). After 2 days of growth, spot the cells onto a SC-Leu-Trp plate (E) and use to replica plate for the phenotypic assays (F).

DB vectors, whereas a strong positive control would be the full-length Gal4 transcription factor. For a description of a larger set of two-hybrid controls, see Walhout and Vidal (2001).

Solutions and Materials

You need SC-Leu-Trp plates, 3-AT plates, 5-FOA plates, and YPD plates. Note that 5-FOA selects against URA3-positive strains, i.e., positive interactors will not grow on 5-FOA. You also need circular nitrocellulose filters.

Steps

1. Get the 96-well plates from Section III,D. Spot 5 μl of culture onto a SC-Leu-Trp plate. Grow at 30°C for 1 to 2 days.

2. You will use three different plates to assay the three two-hybrid reporters: YPD for lacZ, 3-AT for HIS3, and 5-FOA for URA3. The 5-FOA and 3-AT plates can be used as is, but the YPD plate needs a nitrocellulose filter placed on it prior to replica plating. This is necessary to remove the yeast from the YPD plate and perform the β-Gal filter lift assay (see Section III,F). If there are bubbles between the filter and the agar or if the filter is misaligned, use forceps to move the filter or remove the bubble.

3. Using replica velvets and a replica block, replica plate the yeast from the SC–Leu-Trp growth plate to the YPD/filter, 3-AT, and 5-FOA plates. Use the same velvet for each of the assay plates; there should be enough yeast on it for all three plates.

4. Replica clean the 3-AT and 5-FOA plates as follows. Use a clean velvet to remove excess yeast from the 3-AT plate (press firmly but not harshly). Repeat with a clean velvet until there is no longer any visible sign of yeast on the plate. Then repeat one more time (we typically clean with four velvets). Repeat the procedure with the 5-FOA plate. Replica cleaning is necessary to decrease the background growth and to ensure that you start out with comparable amounts of cells in each spot; it is not necessary to perform on the YPD/filter plates.

5. Culture all three assay plates at 30°C. After 1 day the YPD plate should have large spots of yeast on the filter. Take the YPD plate to Section III,F to perform β-Gal filter lift assays.

6. Examine 3-AT and 5-FOA plate. If the negative control shows growth after 1 day, replica clean again.

7. When the controls on the 3-AT and 5-FOA plates have grown to the appropriate levels (it normally takes 2 days for 5-FOA and 3 days for 3-AT), remove plates and score the results. (See Fig. 19.6 for an example of what a 3-AT assay plate can look like.)

8. Score the 3-AT and 5-FOA plates along with the β-Gal filters from Section III,F. Any strain that passes at least two of the three tests is considered positive. Consolidate all of the positives into fresh 96-well plates.

9. Grow plates for 2 days and make glycerol stocks of the strains.

F. β-Gal Filter Lift Assay

Solutions

1. *Z buffer*: To make a liter, add 16.1 g $Na_2HPO_4 \cdot 7H_2O$, 5.5 g $NaH_2PO_4 \cdot H_2O$, 0.75 g KCl, and 0.246 g $MgSO_4 \cdot 7H_2O$. Autoclave to sterilize.

2. *4% X-Gal*: Dissolve 40 mg in 1 ml of *N,N*-dimethylformamide. Store at −20°C wrapped with foil.

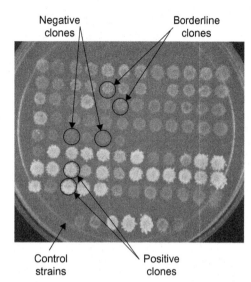

FIGURE 19.6 An example of a 3-AT assay plate. The bottom seven patches are a range of controls. A range of strengths of interactions can be seen by the strength of growth on the spots. Clones that are indicative of positive and negative interactions have been circled as well as a couple of clones that are on the border between being positive and negative.

3. *β-Gal solution*: For each plate, prepare 5 ml of Z buffer with 120 µl of 4% X-Gal and 13 µl of 2-mercaptoethanol. Make fresh each time.

Steps

1. Retrieve the YPD/filter plates from Section III,E.

2. Make up the β-Gal solution according to the number of plates that are needed to assay.

3. For each for each plate to be assayed, get one empty 15-cm petri plate. Put two pieces of Whatman filter paper in the plate. Add 5 ml of β-Gal solution to each plate. Let the paper soak up the solution and make sure there are no bubbles under the Whatman paper.

4. Remove the nitrocellulose filter from the YPD (with yeast on the filter) and place in liquid nitrogen for at least 30 s. This lyses the cells.

5. Remove the filter from liquid nitrogen and allow it to thaw in air (this should take ~30 s). Once the filter is flexible again, place in a petri plate with β-Gal solution-soaked Whatman paper. Use forceps to remove any bubbles that may be under the filter.
6. Put the β-Gal assay plates at 37°C overnight. The next day you can read the results with positives being blue; the stronger the positive, the stronger the blue.
7. Remove the Whatman filter paper. The filters can be stored dry for at least 6 months.

G. Yeast PCR to Identify Preys

Once you have identified the clones that pass the phenotypic assays, it is necessary to isolate prey DNA from the yeast clones and identify them by sequencing. Prey DNA can be obtained by polymerase chain reaction (PCR) using the universal primers on the vector (typically we make primers to the activation domain sequence and the termination sequence; the primer sequences that we use are listed later).

Solutions and Media

1. SC-Leu-Trp plates, YPD plates.
2. Lysis solution: To make 1 ml, add 2.5 mg zymolyase to 0.1 M sodium phosphate buffer (pH 7.4). The solution can be stored at −20°C.

Primer sequences: Activation domain –CGC GTT TGG AAT CAC TAC AGG G
Termination sequence –GGA GAC TTG ACC AAA CCT CTG GCG

Steps

1. Spot 5 μl of the positives from the 96-well plate culture or the glycerol stocks made in Section III,E, steps 8 and 9 onto a SC-Leu-Trp plate. Culture at 30°C for 1 day.

2. Replica plate to YPD. Culture at 30°C for 1 day.
3. Add 15 μl of lysis solution to each well in a 96-well PCR plate.
4. Using a multichannel pipette, scrape some of the yeast cells off the YPD plate (one row at a time as in Section III,D, step 4) and resuspend into the 15-μl lysis solution.
5. Put the yeast at 37°C for 5 min then at 95°C for 5 min (this can be done in a PCR machine).
6. Set up the PCR plate using a final volume of 50 μl for each reaction.
7. Make a 1:10 dilution of the yeast lysis from step 5. Add 5 μl to the PCR plate.
8. Perform PCR using 5-min extension times to make sure that the large ORF inserts are isolated.
9. The PCR products can be sent out for sequencing after confirming it works. Note that most PCR products need to be purified from the primers to be sequenced properly. We use the Millipore MultiScreen PCR plates (Fisher, Cat. No. MANU-030-10) to purify PCR products.

H. Retest by Gap Repair

Sometimes a positive isolate will occur not because the two proteins interact, but through a spontaneous self-activation mutation that makes the growth independent of the two plasmids. This makes it important to isolate the AD-ORF from the strain and retest it in the bait strain. The easiest way to do this is by gap repair (Orr-Weaver et al., 1983). This involves introducing into yeast both a vector cut at the insert sites and the insert itself. If there is some homology between the insert and the cut vector, the yeast will repair it by ligating the insert into the vector.

The protocol is essentially the same as the method listed in Section III,B with a few exceptions.

Step 1. Instead of starting a culture with untransformed MaV203 at the beginning, start a culture with MaV203 transformed with the DB-ORF of interest.

Step 9. Instead of using DB-ORF DNA, use 25 ng of linearized AD plasmid DNA and 2.5 μl of the PCR product from Section III,G. The AD plasmid should be cut in the linker region between the AD sequence and the termination sequence (in our plasmids, we use SmaI) as that way the PCR product can have homology to the AD sequence and the termination sequence and gap repair can be efficient.

Step 15. Select on -Leu, -Trp plates instead of -Leu.

Step 17. Perform phenotypic assays as in Section III,E, skipping step 1 and only using the 3-AT and β-Gal assays.

IV. COMMENT

Notes on High-Throughput Two Hybrid

The simplicity of two-hybrid screening allows for the screening of large numbers of baits of interest at a time. This allows a scientist to study a greater number of genes, but it requires the researcher to be more organized from the start. Here are several points to remember when screening large numbers.

1. Work in 96-well formats from the beginning. This will allow you to transform all the baits that you want to study in 1 day.

2. When transforming the bait strains with the library, it is still necessary to do each transformation individually, but you can do six or more in a day.

3. When picking the colonies, try to patch them on a 96-well grid so that it will be easier to transfer your positives to 96-well plates in the future.

4. Try to keep track of data such as gene names and strength of interactions in the phenotypic assay in a spreadsheet such as Microsoft Excel. This can allow you to monitor your positives easily and keep track of which ones pass all the phenotypic tests.

The strength of the two-hybrid system is its ease of use. It can identify many potential interactors and maybe even bring interaction networks to light. It is ultimately, however, a first-pass prediction of interactions and should be thought of as a jumping off point to other more detailed studies.

References

Boulton, S. J., Gartner, A., et al. (2002). Combined functional genomic maps of the C. elegans DNA damage response. Science 295(5552), 127–131.

Davy, A., Bello, P., et al. (2001). A protein-protein interaction map of the Caenorhabditis elegans 26S proteasome. EMBO Rep. 2(9), 821–828.

Drees, B. L., Sundin, B., et al. (2001). A protein interaction map for cell polarity development. J. Cell Biol. 154(3), 549–571.

Du, W., Vidal, M., et al. (1996). RBF, a novel RB-related gene that regulates E2F activity and interacts with cyclin E in Drosophila. Genes Dev. 10(10), 1206–1218.

Durfee, T., Becherer, K., et al. (1993). The retinoblastoma protein associates with the protein phosphatase type 1 catalytic subunit. Genes Dev. 7(4), 555–569.

Fields, S., and Song, O. (1989). A novel genetic system to detect protein-protein interactions. Nature 340(6230), 245–246.

Hartwell, L. H., Hopfield, J. J. et al. (1999). From molecular to modular cell biology. Nature 402(6761 Suppl.), C47–C52.

Ito, T., Tashiro, K., et al. (2000). Toward a protein-protein interaction map of the budding yeast: A comprehensive system to examine two-hybrid interactions in all possible combinations between the yeast proteins. Proc. Natl. Acad. Sci. USA 97(3), 1143–1147.

Orr-Weaver, T. L., Szostak, J. W., et al. (1983). Genetic applications of yeast transformation with linear and gapped plasmids. Methods Enzymol. 101, 228–245.

Reboul, J., Vaglio, P., et al. (2003). C. elegans ORFeome version 1.1: Experimental verification of the genome annotation and resource for proteome-scale protein expression. Nature Genet. 34(1), 35–41.

Uetz, P., Giot, L., et al. (2000). A comprehensive analysis of protein-protein interactions in Saccharomyces cerevisiae. Nature 403(6770), 601–603.

Vidal, M., Brachmann, R., *et al.* (1996). Reverse two-hybrid and one-hybrid systems to detect dissociation of protein-protein and DNA-protein interactions. *Proc. Natl. Acad. Sci. USA* **93**(19), 10315–10320.

Walhout, A. J., Sordella, R., *et al.* (2000). Protein interaction mapping in *C. elegans* using proteins involved in vulval development. *Science* **287**(5450), 116–122.

Walhout, A. J., Temple, G. F., *et al.* (2000). GATEWAY recombinational cloning: Application to the cloning of large numbers of open reading frames or ORFeomes. *Methods Enzymol* **328**, 575–592.

Walhout, A. J., and Vidal, M. (2001). High-throughput yeast two-hybrid assays for large-scale protein interaction mapping. *Methods* **24**(3), 297–306.

Affinity Electrophoresis for Studies of Biospecific Interactions: High-Resolution Two-Dimensional Affinity Electrophoresis for Separation of Hapten-Specific Polyclonal Antibodies into Monoclonal Antibodies in Murine Blood Plasma

*Kazuyuki Nakamura, Masanori Fujimoto,
Yasuhiro Kuramitsu, and Kazusuke Takeo*

I. INTRODUCTION

Affinity electrophoresis (AEP) was developed as a novel technique for separation of biomolecules by biospecific interactions with their ligands in an electric field (Nakamura, 1959). The techniques of rocket immunoelectrophoresis (Laurell, 1966) and crossed immunoelectrophoresis (Svenson and Axelsen, 1972) are based on the same principle. AEP has been applied not only for separation of a tiny amount of those biomolecules, but also for determination of dissociation constants (K_d) of those interactions (Takeo and Nakamura, 1972; Bog-Hansen, 1973; Horejsi and Kocourek, 1974; Caron *et al.*, 1975; Takeo and Kabat, 1978; Nakamura *et al.*, 1980; Shimura and Kasai, 1982) at different pH (Ek *et al.*, 1980; Mimura *et al.*, 1992) and temperatures for calculation of thermodynamic parameters (Tanaka *et al.*, 1986; Kashiwagi *et al.*, 1991).

Two-dimensional affinity electrophroesis (2DAEP), which was newly developed by a combination of isoelectric focusing (IEF) with AEP (Takeo *et al.*, 1983), has been used for studies of the immune response *in vivo*. Hapten-specific polyclonal IgG antibodies, which were produced by immunization of rabbits and mice with the hapten of dinitrophenyl (DNP)- or fluorescein isothiocyanate (FITC)-conjugated protein carriers (Takeo *et al.*, 1992; Nakamura *et al.*, 1993), were separated into a large number of groups of IgG spots by 2DAEP, and each of the groups showed an identical affinity to the hapten but a different isoelectric point (pI) as in the case of monoclonal antibodies specific for the hapten. Diversification, affinity maturation, and subclass switching of the hapten-specific antibodies *in vivo* were evidenced in the course of immunization of a single mouse (Nakamura and Takeo, 1998). This article describes the procedures for 2DAEP used for separation of anti-DNP antibodies in murine blood plasma.

II. MATERIALS AND INSTRUMENTATIONS

Acrylamide (Cat. No. 00809-85), N, N'-methylenebisacrylamide (Bis) (Cat. No. 22402-02), and N, N, N', N'-tetramethylethylenediamine (TEMED)(Cat. No. 33401-72) are from Nacalai tesque (Kyoto, Japan). Carrier ampholytes, Parmalite pH 4–6.5 (Cat. No. 17-0452-01), Pharmalite pH 6.5–9 (Cat. No. 17-0454-01), and Pharmalite pH 5–8 (Cat. No. 17-0453-01) are from Amersham Biosciences (Little Chalfront, Buckinghamshire, UK). Chicken albumin (Cat. No. A-5503), dinitrofluorobenzene (DNFB)(Cat. No. D1529), dinitrophenyl glycine (Cat. No. 9504), β-alanine (Cat. No. A7752), l-lysine (Cat. No. L5501), human γ-globulins (HGG) (Cat. No. G4386), Tris (Cat. No. T6066), and 4-chloro-1-naphthol (Cat. No. C8890) are from Sigma (St. Louis, Mo). Sodium hydroxide (Cat. No. 28-2940), glycine (Cat. No. 12-1210-5), glycerol (Cat. No. 12-1120-5), urea (Cat. No. 32-0280-5), ammonium peroxodisulphate (APS) (Cat. No. 01-4910-2), potassium hydroxide (Cat. No. 24-4670-5), sucrose (Cat. No. 28-0010-5), methylene blue trihydrate (Cat. No. 19-3200-2), riboflavin (Cat. No. R4500), hydrochloric acid (Cat. No. 13-1700-5), sodium chloride(Cat. No. 28-2270-5), Tween 20 (Cat. No. 30-5450-5), and methanol (Cat. No. 19-2410-8) are from Sigma-Aldrich-Japan (Tokyo, Japan). l-Glutamic acid (Cat. No. 074-00505), hydrogen peroxide (Cat. No. 086-07445), and acetic acid (Cat. No. 012-00245) are from Wako (Osaka, Japan). POD-conjugated rabbit IgG fraction to goat IgG Fc (Cat. No. 55360) and goat antihuman γ-globulin antisera (Cat. No. 55074) are from ICN Pharmaceuticals, Inc. (Cappel Products). Other chemicals are of the highest available purity obtained from various sources.

A. Preparation of Antihapten Antisera

A BALB/c mouse (female, 2 months old) was immunized by an intraperitoneal injection of 50 μg of an antigen, hapten-conjugated protein carrier of DNP-conjugated chicken serum albumin (DNP-CSA) in an emulsion with an equal volume of Freund's complete adjuvant as the primary immunization. The second immunization was performed by an intraperitoneal injection of 50 μg of the antigen 2 weeks after the primary immunization and a boosting was made by an intraperitoneal injection of 100 μg of the antigen weeks after the second immunization. In the course of immunization, 150 μl of whole blood was taken by a puncture of the veniplex lining of an eyeball of a deeply anesthetized mouse every week. The blood was allowed to stand at 37°C for 2 h, and the blood clot was removed by centrifugation at 1000 g for 5 min to yield 70 μl of clear antisera. Antisera were stored at 4°C by adding 0.1% sodium azide and were submitted to 2DAEP.

B. Preparation of Water-Soluble DNP-Conjugated Noncross-Linked Acrylamide–Allylamine Copolymer (DNP-PA) as an Affinity Ligand

Acrylamide monomer (40 g) and 4 g of allylamine were dissolved with 400 ml of distilled water in a 500-ml Erlenmeyer flask, followed by the addition of 0.8 ml of TEMED and 50 ml of 0.8% freshly prepared ammonium peroxodisulphate solution with gentle mixing. A small volume of distilled water was overlaid on top of the solution to shut off oxygen and was allow to stand overnight at 30°C. The solution was then extensively dialyzed against tap water for 3 days and distilled water with four changes a day for 2 days. Sodium bicarbonate (4.5 g) was added to the dialyzed solution, and 4.8 g of dinitrofluorobenzene in 15 ml of acetone was mixed by dropwise addition. The mixture was gently stirred overnight in the dark at 25°C using an evaporator without sucking.

The reaction mixture was then dialyzed against tap water for 5 days and distilled water

with four changes a day for 2 days. The dialyzed solution was concentrated *in vacuo* until the DNP concentration reached 5–10 mM for storage at 4°C in the dark.

The concentration of DNP residue in DNP-conjugated noncross-linked acrylamide–allyamine copolymer (Fig. 20.1) was calculated from the absorbance at 360 nm in 0.1 M sodium hydroxide, using DNP-glycine as a standard (molar absorption at 360 nm was 17530 liter/mol/cm).

III. PROCEDURES

A. Preparation and Running of First-Dimension Gels (IEF)

Solutions

1. *Working solutions for capillary IEF polyacrylamide gels*: To make 6.0 ml of IEF gel solution, prepare the working solutions as shown in Table 20.1
2. *Running buffer solution for anode*: To make 1 liter of 40 mM l-glutamic acid solution,

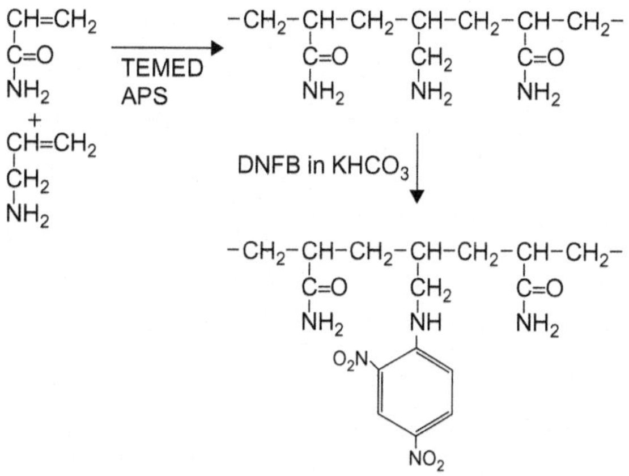

FIGURE 20.1 Preparation of water-soluble dinitrophenyl-conjugated noncross-linked acrylamide–allyamine copolymer (DNP-PA). TEMED, N, N, N′, N′-tetramethylethylenediamine; APS, ammonium peroxodisulphate; DNFB, dinitrofluorobenze.

TABLE 20.1   Preparation of Working Solutions for Capillary IEF Gels

Stock solutions and reagent	Volume and weight
Acrylamide solution	0.6 ml
Acrylamide 48.5 g	
Bis 1.5 g	
Dissolved in distilled water to 100 ml	
Carrier ampholyte	
Pharmalite	
pH 4–6.5	0.0094 ml
pH 6.5–9	0.0094 ml
pH 5–8	0.0094 ml
Glycerol solution	0.5 ml
Glycerol 40.0 g	
Dissolved in 100 ml distilled water	
L-Lysine solution	0.8 ml
L-Lysine 1.58 g	
Dissolved in distilled water to 100 ml	
Distilled water	0.98 ml
Urea	2.162 g
APS solution[a]	0.143 ml
Ammonium peroxodisulphate 0.21 g	
dissolved in distilled water to 10 ml	
TEMED	6 μl
Total volume	6.0 ml

[a] APS solution is prepared freshly before mixing with the deaerated acrylamide gel solution.

dissolve 5.95 g of L-glutamic acid in distilled water and complete to 1 liter

3. *Running buffer solution for cathode*: To make 1 liter of 1 M NaOH, dissolve 40.0 g sodium hydroxide in distilled water and complete to 1 liter

4. *Sample solution*: To make 0.1 ml sample solution, mix 0.027 ml of antisera or purified antibodies with 0.055 ml of phosphate-buffered saline, pH 7.4 (PBS), and add 0.03 g of sucrose

Steps

1. Mark the glass capillary (1.2 mm in inner diameter and 8.5 cm in height, Cat. No. 2-000-100, Drummond Scientific Co., Broomall, PA) at 7.5 cm from the bottom.

2. Prepare the gel solution by mixing 0.6 ml of the acrylamide solution, 0.094 ml of Pharmalite pH 4–6.5, 0.094 ml of Pharmalite pH 6.5–9, 0.188 ml of Pharmalite 5–8, 1.5 ml of 40% glycerol solution, 0.8 ml of L-lysine solution with 0.98 ml distilled water, and add 2.162 g of urea to yield 5.867 ml of gel solution. Deaerate the gel solution and mix with 0.143 ml of APS solution and 6 μl of TEMED just before preparation of gels.

3. Pour two-thirds of the gel solution into a plastic cylindrical chamber (Fig. 20.2: 1.2 cm in inner diameter and 8.0 cm in height) and immerse the glass capillaries in the solution, avoiding the formation of air bubbles in the capillaries, which are tied with a rubber band.

4. Pour the residual solution into the chamber until the meniscus in the capillaries reaches the marker 7.5 cm from the bottom.

5. Stand the chamber at 25°C for 10 min and find a new surface of top of polyacrylamide gel, which is formed 0.6 mm beneath the original meniscus.

6. Remove the rubber band from the capillaries and seal the top of capillaries in the chamber with Parafilm tightly to avoid drying up during storage at 4°C. Use the capillary gels within 4 weeks.

7. Take the capillary from the chamber and wipe up gel crumbs clinging to the outer surface of the capillary.

8. Pour 10 μl of the sample solution onto the top of the capillary by a microsyringe and overlay the running buffer of 40 mM

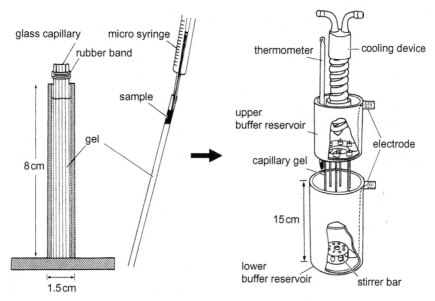

FIGURE 20.2 Preparation of capillary gels and apparatus for first-dimension isoelectric focusing (IEF).

L-glutamic acid (for anode), filling up the capillary.

9. Insert the capillary into a hole of a cylindrical rubber connector and set the connector to the upper buffer reservoir (Fig. 20.2) tightly. After setting six capillaries and a thermometer with the rubber connector to the upper buffer reservoir (for anode) and avoid air bubbles on the bottom of gels with a small volume of the running buffer of 1 M NaOH for cathode.

10. Place the upper reservoir on the lower buffer reservoir (for cathode), which is filled with the running buffer of 1 M NaOH, and place the cooling device in the middle of the apparatus (Fig. 20.2, right).

11. Run the electrophoresis using a constant voltage power supply (Cat. No. 2197-010, Pharmacia LKB, Germany) with stepwise elevation of the voltage from 250 V for 15 min to 500 V for 15 min, 1000 V for 15 min, and 2000 V for 2 h and keep the temperature at 15°C using the cooling device.

12. Detach the capillaries from the apparatus and subject to second-dimension gels (AEP).

B. Preparation and Running Second-Dimension Gels (AEP)

Solutions

1. *Working solutions for AEP polyacrylamide gels*: To make 8.0 ml of the AEP gel solution, prepare the working solutions as shown in Tables 20.2 and 20.3 for separating gel and stacking gel in Reisfeld's buffer system, respectively.

2. *Running buffer solution(Reisfeld, 1962)*: To make 500 ml of running buffer, pH 4.5, mix 100 ml of the stock solution (dissolve 31.2 g of β-alanine and 8.0 ml acetic acid with distilled water to make 1 liter) and 400 ml of distilled water. Check the pH before use.

3. *Stacking solution (Reisfeld, 1962)*: To make 12 ml of the stacking solution, prepare the

TABLE 20.2　Preparation of Working Solutions for AEP Separating Gels[a]

Stock solution	Volume of working solution for a gel	
	Control gel	Affinity gel
Acrylamide solution	1.0 ml	1.0 ml
Acrylamide 40.0 g		
Bis 1.067 g		
Dissolved in distilled water to 100 ml		
Gel buffer solution	2.0 ml	2.0 ml
1 M KOH 48.0 ml		
Acetic acid 17.2 ml		
TEMED 4.0 ml		
Dissolved in distilled water to 100 ml		
DNP-PA solution		
DNP-PA (at any concentration)	—	2.73 ml
Distilled water	2.73 ml	—
Sucrose solution	2.0 ml	2.0 ml
Sucrose 40.0 g		
Dissolved in 100 ml distilled water		
APS solution[b]	0.27 ml	0.27 ml
Ammonium peroxodisulphate 0.21 g		
Dissolved in distilled water to 10 ml		
Total volume	8.0 ml	8.0 ml

[a] Polyacrylamide gels are 5.13% in T, 2.59% in C, pH 4.3.

[b] APS solution is prepared freshly before mixing with the deaerated acrylamide gel solution.

working solutions as shown in Table 20.4. Check the pH before use.

Steps

All steps from 1 to 10 must be finished before the completion of the first-dimension IEF.

1. Mark the slab gel cassette (inner space; 100 mm in height, 85 mm in width, 1 mm in thickness, Cat. Nos. SE-400 and SE-401, Marysol, Tokyo, Japan) at 7.5 cm from the bottom and seal the bottom and both sides of the cassette.

2. Stand the cassette at 37°C in a dry incubator until gel preparation.

3. Prepare the AEP separating gel solution by mixing 1.0 ml of acrylamide solution, 2.0 ml of buffer solution, 2.73 ml of DNP-PA solution (substitute with 2.73 ml of distilled water for control gel solution), and 2.0 ml of 40% (w/v) sucrose solution to yield 7.73 ml of gel solution as shown in Table 20.2.

TABLE 20.3 Preparation of Working Solutions for AEP Stacking Gel[a]

Stock solution	Volume of solution for four gels
Acrylamide solution	1.5 ml
Acrylamide	10.0 g
Bis	2.5 g
Dissolved in distilled water to 100 ml	
Gel buffer solution	1.5 ml
1 M KOH	48.0 ml
Acetic acid	2.87 ml
TEMED	0.46 ml
Dissolved in distilled water to 100 ml	
Sucrose solution	1.5 ml
Sucrose	40.0 g
Dissolved in 100 ml distilled water	
Distilled water	0.75 ml
Riboflavin solution[b]	0.75 ml
Riboflavin	4.0 mg
Dissolved in 100 ml distilled water	
Total volume	6.0 ml

[a] Polyacrylamide gels are 3.13% in T, 20.0% in C, pH 6.7.

[b] The solution should be prepared freshly or kept at 4°C in a brown bottle to use within a few weeks before mixing with the deaerated gel solution.

4. Deaerate the gel solution and mix with 0.27 ml of APS solution well just before preparation of the gel.
5. Pour the gel solution into the cassette until the meniscus reaches the marker 7.5 cm from the bottom.
6. Overlay the DNP-PA solution of the same concentration as in the gel (substitute with distilled water for control gel) on top of the gel solution carefully with a syringe to avoid disturbances to the gel solution meniscus.
7. Stand the cassette at 37°C in a dry incubator to complete gelification within 30–40 min and find a new surface of gel beneath the solution.
8. Discard the residual solution on top of the gel and wash with 1 ml of the stacking gel solution, which has been prepared as in Table 20.3, to remove residual DNP-PA on top of the gel. Shade the part cassette for the separating gel by wrapping with aluminum foil.
9. Place a comb in the cassette to prepare the stacking gel in 1 cm height on the top of the separating gel and pour the stacking gel solution, avoiding the formation of air bubbles.
10. Illuminate the stacking gel solution with a fluorescent lamp to complete gelification the stacking gel in 20–30 min.
11. Remove the comb from the cassette and find the stacking gel as shown in Fig. 20.3 (left). The narrow wells in both sides of the stacking gel are for marker proteins, if necessary.
12. Remove the seal of the bottom of the gel cassette and place two sets of the gel cassette onto the apparatus for AEP as shown in Fig. v20.3 (right), to fixing the cassette with clips.
13. Fill the lower buffer reservoir with the running buffer solution carefully, avoiding the formation of air bubbles at the bottom of the gel.
14. Fill the middle well of the stacking gel with the stacking, solution which has been prepared as described in Table 20.4.
15. Place the IEF capillary gel by pushing out from the glass capillary with an expeller of a syringe filled with the stacking solution as shown in Fig. 20.3 (left) and tightly fit on the top of the stacking gel.

TABLE 20.4 Preparation of Stacking Solution for AEP Gel

Stock solution	Volume of solution for four gels
Buffer solution	3.0 ml
1 M KOH	48.0 ml
Acetic acid	2.87 ml
Dissolved in distilled water to 100 ml	
Sucrose solution	1.5 ml
Sucrose	40.0 g
Dissolved in 100 ml distilled water	
Distilled water	3.7 ml
Methylene blue solution	3.8 ml
Methylene blue	50 mg
Dissolved in 100 ml distilled water	
Total volume	12.0 ml

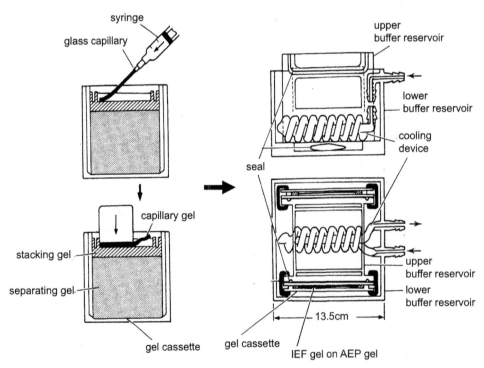

FIGURE 20.3 Preparation of affinity gels and apparatus for second-dimension affinity electrophoresis (AEP). The IEF capillary gel is put on top of the stacking gel of AEP tightly with a metal plate as indicated by the fine arrow (left). Water at the desired temperature circulates in the glass cooling device, which is connected to a thermostat apparatus.

16. Fill the upper buffer reservoir with the running buffer and run the electrophoresis at 50 V for 15 min and 80 V for 2.5 h until the methylene blue reaches the bottom of the separating gel.
17. Using a water circulating cooling device (Cat. No. LKB 2219-001, Pharmacia LKB, Germany) keep the temperature during electrophoresis at the desired temperature.

C. Immunoblotting

Solution

Blotting buffer solution: To make 4 liter of electrode buffer solution of 50 mM glycine, pH 2.5, mix 200 ml of 1 M glycine solution (dissolve 75.07 g of glycine with distilled water to 1 liter) with 100 ml of 1 M HCl solution and add distilled water to 4 liters.

Steps

Steps 1–4 must be completed before completion of the second electrophoresis of AEP.

1. Immerse two sheets of nitrocellulose (NC) membrane (Millipore, pore size 0.22 μm, 100 × 75 mm), four sheets of Whatman 3 MM filter paper (Cat. No. 3030917) in 100 × 80 mm, and two pieces of Scotch-Brite 3 M nylon sponge pad in 100 × 85 mm, into 1 liter of the blotting buffer solution and deaerate extensively with suction by an aspirator for 1 h.
2. Pour the deaerate blotting buffer into the buffer reservoir of the blotting apparatus with carbon electrodes (Fig. 20.4, left).
3. Immerse the cassette holder into 3 liters of the blotting buffer solution in a tray (32 × 23 × 6 cm).
4. Put a sheet of the Whatman 3 MM filter paper on a window of the cassette holder and lay the NC membrane on the filter paper to avoid leaving air bubbles between the sheets.
5. Remove the AEP gel cassette from the upper buffer reservoir (Fig. 20.3) immediately after the completion of AEP and leave the gel on the either side of the cassette plate.
6. Make a small hole with a metal punch (2 mm in diameter) at the corner of the bottom of the AEP gel on the anodic side of IEF.

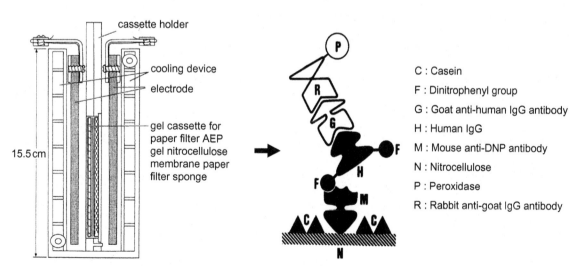

C : Casein
F : Dinitrophenyl group
G : Goat anti-human IgG antibody
H : Human IgG
M : Mouse anti-DNP antibody
N : Nitrocellulose
P : Peroxidase
R : Rabbit anti-goat IgG antibody

FIGURE 20.4 Blotting apparatus and a scheme for immunostaining of mouse anti-DNP antibodies on the NC membrane.

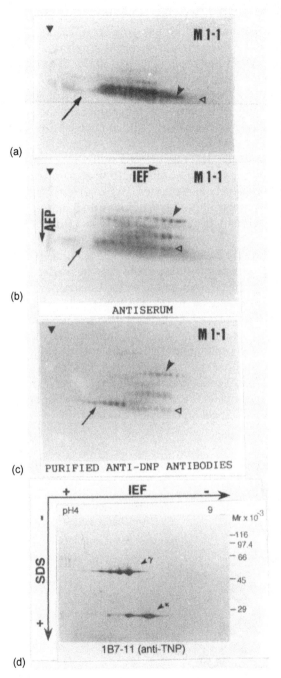

(a)

(b)

ANTISERUM

(c)

PURIFIED ANTI-DNP ANTIBODIES

(d)

FIGURE 20.5 Separations of anti-DNP antibodies in mouse antisera by 2DAEP and of a mouse monoclonal antibody specific to the trinitrophenyl (TNP) group by 2DE. The 2DAEP patterns of anti-DNP antibodies in the absence of affinity ligand (a) and the presence of 0.1 mM DNP-PA (b) are shown. (c) The 2DAEP patterns of 2 μg of anti-DNP antibodies, which were purified from mouse antisera by affinity chromatography with DNP-conjugated lysine Sepharose in the presence of 0.1 mM DNP-PA. Arrows indicate the position of serum albumin which interferes with the blotting of anti-DNP antibodies on the NC membrane. Arrowheads indicate a group of IgG spots with an identical affinity to the ligand of DNP but different isoelectric points (pI), as in the case of the monoclonal antibody. (d) The 2DE patterns of the monoclonal antibody (1B7-11) specific to TNP separated by a combination of IEF and SDS–polyacrylamide gel electrophoresis. The γ and κ indicate the IgG heavy chain and light chain, respectively. Chains show the molecular heterogeneity with different pI values.

7. Release the AEP gel from the cassette plate carefully using a rectangular metal spatular of (85 × 85 mm) and lay on the NC membrane carefully to avoid leaving air bubbles.

8. Overlay another sheet of the Whatman 3 MM filter paper on the AEP gel carefully and put on the sheet of nylon sponge pad.

9. Close the cassette holder and place the holder in the middle of the electrodes of the blotting apparatus as shown in Fig. 20.4.

10. Fill the buffer reservoir with the blotting buffer, if necessary, and run the electrophoresis at 5 V for 2 h.

11. Keep the temperature at 6°C using the cooling device.

12. After completion of blotting, remove the NC membrane from the AEP gel and immerse the NC membrane into Tris-buffered saline, pH 7.5 (TBS), to wash out the blotting buffer overnight at 4°C.

13. Start steps for immunostaining of mouse anti-DNP antibodies blotted on the NC membrane as shown in Table 20.5.

14. Carry out all of the procedures for immunostaining in Tupperware (10 × 14 × 3 cm) at room temperature.

15. Take photographs of the patterns of the anti-DNP antibodies spots on the NC

TABLE 20.5 Procedure for Immunostaining of Anti-DNP Antibodies Blotted on NC Membrane[a]

No.	Step	Solution	Volume (ml) per NC membrane	Incubation time
1.	Washing	TBS[b]	100	Overnight
2.	Blocking	Skim milk[c]	20	1 h
3.	Washing	TBS	50	30 s
4.	Reaction with DNP–conjugate	DNP-conjugated HGG in 1% skim milk[d]	10	2 h
5.	Washing (repeat four times)	Tween-TBS[e]	50	10 min
6.	Reaction with anti-HGG antibody	Goat anti-HGG antisera in 1% skim milk[f]	10	2 h
7.	Washing (repeat four times)	Tween-TBS	50	10 min
8.	Reaction with POD-conjugated antibody	POD-conjugated rabbit antigoat IgG antibodies in 1% skim milk[g]	10	2 h
9.	Washing (repeat four times)	Tween-TBS	50	10 min
10.	Visualization	POD substrate solution[h]	20	5–15 min
11.	Washing (repeat four times)	Distilled water	50	5 min

[a] Carry out all steps of the procedure at room temperature with gentle shaking.
[b] Tris-buffered saline: dissolve $2 M$ Tris 50 ml, $1 M$ HCl 81 ml, and NaCl 43.5 g in distilled water to 5 liters.
[c] Dissolve 5 g of skim milk in 100 ml of TBS.
[d] Prepare DNP-conjugated human γ-globulin (HGG) (0.1 mg/ml) in 1% skim milk of TBS.
[e] Dissolve 50 mg of Tween 20 in 100 ml of TBS.
[f] Dissolve 0.08 ml of goat anti-HGG antisera in 1% skim milk of TBS.
[g] Dissolve 0.02 ml of peroxidase (POD)-conjugated rabbit antigoat IgG antibody solution in 1% skim milk of TBS.
[h] Dissolve 0.06 g of 4-chloro-1-naphthol in 20 ml of cold methanol and mix with 100 ml of TBS and 0.06 ml of 30% of hydrogen peroxide just before visualization.

membrane as shown in Fig. 20.5 soon after immunostaining and store the NC membrane by wrapping with aluminum foil at 4°C if necessary.

IV. COMMENTS

Dissociation constants for interactions between each group of anti-DNP antibodies in mouse antisera and the DNP group can be determined using gels containing a series of different concentrations of DNP-PA for second-dimension AEP. The migration distance of an anti-DNP antibody is decreased by increasing the concentration of DNP-PA in the gel, and the dissociation constant (K_d) can be calculated by the equation $1/r = (1/R_o)[1 + (c/K_d)]$, where r is the relative migration distance of the antibody in the presence of DNP-PA, R_o is the relative migration distance of the antibody in the absence of DNP-PA, and c is the concentration of DNP-PA. To obtain reproducible results, the voltage of the electric field and the temperature of the gels should be kept constant during AEP.

References

Bog-Hansen, T. C. (1973). *Anal. Biochem.* **56**, 480–488.

Caron, M., *et al.* (1975). *J. Chromatogr.* **103**, 160–165.

Ek, K., and Righetti, P. J. (1980). *Electrophoresis* **1**, 137–140.

Horejsi, V., and Kocourek, J. (1974). *Biochim. Biophys. Acta* **336**, 338–343.

Kashiwagi, S., *et al.* (1991). *Electrophoresis* **12**, 420–424.

Laurell, C.-B. (1966). *Anal. Biochem.* **15**, 45–52.

Mimura, Y., *et al.* (1992). *J. Chromatogr.* **597**, 345–350.

Nakamura, K., *et al.* (1980). *J. Chromatogr.* **192**, 351–362.

Nakamura, K., *et al.* (1993). *Electrophoresis* **14**, 81–87.

Nakamura, K., and Takeo, K. (1998). *J. Chromatogr. B.* **715**, 125–136.

Nakamura, S. (1966). "Cross Electrophoresis." Igaku Shoin, Tokyo and Elsevier, Amsterdam.

Nakamura, S., *et al.* (1959). *Nature (London)* **184**, 638–639.

Reisfeld, R. A., *et al.* (1962). *Nature* **195**, 281–283.

Shimura, K., and Kasai, K. (1982). *J. Biochem.* **92**, 1615–1622.

Svenson, P. J., and Axelson, N. H. (1972). *J. Immunol. Methods* **1**, 169–172.

Takeo, K. (1987). Affinity electrophoresis. *In* "*Advances in Electrophoresis*" (A. Chrambach, M.J. Dunn, and B. J. Radola, eds.), pp. 229–279. VCH, Weinheim.

Takeo, K., and Kabat, E. A. (1978). *J. Immunol.* **121**, 2305–2310.

Takeo, K., and Nakamura, S. (1972). *Arch. Biochem. Biophys.* **153**, 1–7.

Takeo, K., *et al.* (1983). *In* "Electrophoresis 82" (D. Stathakos, ed.), pp. 277–283. Walter de Gruyter, Berlin.

Takeo, K., *et al.* (1992). *J. Chromatogr.* **597**, 365–376.

Tanaka, T., *et al.* (1986). *Electrophoresis* **7**, 204–209.

Methods in Protein Ubiquitination

Aaron Ciechanover

I. INTRODUCTION

Degradation of a protein via the ubiquitin-proteasome (UPS) pathway involves two discrete and successive steps: **(a)** tagging of the substrate by covalent attachment of multiple ubiquitin molecules and **(b)** degradation of the tagged protein by the 26S proteasome complex with release of free and reusable ubiquitin. This last process is mediated by ubiquitin recycling enzymes (isopeptidases; deubiquitinating enzymes; DUBs)(for a scheme of the UPS, see Fig. 21.1).

Conjugation of ubiquitin, a highly conserved 76 residue polypeptide, to the protein substrate proceeds via a three-step cascade mechanism. Initially, the ubiquitin-activating enzyme, E1, activates ubiquitin in an ATP-requiring reaction to generate a high-energy thiol ester intermediate, E1-S~ubiquitin. One of several E2 enzymes (ubiquitin-carrier proteins or **ub**iquitin-conjugating enzymes, UBCs)

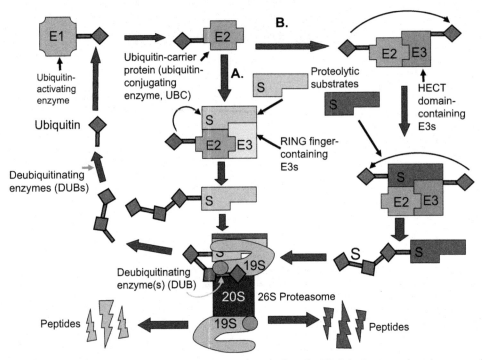

FIGURE 21.1 Ubiquitin is first activated to a high-energy intermediate by E1. It is then transferred to a member of the E2 family of enzymes. From E2 it can be transferred directly to the substrate (S; light) that is bound specifically to a member of the ubiquitin ligase family of proteins, E3. This occurs when the E3 belongs to the RING finger family of ligases (A). In the case of an HECT domain-containing ligase (B), the activated ubiquitin is transferred first to the E3 before it is conjugated to the E3-bound substrate (S; dark). Additional ubiquitin moieties are added successively to the previously conjugated moiety to generate a polyubiquitin chain. The polyubiquitinated substrate binds to the 26S proteasome complex (constituted of 19S and 20S subcomplexes): the substrate is degraded to short peptides, and free and reusable ubiquitin is released via the activity of deubiquitinating enzymes. These enzymes are both proteasomal and soluble.

transfers the activated ubiquitin from E1 via an additional high-energy thiol ester intermediate, E2-S~ubiquitin, to the substrate that is specifically bound to a member of the ubiquitin-protein ligase family, E3. There are several families of E3 enzymes. Members of the RING finger-containing E3s, the largest family of E3s, catalyze direct transfer of the activated ubiquitin from E2 to the E3-bound substrate. For HECT (**H**omologous to the **E**6-AP **C T**erminus) domain E3s, the ubiquitin is transferred from the E2 to an active site Cys residue on the E3 to generate a third high-energy thiol ester intermediate, ubiquitin-S~E3, prior to its transfer to the ligase-bound substrate.

E3s catalyze the last step in the conjugation process: covalent attachment of ubiquitin to the substrate. Ubiquitin is generally transferred to an ε-NH$_2$ group of an internal lysine residue in the substrate to generate a covalent isopeptide bond. In some cases however, ubiquitin is conjugated to the N-terminal amino group of the substrate. By successively adding activated ubiquitin moieties to internal lysine residues on the previously conjugated ubiquitin molecule, a polyubiquitin chain is synthesized. The chain is recognized by the downstream 26S proteasome complex. Thus, E3s play a key role in the ubiquitin-mediated proteolytic cascade, as they serve as the specific substrate-recognition elements of the system. Approximately 1,000 different E3s have been identified in the human genome based on specific, commonly shared structural motifs. A single modification by ubiquitin or by ubiquitin-like proteins (UBLs), such as the Small Ubiquitin MOdifier (SUMO) or NEDD8, serves other, nonproteolytic purposes, such as routing cellular proteins to subcellular destinations or to the lysosome/vacule for degradation. UBLs are also conjugated via their C-terminal residue to an internal lysine residue in the acceptor protein. The specific enzymes that catalyze modification by ubiquitin-like proteins are somewhat different from those involved in conjugation of ubiquitin, although they utilize a similar mechanism. SUMO, for example, is conjugated by a heterodimeric E1, Aosl•Uba2, and the E2conjugating enzyme Ubc9. Although Ubc9 can recognize the SUMOylation motif and transfer SUMO to certain substrates, for other proteins, specific E3 enzymes have been described (for reviews on poly- and oligoubiquitination and on UBLs, see Pickart, 2001; Weissman, 2001; Glickman and Cicchanover, 2002; Schwartz and Hochstrasser, 2003; Hicke and Dunn, 2003; Huang *et al.*, 2004).

Degradation of the polyubiquitinated substrates is carried out by the 26S proteasome that does not recognize, in the vast majority of cases, nonmodified substrates. The proteasome is a multicatalytic protease that degrades polyubiquitinated proteins to short peptides. It is composed of two subcomplexes: a 20S core particle (CP) that carries the catalytic activity and a regulatory 19S regulatory particle (RP). The 20S CP is a barrel-shaped structure composed of four stacked rings, two identical outer α rings, and two identical inner β rings. The eukaryotic α and β rings are composed each of seven distinct subunits, giving the 20S complex the general structure of $\alpha_{1-7}\beta_{1-7}\beta_{1-7}\alpha_{1-7}$. The catalytic sites are localized to some of the β subunits. Each extremity of the 20S barrel can be capped by a 19S RP. One important function of the 19S

RP is to recognize ubiquitinated proteins and other potential substrates of the proteasome. A ubiquitin-binding subunit of the 19S RP has indeed been identified; however, its biological function and mode of action have remained enigmatic. A second function of the 19S RP is to open an orifice in the α ring that will allow entry of the substrate into the proteolytic chamber. Also, because a folded protein would not be able to fit through the narrow proteasomal channel, it is assumed that the 19S particle unfolds substrates and inserts them into the 20S CP. Both the channel opening function and the unfolding of the substrate require metabolic energy, and indeed, the 19S RP contains six different ATPase subunits. Following degradation of the substrate, short peptides derived from the substrate are released, as well as reusable ubiquitin. These peptides are further degraded into free amino acids by cytosolic amino- and carboxypeptidases. A small fraction of the peptides is transported across the ER membrane, binds to the MHC class I complex, and is carried to the cell surface to be presented to cytotoxic T cells. In case the peptides are derived from a "nonself" antigen, the T cell lyses the presenting cell. Proteasomal degradation is not always complete. In some cases, the proteasome processes the ubiquitinated substrate in a limited manner, releasing a truncated product. In the case of the NF-κB transcriptional regulator, an active subunit (p50 or p52) is thus released from a longer inactive precursor (p105 or p100). For reviews on the proteasome, see Zwickl *et al.* (2002), Gröll and Huber (2003), Adams (2003), and Glickman and Adir (2004).

A major, yet unresolved problem is how the UPS achieves its high specificity and selectivity toward its innumerable substrates. Why are certain proteins extremely stable in the cell, whereas others are extremely short-lived? Why are certain proteins degraded only at a particular time point during the cell cycle or only following specific extracellular stimuli, yet they are stable under most other conditions? It appears that specificity of the ubiquitin system is determined by two distinct and unrelated groups of proteins: (i) E3s and (ii) modifying enzymes and ancillary proteins. Within the ubiquitin system, substrates must be recognized and bind to a specific E3 as a prerequisite to their ubiquitination. In most cases, however, the substrates are not recognized in a constitutive manner and they must undergo a posttranslational modification such as specific phosphorylation or oxidation that renders them susceptible for recognition. In some other cases the target proteins are not recognized directly by the E3, and their recognition depends on association with ancillary proteins such as molecular chaperones that act as recognition elements in *trans* and serve as a link to the appropriate ligase. Other proteins, such as certain transcription factors, have to dissociate from the specific DNA sequence to which they bind in order to be recognized by the system. Stability of yet other proteins depends on oligomerization. Thus, in addition to E3s, modifying enzymes, such as kinases, ancillary proteins, or DNA sequences to which substrates bind, also play an important role in the recognition process. In some instances, it is the E3 that must be switched on by undergoing posttranslational modification in order to yield an active form that recognizes the target substrate.

II. MATERIALS AND INSTRUMENTATION

ATP (adenosine-5'-triphosphate disodium salt)(Sigma A-7699)

ATP-γ-S (adenosine-5'-[γ-thio]triphosphate tetralithium salt)(Sigma A-1388)

Chloramine-T (Sigma C-9887)

Creatine phosphokinase (CPK; from rabbit muscle)(Sigma C-3755)

Creatine phosphate (CP; phosphocreatine, disodium salt) (Sigma P-7936)

Cultured cells (HeLa, 293, Cos; E1 WT and its mutant cells—CHO-E36 and CHO-ts20) for preparation of cell extract and for pulse–chase labeling and immunoprecipitation experiments

Cycloheximide (Sigma C-1988)

2-Deoxyglucose (2-DOG; Sigma D-6134)

Diethylaminoethyl cellulose (DEAE; DE52; Whatman 4057-050)

2,4-Dinitrophenol (DNP; Aldrich D198501)

1,4-Dithiothreitol (DTT)(Sigma D-8255)

Dulbecco's modified Eagle medium (DMEM) with high glucose, l-glutamine, and sodium pyruvate (GIBCO 11995-065)

DMEM with high glucose, without L-glutamine, sodium pyruvate, and L-methionine (Biological Industries, Beit Ha'emek, Israel 01-054-1)

Epoxomicin (BIOMOL International L.P. PI-127)

N-Ethylmaleimide (NEM; Sigma E-1271)

Foetal bovine serum (certified, heat inactivated; US)(GIBCO 10082-147)

Foetal calf serum, dialyzed (Biological Industries, Beit Ha'emek, Israel 04-011-1)

Anti-HA antibody (Roche; clone 3F10; 1867423)

HEPES buffer (Sigma H-4034)

Hexokinase (~1,500 units/ml as ammonium sulfate slurry, Roche 1426362)

Anti-6X-His antibody (Qiagen; RGS; 34610)

[^{125}I]Na (350–600 mCi/ml; Amersham Biosciences IMS 300)

clasto-Lactacystin β-lactone (BIOMOL International L.P. PI-108)

Z-Leu-Leu-Leu-H (MG132; BIOMOL International L.P. PI-102)

L-[^{35}S]methionine (>1000 Ci/mmol; *in vitro* translation grade; Amersham biosciences SJ1515)

L-[^{35}S]methionine (×1000 Ci/mmol; cell labeling grade; Amersham biosciences SJ1015)

Methylated ubiquitin (MeUb; BIOMOL International L.P. UW8555)

Anti-Myc antibody (Roche; clone 9E10; 1667149)

New Zealand white rabbits (preferably females) of ~2 kg body weight (2-3 months old; for preparation of reticulocyte lysate)

Phenylhydrazine-HCl (Sigma P-6926)

Ribonuclease inhibitor (recombinant RNasin; Promega N2511)

Salts (KCl, NaCl, MgCl$_2$, NaF, NaN$_3$, ammonium sulfate, NaPi, KPi) (various suppliers; of highest analytical grade)

Sodium bisulfite (sodium metabisulfite; Sigma S-9000)

Transcription-translation kit (rabbit reticulocyte lysate based; L4600, L4610, or L4950, dependent on the RNA polymerase promoter in the vector, SP6, T3 or T7, respectively; TNT; Promega)

Transcription-translation kit (wheat germ extract based; L4120, L4130, or L4140, dependent on the RNA polymerase promoter in the vector, T3, SP6, or T7, respectively; TNT; Promega)

Trichloroacetic acid (TCA; 100% solution; Sigma T-0699)

Tris buffer (Trizma base; Tris base; THAM; Trometamol; 2-Amino-2-(hydroxymethyl)-1,3-propanediol; Sigma T-1378)

Ubiquitin (Sigma U-6253)

Ubiquitin (His-tagged; BIOMOL International L.P. UW8610)

Ubiquitin aldehyde (UbAl; BIOMOL International L.P. UW8450)

Ubiquitin, antibody (BIOMOL International L.P. PW8810)

Water (nuclease free; Promega P1193)

Dialysis tubing (SnakeSkin, Pierce, 68100)

Desalting column (PD-10, Amersham Biosciences 17-0851-01)

Nitrogen cavitation bomb (Parr Instrument Company)

III. PROCEDURES

A. Preparation of Cell Extracts for Monitoring Conjugation and Degradation

To conjugate or degrade a protein substrate *in vitro*, one has to utilize the appropriate cell extract. Rabbit reticulocyte lysate contains the enzymes required for degradation of most proteins and can be therefore used in most cases. Reticulocytes have several advantages. They contain a relatively small number of proteins and do not have lysosomes from which proteases can leak during preparation of the extract. Unlike cultured cells lysates, one can obtain reticulocyte lysate in a relatively large amount. Also, the lack of requirement for tissue culture media and sera makes this lysate significantly less expensive then its nucleated cultured cells counterpart. All these attributes make this lysate an ideal extract in which one can test conjugation and proteolysis of the studied protein. For monitoring conjugation and degradation of labeled proteins in the crude extract, it is not necessary to deplete ATP from the cells prior to the preparation of the extract. This will be necessary, however, in order to reconstitute the cell-free proteolytic system and to monitor dependence of the proteolytic process on the addition of exogenous ubiquitin. It will also be necessary in order to monitor conjugation of labeled or tagged ubiquitin to different substrates. Depletion of ATP from cells leads to deubiquitination of most proteins. Once such an ATP-depleted lysate is fractionated over the anion-exchange resin diethylaminoehtyl (DEAE)-cellulose, ubiquitin is eluted in the unadsorbed, flow-through material (fraction I) that also contains several E2 enzymes. Fraction II, the high salt eluate, contains E1, the remaining E2s, all the E3s, and the 26S proteasome.

1. *Preparation of Reticulocyte Lysate*

Steps

1. Inject rabbits subcutaneously with 10 mg/kg of phenylhydrazine (freshly dissolved in phosphate-buffered saline, PBS) on days 1, 2, 4, and 6.
2. Bleed the rabbits from the ear artery or vein or from the heart (following anaesthesia) on day 8. Induction of reticulocytosis

is dramatic, and more than 90% of the circulating red blood cells are reticulocytes as determined by methylene blue or brilliant cresyl blue staining.

3. Wash the cells three times with ice-cold PBS and, using a Pasteur pipette, aspirate carefully the thin layer of white blood cells ("buffy coat") that overlays the pelleted red blood cells.

4. Lyse the cells in 1.6 volumes (of pelleted cells volume) of ice-cold $H_2O \times 2$ (double distilled water) containing $1\,mM$ DTT (diluted from $1M$ solution).

5. Centrifuge at $80,000\,g$ for $1\,h$ at $4°C$ to remove particulate material.

6. Collect the supernatant and freeze in aliquots at $-70°C$.

7. To deplete ATP, wash cells twice in PBS and resuspend in 1 volume of Krebs Ringer phosphate buffer ($120\,mM$ NaCl, $4.8\,mM$ KCl, $1\,mM$ CaCl$_2$, $1.2\,mM$ MgSO$_4$, and $16.5\,mM$ NaPi, pH 7.4) containing $20\,mM$ 2-deoxyglucose and $0.2\,mM$ 2,4-dinitrophenol. Following incubation accompanied by gentle shaking for $90\,min$ at $37°C$, wash cells twice in PBS, lyse, and centrifuge as described previously.

2. Preparation of Extract from Cultured Cells

All procedures are carried out at $4°C$.

Steps

1. Wash cells three times in HEPES ($20\,mM$ pH 7.5)-saline buffer and resuspend to a concentration of 10^7–10^8/ml in $20\,mM$ HEPES, pH 7.5, that contains also $1\,mM$ DTT.

2. Cavitate cells in a high-pressure nitrogen chamber. For HeLa cells, the best conditions are $1000\,psi$ for $30\,min$. However, these conditions may vary among different cell species. For example, it may be necessary

to repeat the pressure cycle twice. Make sure that most of the cells are disrupted by visualizing the suspension in a light microscope before and after cavitation. Following disruption, one should observe intact nuclei (that are not broken) and cell debris.

3. Centrifuge the homogenate successively at $3,000\,g$ and $10,000\,g$ for $15\,min$ each time and then at $80,000\,g$ for $60\,min$. Collect and freeze the supernatant at $-70°C$.

4. To deplete ATP, wash cells twice in HEPES-saline buffer and resuspend in Krebs Ringer phosphate buffer (to a density of 10^7 cells/ml) in the presence of 2-deoxyglucose, 2,4-dinitrophenol (as described earlier), $20\,mM$ NaF, and $10\,mM$ of NaN$_3$. Following incubation for $60\,min$ at $37°C$, wash cells twice in HEPES-saline, resuspend in HEPES-DTT, lyse, and centrifuge as described earlier.

B. Fractionation of Cell Extract to Fraction I and Fraction II

As described earlier, fractionation of the lysate into fraction I and fraction II separates ubiquitin from many of the other components of the system, thus enabling one to monitor the dependence of conjugation and degradation upon the addition of exogenous ubiquitin and certain E2 enzymes. To fractionate the lysate, ATP-depleted lysate is resolved on a DEAE-cellulose column. In the ATP-depleted lysate, all the ubiquitin is free: it was released from conjugates by isopeptidases during incubation in the presence of the glycolysis and respiration inhibitors. In the absence of ATP, reconjugation cannot occur. Under these conditions, ubiquitin is resolved in fraction I, and fraction II is dependent for its conjugating and proteolytic activities upon the addition of exogenous ubiquitin. In cell extracts from which ATP was not depleted, the ubiquitin that is still conjugated to endogenous protein substrates will adsorb to the anion-exchange resin DEAE (via the protein substrate moiety) and will elute in fraction II. During incubation, the bound ubiquitin will be released by the activity of isopeptidases and will be available for conjugation to other proteins, including the test substrate examined. Therefore, it will be difficult to demonstrate ubiquitin-dependent conjugation and degradation in fraction II that is prepared from an extract from which ATP was not depleted. In addition, the bound ubiquitin fraction, when released, will dilute any added labeled or tagged ubiquitin, thus decreasing the detectable signal in the biosynthesized ubiquitin adducts.

All procedures are carried out at 4°C.

Steps

1. Swell the DEAE resin in 0.3M KPi, pH 7.0, for several hours. Use enough resin to adsorb all the proteins in the extract that can be bound. As a rule, use 0.6 resin volume per volume of reticulocyte lysate, or 1 ml resin/~5 mg protein of nucleated cell extract [in principle, one can use also a chromatographic system such as the AKTÄ FPLC (Amersham Biosciences) with a Mono Q column, although for resolution of large quantities, the DEAE resin procedure is advantageous].
2. Load the resin onto a column and wash with 10 column volumes of a buffer containing 5 mM KPi, pH 7.0, and 1 mM DTT (buffer A).
3. Load the extract. Once all the material is loaded, elute fraction I with buffer A. When resolving reticulocyte lysate, collect only the dark red fraction. When resolving mucleated cell extract, collect only the fractions with the highest absorption at 280 nm. Freeze fraction in aliquots at −70°C.
4. Wash the column extensively with a buffer containing 20 mM KCl in buffer A. When resolving reticulocyte lysate, make sure all the hemoglobin is eluted. When resolving

nucleated cell extract, wash until the absorbency at 280 returns to baseline.
5. Elute fraction II with 2.5 column volumes of a buffer containing 20 mM Tris–HCl, pH 7.2, 1 mM DTT, and 1M KCl.
6. Add to the eluted fraction II ammonium sulfate to saturation (~70 g/Li solution) and swirl on ice for 30 min.
7. Centrifuge at 15,000 rpm for 15 min.
8. Resuspend pellet in 0.2–0.3 of the volume of the original extract in a buffer containing 20 mM Tris–HCl, pH 7.2, and 1 mM DTT. At times, it will be difficult to dissolve all the proteins. This is not essential. They will be dissolved during dialysis.
9. Dialyze against two changes of a large volume of a buffer containing 20 mM Tris–HCl, pH 7.2, and 1 mM DTT. Dialysis should be carried out on ice. Remove particulate material by centrifugation at 15,000 rpm for 15 min. Freeze in aliquots at −70°C.

C. Labeling of Proteolytic Substrates

In most cases, monitoring the conjugation and/or degradation of a specific protein requires its labeling. The fate of the protein can also be followed via Western blot analysis using specific antibodies directed against the test protein [Western blot analysis will not be described here; however, the conjugation and degradation assays for labeled proteins (described later) can be applied in an almost identical manner for unlabeled proteins, using immune detection]. Two methods of labeling have proven to be useful: iodination and biosynthetic incorporation of [^{35}S]methionine. Iodination is utilized mostly when a purified recombinant/commercial protein is available. The main advantage of the method is the high specific radioactivity that can be attained. The disadvantage of the method is that one needs a pure protein. Also, during iodination, unless it is carried out using the Bolton–Hunter reagent, the protein can be damaged from the chloramine T used to oxidize the iodide. In addition, during storage, the labeled substrate may be subjected to radiochemical damage. A different method of labeling utilizes incorporation of ^{35}S-labeled methionine to a protein that is synthesized in a cell-free system from its corresponding cDNA/mRNA. The generated protein is native; however, the specific activity obtained is relatively low. Also, the labeled protein is contained in the crude extract in which it is synthesized and it is not pure. This extract contains, among other proteins, enzymes of the ubiquitin system that may interfere in the reconstitution of a cell-free system from purified components.

Despite these shortcomings, metabolic labeling of proteins is the most frequently used procedure to label them and follow their fate *in vitro*. To label proteins biosynthetically, one can first synthesize the mRNA on the cDNA template, using the appropriate RNA polymerase. Following digestion of the cDNA, the extracted mRNA can be translated *in vitro* in reticulocyte or wheat germ extracts. Alternatively, one can use a coupled transcription–translation cell-free extract that synthesizes the mRNA and translates it simultaneously. Such systems are available commercially (TNT; Promega). An even more advanced system (Quick TNT; Promega) contains all the necessary components for biosynthesis, except for the template cDNA and the labeling amino acid. Biosynthesis is carried out basically according to the manufacturer's instructions. In principle, it is preferable to use wheat germ extract. This extract lacks many, although not all, of the mammalian E3 enzymes. Therefore, in most cases, a protein synthesized in this extract can be used in experiments in which a cell-free system is reconstituted from purified enzymes and, in particular, when the role of a specific E3 is tested. A protein synthesized in reticulocyte lysate may be "contaminated" in many cases with its cognate E2 and/or E3 enzyme(s). This enzyme(s), which is being carried to the reconstituted system,

may interfere with the examination of the role of an exogenously added E2 and/or E3 in the conjugation/degradation of the translated protein. However, at times, one must use the reticulocyte lysate, as the translation efficiency in the wheat germ extract is extremely low. In that case, the "contaminating" E2 or E3 in the reticulocyte lysate can be inactivated, if necessary, by NEM (10-min incubation at room temperature in a final concentration of 10 mM of freshly prepared solution). Because E1, all known E2s, and some of the E3s (HECT domain-containing) have an essential $-$SH group, the alkylating agent inactivates them. The NEM is then neutralized by the addition of DTT (final concentration of 7.5 mM). It should be noted that this procedure can also denature/inactivate the substrate. Alternatively, the labeled substrate can be immunoprecipitated from the translation mixture, and the system can be reconstituted using the isolated protein. In most cases, however, the substrate can still be utilized and reproduces faithfully the behavior of the native substrate. The cDNA template coding for the test protein should be driven by one of the following RNA polymerase promoters: SP6, T7, or T3.

Radioiodination of Proteins

Steps

1. Add the following reagents in the following order to 1.5-ml microcentrifuge (Eppendorf) tube. The volume of the reaction mixture can vary from 20 to 100 µl.
   a. NaPi, pH 7.5, final concentration of 100 mM
   b. Protein substrate, 10–500 µg
   c. Unlabeled NaI, 50 nmol (use a stock solution of 10 mM in H_2O)
   d. Radiolabeled Na^{125}I, 0.1–2.0 mCi
   e. Chloramine-T solution (10 mg/ml in 10 mM NaPi, pH 7.4, freshly dissolved), 10–50 µg

2. Mix once (vortex) and incubate for 1–2 min at room temperature.

3. Add 20–100 μg Na-metabisulfite solution (20 mg/ml in 10 mM NaPi, pH 7.4, freshly dissolved) and mix. Add two-fold the amount of the added chlovamine T.

4. To remove unreacted radioactive iodine, resolve the mixture over a desalting column (PD10) equilibrated with 10 mM Tris–HCl, pH 7.6, and 150 mM NaCl. Collect fractions (in a fraction collector or manually) of ~10% of column volume each. The radioactive protein is typically eluted in fraction 4 (void volume of the column, which is ~35% of the total volume of the column). To keep a relatively small elution volume, it is recommended not to follow the one step elution procedure suggested by the manufacturer.

5. Store in aliquots at −18°C.

When monitoring conjugation, the *in vitro*-translated substrate can be used without further processing. This is also true in many cases when the degradation of the labeled substrate is followed by monitoring its disappearance in PhosphorImager-analyzed SDS–PA electrophoresed gels. However, as the degradation of certain proteins is not always efficient, it may be difficult to follow with accuracy the disappearance of 10–30% of a labeled protein band in a gel. In this case it will be necessary to monitor the release of radioactive material into a TCA-soluble fraction. For such preparations, in order to decrease background, the excess of unincorporated labeled methionine (or iodine) in the preparation of the translated protein must be first removed. This can be achieved via chromatography over DEAE exactly as described earlier for fractionation of lysate into resin-unadsorbed (fraction I) and adsorbed fractions (fraction II). The vast majority of the labeled proteins will resolve in fraction II, whereas the labeled amino acid (or iodine) will be eluted in fraction I. If the labeled protein is eluted in fraction I, changing

the pH may lead to its adsorption. Alternatively, extensive dialysis of the labeled protein against a solution of 20 mM Tris–HCl, pH 7.6, 150 mM NaCl that also contains 1 mM of unlabeled methionine (or No. I) will also remove efficiently the labeling amino acid.

D. Conjugation of Proteolytic Substrates *in vitro*

To demonstrate that the degradation of a certain protein proceeds in a ubiquitin-dependent manner, it is essential to demonstrate the intermediates in the process, ubiquitin–protein adducts. Typically, incubation of the labeled protein in a complete cell extract supplemented with ubiquitin and in the presence of ATP will lead to the formation of high molecular mass adducts that can be detected following resolution of the mixture in SDS–PAGE. To increase the amount of the adducts generated, one can use two approaches, independently or simultaneously. The nonhydrolyzable ATP analog, adenosine-5'-O-(3-thiotriphosphate), ATP-γ-S, can be used instead of ATP (Johnston and Cohen 1991). The ubiquitin-activating enzyme, E1, can catalyze activation of ubiquitin in the presence of the analog, as it utilizes the α–β high energy bond of the nucleotide that is cleavable also in the analog. In contrast, assembly and activity of the 26S proteasome complex require the β–γ bond, which cannot be cleaved in the analog. Caution should be exercised, however, when utilizing the ATP analog. Often, phosphorylation of the target protein is required in order for the ubiquitin ligase to recognize it (for example, see Yaron *et al.*, 1997). In these cases, the analog cannot substitute for the hydrolyzable native nucleotide, ATP. An additional approach to increase the amount of generated conjugates in a cell-free system is to use ubiquitin aldehyde (UbAl), a specific inhibitor of certain ubiquitin C-terminal hydrolases, isopeptidases (Hershko and Rose, 1987).

Steps

1. Add the following reagents to a 0.5-ml microcentrifuge (Eppendorf) tube. The volume of the reaction mixture can vary from 10.0 to 50.0 μl. Addition of all the reagents should be carried out on ice:

   a. 50 mM Tris–HCl, pH 7.6 (1M stock solution)

   b. 5 mM MgCl$_2$ (1M stock solution)

   c. 2 mM DTT (1M stock solution)

   d. 5.0–30 μl of reticulocyte lysate or 50–200 μg of complete mucleated cell extract protein

   e. 2.5–10 μg ubiquitin (10 mg/ml stock solution in H$_2$O)

   f. 0.5–2.0 μg UbAl (1 mg/ml stock solution in H$_2$O)

   g. ATP and ATP-regenerating system [0.5 mM ATP (0.1M stock solution), 10 mM CP (0.5M stock solution), and 2.5–10 μg CPK (10 mg/ml stock solution in 10 mM Tris, pH 7.2) or 2 mM ATP-γ-S (0.1M stock solution)]

   h. For depletion of endogenous ATP, the system should contain instead of ATP and the regenerating system 10 mM 2-DOG (1M stock solution) and 0.2–1.0 unit hexokinase (ammonium sulfate slurry; centrifuge an aliquot of the slurry and resuspend to the same volume in 20 mM Tris–HCl buffer, pH 7.6. Dilute in the same buffer. Stock solution in the buffer can be stored at 4°C for at least 4 weeks)

   i. For the substrate, use either a labeled protein (25,000–100,000 cpm) or an unlabeled substrate in an amount that is sufficient for detection by Western blot analysis (100–2000 ng)

Typically we prepare three reaction mixtures: (i) one that contains Tris, MgCl$_2$, DTT, ubiquitin, and UbAl, (ii) one that contains ATP, CP, and CPK, and (iii) one that contains 2-DOG and hexokinase.

2. Incubate the mixture for 30 min at 37°C and resolve via SDS–PAGE (7.5–10% acrylamide).

3. Detect high molecular mass conjugates by PhosphorImager analysis (labeled proteins) or via enhanced chemiluminescence (ECL) following Western blot (for unlabeled substrates) using a specific primary antibody against the test protein and a secondary tagged antibody.

There are several ways to demonstrate that the high molecular mass adducts generated are indeed ubiquitin conjugates of the test protein.

a. It is expected that the adducts will not be generated in an ATP-depleted system.

b. Generation of conjugates of the specific substrate should be inhibited reversibly by the addition of increasing amount of MeUb (Hershko and Heller, 1985). This reductively methylated derivative of ubiquitin lacks free amino groups and therefore cannot generate polyubiquitin chains. It serves therefore as a chain terminator in the polyubiquitination reaction, and consequently as an inhibitor in this reaction.

c. Adducts can be precipitated from the reaction mixture with an antibody directed against the test protein and, following SDS–PAGE, can be further detected with an antiubiquitin antibody. Alternatively, the reaction can be carried out in the presence of His-, HA-, or Myc-tagged ubiquitin (His-tagged ubiquitin is available commercially; HA- and Myc-tagged ubiquitins or the bacterial expression cDNA clones that code for them can be obtained from different researchers), and the immunoprecipitate can be detected with an antibody against the appropriate tag.

d. A cell-free system can be reconstituted from purified or isolated components of the ubiquitin system and the formation of conjugates can be followed, dependent upon the addition of these components.

Instead of adding a complete cell extract, it is possible to add fraction II (50–200 µg; derived from ATP-depleted cells) and free or tagged ubiquitin (2.5–10 µg: same amount as added to supplement the complete extract; see earlier discussion). Because fraction II is devoid of ubiquitin, the formation of conjugates that is dependent upon the addition of exogenous ubiquitin will strongly suggest that the high molecular mass derivatives generated slurring the reaction are indeed ubiquitin adducts of the test substrate. Because not all E2 enzymes are present in fraction II, it may be necessary, at times, to add to the reconstituted system purified UbcH5a, b, or c, UbcH7, or UbcH8 (~0.5–2.0 µg; available from BIOMOL International L.P.). In most cases, one of the UbcH5 (typically b or c) enzymes will be able to reconstitute activity and support conjugation.

E. Degradation of Proteolytic Substrates *in vitro*

With several exceptions, cell-free systems for monitoring the degradation of proteolytic substrates are similar to those used for monitoring their conjugation. In the proteolytic assays however, unlike in the conjugation assays, ATP (and not ATP-γ-S) must be used, as activity of the 26S proteasome complex is dependent upon cleavage of the high-energy β–γ bond (see earlier discussion). ATP is added along with ATP-regenerating system as described earlier. Also, UbAl is not added. Following incubation for 2–3 h at 37°C, the reaction mixture is resolved via SDS–PAGE and disappearance of the substrate can be monitored either via PhosphorImager analysis (in case the protein substrate is radioactively labeled) or via Western blot analysis (in case of unlabeled substrate). Control reactions are complete mixtures that have been incubated on ice and/or mixtures that were incubated at

37°C in the absence of ATP. At times, degradation efficiency is low and it is difficult to follow the reduction in the amount of a protein band in gel analysis. In these cases, it is necessary to monitor the appearance of radioactivity in the TCA-soluble fraction. Here, only a radioactive substrate can be used. In moA cases, radioiodinated proteins can be used directly. At times, excess unincorporated iodine must be removed. *In vitro*-translated proteins must first undergo DEAE fractionation or extensive dialysis in order to remove excess unincorporated labeled methionine (see earlier discussion). At the end of the incubation, a carrier protein (10–25 µl of 100 mg/ml solution of BSA) is added, followed by the addition of 0.5 ml of ice-cold TCA (20%). Following mixing, the reaction is incubated on ice for 10 min and centrifuged (5 min at 15,000 g). The supernatant is collected and the radioactivity is determined in either a β-scintillation counter (for methionine) or a γ-counter (for iodine-labeled substrates). Again, control reactions are complete mixtures that have been incubated on ice and/or mixtures that were incubated at 37°C in the absence of ATP.

F. Involvement of Ubiquitin System in Degradation of Proteins *in vivo*: Effect of Specific Proteasomal Inhibitors and Inactivation of E1 on Stability of Proteins in Intact Cells

All the known proteolytic substrates of the ubiquitin system are degraded, following generation of the polyubiquitin chain, by the 26S proteasome. The opposite notion, that all substrates of the 26S proteasome must be ubiquitinated prior to their recognition by the enzyme, is true in all but one established case, that of ornithine decarboxylase (ODC) (Murakami *et al.*, 1992). This enzyme is degraded by the 26S complex without prior ubiquitination. A noncovalent association with a specific binding protein, antizyme, renders ODC susceptible to degradation. The core catalytic subunit of the 26S enzyme is the 20S proteasome complex and inhibition of this complex inhibits all proteolytic activities of the 26S proteasome. To test whether a certain protein is degraded by the proteasome, it is possible to inhibit the enzyme, both *in vitro* and *in vivo*. Inhibition of the proteasome in a cell-free system requires higher concentrations of the inhibitor (two- to five-fold) compared to the concentrations required to inhibit it in cultured cells. Also, as noted earlier, for accumulation of ubiquitin adducts in cell-free systems, it is possible to inhibit the proteasome by using ATP-γ-S, the nonhydrolyzable ATP analog (see earlier discussion). Stabilization of a protein following inhibition of the proteasome is a strong indication that the protein is indeed degraded by this enzyme. Furthermore, inhibition of the 20S proteasome may lead to accumulation of ubiquitin adducts of the test protein that cannot be detected under conditions of rapid degradation when the proteasome is active. Detection of such intermediates serves as strong support for the notion that the protein is degraded by the 26S proteasome complex following its tagging by ubiquitin. The adducts can be detected by probing the specific immunoprecipitate that was resolved via SDS–PAGE and blotted onto the membrane with an anti-ubiquitin antibody. Alternatively, to increase the sensitivity of the signal, the cells can be transfected with a cDNA coding for HA-, His-, or Myc-tagged ubiquitin, and the immunoprecipitate can be detected with the appropriate anti-tag antibody.

Determination of Stability (Half-Life; $t_{1/2}$) of a Protein in Cells; Effect of Proteasome Inhibitors

Steps

1. Wash cells twice in a methionine-free medium at 4°C.
2. Add methionine-free medium that contains dialyzed serum (serum is added in the concentration used for growing the cells).
3. Incubate for 30 min (to remove endogenous methionine), remove the medium (by aspiration for adherent cells and following centrifugation at 800 g for 10 min for cells in suspension), and add fresh methionine-free medium with dialyzed serum. To save on labeled methionine, for adherent cells add medium to barely cover the cells (1–1.5 ml to a 60-mm dish). For cells in suspension, resuspend cells to 2 × 10^6/ml.
4. Add labeled methionine (50–250 μCi/ml) and continue the incubation for 1 h (pulse).
5. Add the inhibitor to the experimental dishes. Lactacystin or its lactone inhibitor (which penetrates cells better) or epoxomicin should be added to a final concentration of 5–20 μM, whereas MG132 should be added to a final concentration of 20–50 μM. The inhibitor should be added for 0.5 h (the last 30 min of the labeling period).
6. Remove the labeling medium.
7. Add ice-cold complete medium that contains, in addition to the inhibitor, also 2 mM of unlabeled methionine, and wash the cells twice. The complete medium should contain also 10% untreated complete FBS.
8. Add prewarmed complete medium (that contains the inhibitor and 2 mM of unlabeled methionine) and continue the incubation for the desired time periods (chase).
9. Withdraw samples at various time points and monitor degradation/stabilization of the target protein by immunoprecipitation, SDS–PAGE, and PhosphorImaging analysis. High molecular mass conjugates of the labeled protein should be precipitated by a specific antibody directed against the target protein under study. To avoid proteolysis of the conjugates by ubiquitin C-terminal hydrolases, it is recommended to dissolve the cells in a detergent-containing lysis

buffer at 100°C. Also, the buffer should contain 10 mM NEM to inhibit ubiquitin hydrolases. The NEM can be later neutralized by DTT (7.5 mµ) or β-mercaptoethenol (15 mµ).

Instead of using pulse–chase labeling and immunoprecipitation, one can use cycloheximide (20–100 µg/ml diluted from 20 to 100 mg/ml freshly water-dissolved solution) to stop protein synthesis and follow its degradation via Western blot analysis. The advantage of this approach is that it does not necessitate the use of radioactive material and immunoprecipitation, and one can load a whole cell extract onto the gel. Utilization of the proteasome inhibitors in this system is similar to that described previously for the pulse–chase labeling approach. The disadvantage of the method is the potential interference of the drug in the proteolytic process. Thus, if cycloheximide inhibits the synthesis of a short-lived ubiquitin ligase, E3, involved in the process, the test protein can be stabilized or further destabilized, dependent on the role of the ligase in its degradation.

A complementary approach to the utilization of proteasome inhibitors, stabilization of the protein and accumulation of ubiquitin adducts, is the use of cells that harbor a temperature-sensitive mutation in the ubiquitin-activating enzyme E1, the first enzyme in the ubiquitin proteolytic cascade. At the nonpermissive temperature, the cells fail to conjugate the target proteins, which are consequently stabilized. Such cells can be, for example, the CHO-E36 (WT) and CHO-ts20 (E1 ts mutant) (used, for example, in Aviel et al., 2000). The experimental approach used with these cells can be either pulse–chase labeling and immunoprecipitation or a cycloheximide chase.

Acknowledgments

Research in the laboratory of A.C. is supported by grants from Prostate Cancer Foundation (PCF) Israel—Centers of Excellence Program, the Israel Science Foundation founded by the Israeli Academy of Sciences and Humanities—Centers of Excellence Program, and the Foundation for Promotion of Research in the Technion. A.C. is an Israel Cancer Research Fund (ICRF) Professor. Infrastructural equipment was purchased with support of the Wolfson Charitable Fund Center of Excellence for studies on *Turnover of Cellular Proteins and its Implications to Human Diseases*.

References

Adams, J. (2003). The proteasome: Structure, function, and role in the cell. *Cancer Treat. Rev.* **29** (Suppl. 1), 3–9.

Aviel, S., Winberg, G., Massucci, M., and Ciechanover, A. (2000). Degradation of the Epstein–Barr virus latent membrane protein 1 (LMP1) by the ubiquitin-proteasome pathway: Targeting via ubiquitination of the N-terminal residue. *J. Biol Chem.* **275**, 23491–23499.

Glickman, M. H., and Adir, N. (2004). The proteasome and the delicate balance between destruction and rescue. *PLoS Biol.* **2**, E13.

Glickman, M. H., and Ciechanover, A. (2002). The ubiquitin proteasome pathway: Destruction for the sake of construction. *Physiol. Rev.* **82**, 373–428.

Gröll, M., and Huber, R. (2003). Substrate access and processing by the 20S proteasome core particle. *Int. J. Biochem. Cell Biol.* **35**, 606–616.

Hershko, A., and Heller, H. (1985). Occurrence of a polyubiquitin structure in ubiquitin-protein conjugates. *Biochem. Biophys. Res. Commun.* **128**, 1079–1086.

Hershko, A., and Rose, I. A. (1987). Ubiquitin-aldehyde: A general inhibitor of ubiquitin-recycling processes. *Proc. Natl. Acad. Sci. USA* **84**, 1829–1833.

Hicke, L., and Dunn, R. (2003). Regulation of membrane protein transport by ubiquitin and ubiquitin-binding proteins. *Annu. Rev. Cell Dev. Bio.* **19**, 141–172.

Huang, D. T., Walden, H., Duda, D., and Schulman, B. A. (2004). Ubiquitin-like protein activation. *Oncogene* **23**, 1958–1971.

Johnston, N. L., and Cohen, R. E. (1991). Uncoupling ubiquitin-protein conjugation from ubiquitin-dependent proteolysis by use of β, γ-nonhydrolyzable ATP analogues. *Biochemistry* **30**, 7514–7522.

Murakami, Y., Matsufuji, S., Kameji, T., Hayashi, S.-I., Igarashi, K., Tamura, T., Tanaka, K., and Ichihara, A. (1992). Ornithine decarboxylase is degraded by the 26S proteasome without ubiquitination. *Nature* **380**, 597–599.

Pickart, C. M. (2001). Mechanisms of ubiquitination. *Annu. Rev. Biochem.* **70**, 503–533.

Schwartz, D. C., and Hochstrasser, M. (2003). A superfamily of protein tags: Ubiquitin, SUMO and related modifiers. *Trends Biochem. Sci.* **28**, 321–328.

Weissman, A. M. (2001). Themes and variations on ubiquitylation. *Nature Rev. Cell Mol. Biol.* **2**, 169–179.

Yaron, A., Gonen, H., Alkalay, I., Hatzubai, A., Jung, S., Beyth, S., Mercurio, F., Manning, A. M., Ciechanover, A., and Ben-Neriah, Y. (1997). Inhibition of NF-κB cellular function via specific targeting of the IκBα-ubiquitin ligase. *EMBO J.* **16**, 6486–6494.

Zwickl, P., Seemüller, E., Kapelari, B., and Baumeister W. (2001). The proteasome: A supramolecular assembly designed for controlled proteolysis. *Adv. Protein Chem.* **59**, 187–222.

III.   PROTEIN–PROTEIN AND PROTEIN–SMALL MOLECULE INTERACTIONS

PROTEIN–DNA INTERACTIONS

22

Gel Mobility Shift Assay

Peter L. Molloy

I. INTRODUCTION

Because they are conceptually simple and also relatively straightforward to perform practically, gel mobility shift assays (otherwise known as gel retardation or electrophoretic mobility shift assays) have become one of the most widely used techniques in molecular and cell biology. They provide a key point of entry for identification of protein–DNA interactions important for regulation of gene expression. The discussion in this article focuses on DNA-binding proteins in relation to transcriptional control, but similar principles apply to the use of gel mobility shift assays to study protein/DNA interactions in other processes (replication, recombination, and repair), as well as proteins involved in RNA metabolism. The assay relies on the increased molecular size and decreased charge:mass ratio of a protein–DNA complex compared to free DNA and the observation that many protein–DNA complexes are stable through electrophoresis. Specific protein–DNA complexes can therefore be readily distinguished from free DNA by their slower mobility during electrophoresis. The use of labeled DNA probes in gel mobility shift assays enables visualisation of the specific complexes even in complex protein mixtures. The format of the assays allows both characterisation of DNA sequence requirements for protein binding and characterisation or identification of the proteins involved in complex formation, linking back to the cellular regulatory networks controlling gene expression.

II. MATERIALS AND INSTRUMENTATION

Electrophoresis requires a power supply capable of supplying approximately 30 mA at 200 to 300 V. Conditions described are for a vertical gel apparatus with 20 × 20-cm plates and 0.75- to 1-mm spacers. Conditions can be scaled down for a minigel apparatus, e.g., Bio-Rad 8 × 7.3 cm.

General chemicals should all be analytical reagent grade. Solutions for DNA-binding reactions should be prepared using nuclease-free water and reagents. HEPES (Cat. No. H4034), Nonidet P-40 (identical to Igepal CA-630, Cat. No. I8896), and dithiothreitol (DTT, Cat. No. D5545) can be purchased from Sigma Chemical Co. Nonspecific competitor polynucleotides, poly(dI-dC) poly(dI-dC), poly(dG-dC) poly(dG-dC), and poly(dA-dT) poly(dA-dT) are available from Amersham Biosciences, (Cat. Nos. 27-7880-02, 27-7910-02, and 27-7870-02, respectively). Nuclease-free bovine serum albumin (BSA) can be purchased from Promega. Restriction enzymes, Klenow fragment of DNA polymerase 1, T4 polynucleotide kinase, and premixed acry-lamide solutions are available from a number of suppliers.

Radioactive nucleotides [α-^{32}P]dATP or dCTP (>3000 Ci/mmol, 10 mCi/ml) and [γ-^{32}P]ATP (>3000 Ci/mmol, 10 mCi/ml) can be purchased from Amersham Biosciences. Reagents for nonradioactive labeling of DNA probes are also available, e.g., digoxygenin-labeling kit from Roche Applied Science and chemiluminescent kit from Pierce. Direct postelectrophoresis

staining for DNA and protein components of complexes using sensitive dyes are now also possible (Jing et al., 2002)

Nuclear extracts and related reagents, including specific oligonucleotide sets, are available from Promega, and a kit for nuclear extract preparation can be purchased Pierce. A wide range of antibodies targeted to transcription factors and other DNA-binding proteins is available from Santa Cruz Biotechnology, Inc. and also Chemicon; both catalogues indicate which antibodies are suitable for "supershifting." A range of individual protease inhibitors and prepared cocktails are available from Roche Applied Science and Calbiochem.

III. PROCEDURES

A. Preparation of Probes and Competitor DNAs

Procedures described in this article utilise ^{32}P-labeled probes that have traditionally been used in gel shift assays to allow ready visualisation and potential quantification by autoradiography or phosphorimaging. It is important to use proper shielding (e.g., Perspex screens) to limit exposure to ^{32}P radiation. There are also a number of nonradioactive methods available for visualisation of complexes, including fluorescent probes and biotinylated or digoxygenin-labeled probes. Nonradioactively labeled probes have advantages in safety and length of storage but also have disadvantages, such as the greater number of handling steps needed for visualisation, sensitivity levels, and linearity of signals. Probes are normally prepared by either restriction enzyme digestion and labeling or by labeling of oligonucleotides.

1. Restriction Digestion

Labeling of restriction digests of 1 to 2 μg of plasmid DNA should provide sufficient DNA probe to perform 50 to 100 binding reactions (for a 3-kb plasmid, 2 µg yields about 1pmol of fragments). To minimise the effect of nonspecific binding to the probe and the number of potential binding sites, restriction fragments should be relatively short (ideally less than 100bp). If possible, it is best to have binding sites located centrally within the fragment; because affinity is often enhanced by nonspecific interaction of proteins with DNA surrounding their specific binding site, a fragment length of at least 30 to 40bp is advisable. Restriction enzymes that leave 5′ overhangs are most convenient as they can readily be filled in and radiolabeled using the Klenow fragment of DNA polymerase I and an appropriate $[\alpha\text{-}^{32}P]dNTP$ in the presence of other unlabeled deoxynucleotides.

For end labeling, we digest 2 µg of plasmid DNA in a 20-µl reaction in the recommended enzyme buffer. To this is added 20 µCi of suitable ^{32}P-labeled deoxynucleotide for end labeling along with the other three unlabeled dNTPs to a concentration of 100 µM each and 1 unit of Klenow fragment of DNA polymerase I. After a 15-min incubation at room temperature, the fourth unlabeled dNTP is added to 100 µM and incubation is continued for 5min. Chasing the reaction with the unlabeled nucleotide is important, as single-stranded ends can provide avid binding sites for some proteins.

For optimal gel shift results it is important to gel purify restriction fragment probes. Depending on their size, probes can be separated on 5 to 10% acrylamide gels. For digests of 1–2 µg of DNA, load digest in a 2.5-cm-wide well on a 1-mm-thick gel in TBE buffer and electrophorese at 10V/cm until the bromphenol blue dye is near the bottom of the gel. For probes 50bp or less, we routinely run gels and elute small DNA fragments in the cold room. This avoids DNA melting, as single-strand DNAs can produce artefactual results. After the gel apparatus is dismantled, leaving the gel on one of the glass plates, it is covered with plastic wrap and exposed to X-ray film for 2 to 5 min, marking the film for alignment with the gel. The position of the labeled band is identified from the autoradiograph, and the gel slice is excised with a scalpel blade. Fragments can be recovered by elution overnight on a rocking platform or shaker. Depending on its concentration, DNA probes may be used directly or ethanol precipitated and resuspended in TE buffer. Storage of radiolabeled fragments in 0.5mM DTT or 1mM β-mercaptoethanol is recommended to limit radiolytic breakdown.

Unlabeled competitor DNA fragments are prepared similarly except that the quantity of DNA is increased to 5 µg, restriction digestion is done in 50 µl, and all four deoxynucleoside triphosphates are added for end filling. A thicker, 2-mm gel should be used and the gel stained lightly with ethidium bromide (soaking in 0.5 µg/ml solution for 5 to 10min) for fragment visualisation and isolation. The unlabeled competitor needs to be concentrated by ethanol precipitation.

2. Oligonucleotide Probes

Once a target region for protein binding within a gene regulatory region is identified, oligonucleotide probes provide a powerful way to characterise DNA sequences responsible for DNA–protein complex formation. Sets of mutations within the putative binding site can be prepared and assayed readily and binding and migration of complexes can be compared with complexes formed on binding sites for well-characterised proteins. Probes may be prepared by end filling using deoxynucleoside triphosphates (A and B in Fig. 22.1) or by kinasing one or both strands. For methods A and C, two complementary 25–30 base oligonucleotides must be prepared for each sequence to be studied. For method B, a single 12-base primer adjacent to the binding sequence to be studied can be used to prime on separate oligonucleotides containing variants of the target sequence.

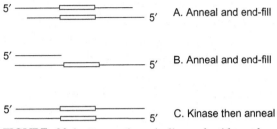

FIGURE 22.1 **Preparation of oligonucleotide probes.** Three methods as described in the text for preparation of oligonucleotide probes are shown. The boxed region indicates sequences required for protein–DNA binding. (A and B) Oligonucleotides are extended from 3′ ends to flush-ended double-stranded oligonucleotides. (C) Fully two complementary oligonucleotides are annealed after kinase labeling.

a. End-Filling Reactions. Mix 2pmol of each oligonucleotide (for method A) or 2pmol of the long oligonucleotide and 10pmol of primer for method B in 10μl of medium salt restriction enzyme buffer, warm to 60°C, and allow to cool slowly to room temperature. Adjust the reaction volume to 20 μl with additional 10× restriction enzyme buffer, 20μCi of ^{32}P-labeled deoxynucleotide, the other three unlabeled dNTPs to a concentration of 100μM, and 1 unit of Klenow fragment of DNA polymerase I. Incubation and chase with unlabeled dNTP are as for labeling restriction fragments.

b. Kinase Reactions. Incubate oligonucleotides for 30min at 37°C in 20 μl final volume reactions containing 2pmol of oligonucleotide, 2μL of 10× kinase buffer, 3 μl (30μCi) of [γ-^{32}P]ATP, and 1 μl of polynucleotide kinase. After kinasing one or both complementary oligonucleotides, mix 20-μl reactions containing 2pmol of each and add 5 μl of 100mM MgCl$_2$. Heat the mix to 60°C and allow oligonucleotides to anneal by cooling slowly to room temperature.

After labeling by either method, load reactions on a 10% acrylamide/0.5% bisacrylamide gel run at 4°C in TBE buffer and purify fragments as described earlier. Gel purification is essential to remove any single-stranded oligonucleotide.

B. Protein–DNA-Binding Reactions

1. *Protein Source*

Nucleic acid-binding proteins used in gel shift assays may be either purified proteins (endogenous or expressed recombinant proteins) or contained in relatively crude cellular or nuclear extracts. *In vitro* expression in coupled transcription/translation extracts is a convenient way to produce proteins if the gene has been cloned. For many studies, nuclear extracts from cells or tissues in which a gene is expressed provide the starting point for the identification of transcription factors relevant to the expression of a gene. The nuclear extract preparation is based on the principle that elevated levels of salt release specific DNA-binding proteins from chromatin; it is critical that the salt level remains below that which will begin to dissociate histones that bind DNA strongly and nonspecifically and interfere with gel shift assays. Preparation of nuclear extracts is usually based on the method of Dignam *et al.* (1984), which is applicable to quantities of 10^8 to 10^9 cells in culture. After isolation of a crude nuclear fraction, proteins are extracted in 0.42M NaCl and the extract is dialysed against buffer containing 0.1M KCl. A number of variations of the method have been published that allow for extract preparation from tissue sources (Gorski *et al.*, 1986; Fei *et al.*, 1995) or rapid miniextract preparation from small numbers of cultured cells (e.g., Schreiber *et al.*, 1989). The rapid ammonium sulfate nuclear extract protocol of Slomiany *et al.* (2000) provides a convenient approach for the isolation of high activity extracts from a range of starting sources. When preparing extracts from tissue sources, it is especially important to minimise proteolytic degradation. Early methods included the serine protease inhibitor phenylmethylsulfonyl fluoride, but use of an inhibitor cocktail is recommended to obtain extracts of maximal activity and to avoid potential confusion caused by probes binding to different proportions of intact proteins and proteolytic fragments in extracts from different tissues or cell types.

2. Reaction Setup and Parameters

For each protein–DNA interaction studied it is necessary to optimise the binding conditions. For nuclear extracts and a range of binding proteins, the standard conditions shown in Table 22.1 provide a good starting point. Because the amount of a specific protein in an extract and the amount of nonspecific DNA-binding proteins will vary widely, it is important to initially survey a range of protein and nonspecific DNA competitor

concentrations to identify levels that allow clear distinction of the protein–DNA complexes.

A setup for a typical exploratory experiment is shown in Table 22.2. Reactions should be set up on ice in 1.5- or 0.5-ml microfuge tubes, first adding all components except the nuclear extract and DNA probe. For nuclear protein extracts, a wide range of protein concentrations, e.g., from 2 to 20 µg in a 20-µl reaction, should be assayed. A typical nuclear extract contains 2 to 4 µg of protein per microliter. Levels of purified proteins need to be titrated to determine optimal levels. It can be preferable to add the nuclear extract prior to the probe if the extract contains significant levels of avid, nonspecific

TABLE 22.1   Protein–DNA-Binding Reaction Conditions with Nuclear Extracts

Component[a]	Standard conditions	Range/alternates
HEPES (pH to 7.9 with 2M KOH)	12 mM	Tris–HCl pH range 6.5–8.5
KCl	60 mM	0–200 mM sodium or ammonium salts
EDTA	0.6 mM	0.1 mM EGTA for selective inhibition of Ca-dependent proteases
Glycerol	12%	0–12%
Dithiothreitol (DTT)	1.2 mM	0 to 5 mM. *Note*: Binding of some proteins is redox sensitive, whereas for others binding may be enhanced by higher levels of DTT
NP40	0.1%	Up to 2% Tween or Triton X-100 as alternates
poly(dI-dC):poly(dI-dC)	1 µg/20 µl Rn	200 ng to 4 µg; other synthetic DNA polymers or mixed sequence DNA such as calf thymus or *E. coli*
Nuclease-free BSA	10 µg/20 µl Rn	Up to 20 µg
MgCl$_2$		Up to 10 mM
Labeled DNA probe	10 fmol	
Temperature	30°C	0–37°C
Time	30 min	5–30 min
Protease inhibitors		Use as required. Cocktail of inhibitors or individual inhibitors: AEBSF (Pefabloc), leupeptin, aprotinin, calpain inhibitors I and II, soybean trypsin inhibitor, chymostatin, pepstatin

[a] HEPES, KCl, EDTA, glycerol, and DTT are all contained within binding buffer A, which comprises 20 mM HEPES, pH 7.9, 100 mM KCl, 1 mM EDTA, 2 mM DTT, and 20% glycerol. Prepare by dissolving the HEPES base, KCl, and EDTA and adjusting the pH to 7.9 using 2M KOH. Add glycerol and adjust the volume prior to autoclaving. Buffer without DTT can be stored at room temperature. Buffer with DTT should be stored in aliquots at −20°C.

22. GEL MOBILITY SHIFT ASSAY

TABLE 22.2 Exploration of Binding Conditions

Reaction #	1	2	3	4	5	6	7	8	9	10	11	12
Nuclear extract[a]	0	1	2	4	8	12	4	4	4	4	4	4
Binding buffer A[b]	12	11	10	8	4	0	8	8	8	8	8	8
BSA, 10 mg/ml	1	1	1	1	1	1	1	1	1	1	1	1
poly(dI-dC):poly(dI-dC), 2 mg/ml[c]	1	1	1	1	1	1	0.5	1.5	2	1	1 dAT[e]	1 dGC[e]
NP-40, 2%	1	1	1	1	1	1	1	1	1	1	1	1
MgCl$_2$, 100 mM[d]										1		
H$_2$O	3	3	3	3	3	3	3.5	2.5	2	2		
DNA probe (10 fmol)	2	2	2	2	2	2	2	2	2	2	2	2

[a] Conditions based on nuclear extract being dialysed against binding buffer A.
[b] Composition of buffer is in footnote to Table 22.1.
[c] Polynucleotides are dissolved in 10 mM Tris–HCl, 0.1 mM EDTA, pH 8, and stored in aliquots at −20°C.
[d] MgCl$_2$ should be prepared from fresh solid using sterile, nuclease-free water and filter sterilised.
[e] dAT = poly(dA-dT) poly(dA-dT)poly and dGC = (dG-dC) poly(dG-dC), both prepared as in footnote c.

DNA-binding proteins, but in practice the order of addition does not normally make a significant difference. The important point is that the protein and DNA probe should not come into contact until the final mixing step. Binding reactions are typically incubated at 30°C for 30 min. Binding buffer (4 μl) containing bromphenol blue dye is added to aid in gel loading. Provided the binding buffer contains >5% glycerol, there is no need for addition of a density reagent. At this early stage it is also a good idea to test the separation of complexes on different gel systems (see later), as the buffer type and ionic strength can significantly affect separation and band quality.

The exploratory conditions of Table 22.2 were applied to the binding site for an ets-related protein (GGAA core sequence) found in the N-ras gene promoter (Fig. 22.2). The amount of specific complex formed increases with increasing levels of nuclear extract, but at higher levels of extract, a significant fraction of the labeled probe is trapped in the well. A clear signal with minimal background is achieved with 0.5 to 1 μl of nuclear extract and 1 μg of poly(dI-dC) competitor. Replacement of poly(dI-dC) with either poly(dA-dT) or poly(dG-dC) reveals the

presence of additional complexes, indicating the value of testing alternate nonspecific competitor DNAs.

Table 22.1 indicates a range of possible variations of binding conditions, and the effect of reaction conditions on complex formation should be explored systematically.

a. Time and Temperature. While reactions are often incubated for 30 min, all that is necessary is sufficient time for complex formation to have reached equilibrium, which can often be as short as 5–10 min. Binding reactions are done most commonly at 30°C or room temperature. Lower temperatures sometimes improve binding, e.g., Sp1 binds better at 20°C than at 30°C. Lower temperatures and/or shorter times of incubation can also limit the effect of phosphatases that may be present in some extracts (critical if the active form of a protein is phosphorylated or if probe 5′ is end labeled) (Laniel and Guerin, 1998). Similar considerations apply to the inhibition of nucleases if divalent ions are present in the binding reaction. A characteristic downward smearing of the unbound DNA band is indicative of nuclease activity.

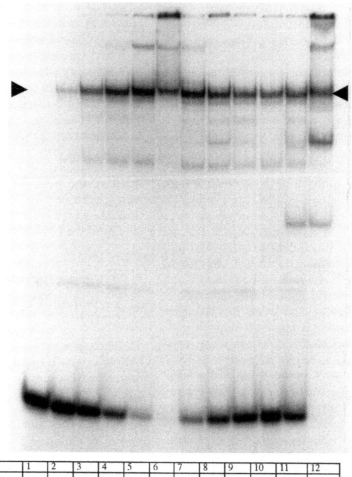

	1	2	3	4	5	6	7	8	9	10	11	12
Nuc. extract (μl)	0	0.2	0.5	1.0	2.0	4.0	1.0	1.0	1.0	1.0	1.0	1.0
poly dI-dC (μg)	1.0	1.0	1.0	1.0	1.0	1.0	0.5	1.5	2.0	1.0		
poly dX-dY, 1μg											dA-dT	dG.dC
MgCl₂, 5 mM	-	-	-	-	-	-	-	-	-	+	-	-

FIGURE 22.2 Analysis of protein–DNA-binding conditions. Binding reactions were set up using a 50 bp fragment containing a binding site for an ets-related protein. Reaction components were as in Tables 22.1 and 22.2; the variation in specific components is indicated in the table beneath the gel lanes. Free and bound complexes were separated on a 5% acrylamide gel(30:1 acrylamide: bisacrylamide) run in TNAE buffer. The major specific complex is indicated by the arrow.

b. Ionic Strength. The relative level of formation of specific and nonspecific complexes is influenced significantly by the concentration of monovalent cation; increased salt concentrations can favour the formation of specific complexes. Most nonspecific protein–DNA interactions are principally ionic and the strength of interaction declines as the salt concentration is increased. For sequence-specific complexes, key interactions involve hydrogen bonding and hydrophobic interactions between bases and amino acids, although the interaction between positively charged amino acids and the phosphate backbone still makes an important contribution

IV. PROTEIN–DNA INTERACTIONS

to overall affinity. When working with crude protein extracts, salt concentrations in the range of 50 to 100 mM are generally optimal, but individual complexes can sometimes be differentiated by their stability at concentrations up to 200 mM. Particularly for purified proteins, when the issue of nonspecific binding by other proteins is not an issue, use of a lower or no added monovalent ion can improve binding affinity.

c. Divalent Ions. While many protein–DNA complexes form in the absence of divalent ions, it is advisable to always evaluate binding in both the presence and the absence of Mg^{2+}, as sometimes binding properties can be affected significantly. An illustrative example is binding of the helix–loop–helix protein USF. In the presence of Mg^{2+}, the rates of both association and dissociation and equilibrium binding are affected significantly and even the sequence specificity of binding is altered (Chodosh *et al.*, 1986; Bendall and Molloy, 1994). In some cases, a specific metal ion can be necessary for DNA binding and as well as inclusion of the metal ion in the binding reaction it may be necessary to omit EDTA from the electrophoresis buffer (Anderson *et al.*, 1990). Examples are the metal response element-binding factor that requires Cd^{2+} and Zn finger proteins, which require low levels of Zn^{2+} for proper protein folding.

d. Nonionic Detergents and BSA. The addition of a nonionic detergent and a carrier protein such as NP-40 and BSA helps minimise protein aggregation and generally results in less smearing of bands and less trapping of the DNA probe in the wells. This is beneficial when working with crude nuclear extracts or to minimise protein denaturation and sticking to tube walls when working with highly purified proteins

e. Nonspecific Competitor DNA. All DNA-binding proteins will bind to a certain extent to both the specific probe and nonspecific competitor DNA. The purpose of the nonspecific competitor is

to bind as much of the nonspecific DNA-binding proteins in an extract with minimal binding of the specific protein being studied. Synthetic copolymers have become the preferred choice for competing nonspecific DNA binding because they are less likely to be able to bind efficiently with sequence-specific DNA-binding proteins. Natural, mixed sequence competitor DNAs such as *Escherichia coli* or salmon sperm DNA can also prove to be very effective. However, competitors that have similarity to the binding site of the protein(s) being studied have the potential to interfere significantly with specific binding. For example, poly(dA-dT) would be a poor choice of competitor for studying binding of the TATA-binding protein.

C. Specificity of Protein–DNA Complex Formation: Characteristics of Target DNA Site

Gel retardation assays can be used effectively both to establish that formation of a complex is dependent on specific DNA-sequences and to define the sequence characteristics of the DNA binding site. This can be achieved through a combination of DNA-binding assays using probes of different sequence and competition experiments using unlabeled competitor DNAs.

1. Competitor DNA

Competition assays are based on the principle that the amount of a sequence-specific complex formed on a labeled probe will be reduced in proportion to the amount of competitor DNA if the competitor DNA has specificity for the same sequence. In contrast, if the competitor lacks the specific binding site the additional DNA will compete generally for binding of proteins in the reaction and will add to the overall level of nonspecific competitor DNA. Competitor DNAs may be either restriction fragments or oligonucleotides (see earlier discussion for preparation), but for comparison between competitors it is important to use DNAs of equivalent size as affinity can increase significantly with increasing fragment size, particularly with oligonucleotides of about 20 bp. Competitor DNAs should minimally include a DNA fragment identical to the probe and a control of equivalent length but unrelated sequence (Fig. 22.3, tracks 4–9). Competitors with specific mutations can be used to identify or confirm the location of bases important for binding; this approach can be extended to a series of mutants to provide a detailed analysis of the relative importance of different bases to binding affinity. If binding of a known protein is suspected, competition with a DNA fragment that contains a known high-affinity binding site surrounded by unrelated DNA can provide good supporting evidence.

For competition assays the level of DNA-binding protein should not be saturating on the labeled probe; binding of 20 to 30% of the labeled probe in the absence of competitor is a good starting point. It is reasonable to use up to 1 pmol of competitor in a 20-μl reaction, allowing, for example, 10-, 50-, and 200-fold ratios if 5 fmol of probe is included in the reaction. For proper comparison of relative binding affinities the probe and competitor should be added together. It is important to remember, especially with impure protein preparations, that binding kinetics are not simple and that the final level of binding seen depends not just on the specific binding parameters of the protein with its specific DNA target, but also the binding of the protein to both specific and nonspecific competitor DNAs and to other proteins in the reaction, as well of the binding of the DNA probe to other proteins.

2. Direct Analysis of Binding to Probes of Different Sequences

Complementary to the aforementioned competitor approach, DNA fragments of different sequence can be used as labeled probes.

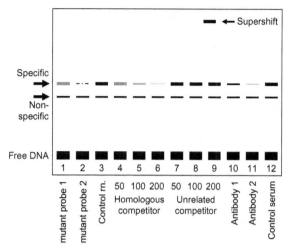

FIGURE 22.3 **Gel mobility shift schematic.** The control binding reaction that produces a specific as well as a nonspecific complex is shown in track 3. Reactions are identical to that of track 3 except for a single factor as indicated below each track. Tracks 3 to 12 all contain the same labeled DNA probe. Tracks 1 and 2 contain labeled mutant probes of different binding affinity. In tracks 4 to 6, competitor unlabeled DNA identical in sequence to the probe has been added at 10-, 50-, and 200-fold excess. In tracks 7–9 an unrelated competitor DNA of the same length as the labeled probe has been added at 10-, 50-, and 200-fold excess. In track 10, an antibody that specifically binds the protein in the specific complex without disrupting the DNA-binding interaction has been added. In track 11, antiserum that binds the protein in the specific complex and disrupts complex formation has been preincubated in the reaction prior to probe addition. In track 12, nonimmune control serum has been added.

Probes covering sequence variants in the same way as for competitor analysis can be run in parallel. Formation of complexes with equivalent migration in the gel provides evidence that both sequences can bind the same protein (Fig. 22.3, tracks 1–3). This approach is well suited to analysis of series of mutations within a fragment and identification of bases critical for binding. An approach that allows relative binding affinities to be estimated is to include a reference-labeled DNA probe of different size in binding reactions; if probe lengths are sufficiently different, it can be feasible to separate complexes formed on the separate probes and quantify binding in relation to the reference probe (Bendall and Molloy, 1994). The principle of reacting proteins with sequence variants can be extended to incubating proteins with populations of oligonucleotides containing segments of random sequence DNA. Complexes can be isolated from gels and successive rounds of PCR and gel retardation used to isolate enriched populations of DNA-binding sites (Oliphant *et al.*, 1989). For studies of a limited number of base positions, bound populations can be studied by direct sequencing (Luo *et al.*, 1997).

D. Characterisation of DNA-Binding Proteins

Gel retardation assays can provide an entry point for identification and characterisation of proteins regulating gene expression. Antibodies to a wide range of characterised transcription factors are available and can be used to determine if a specific protein is contained in a complex (see Fig. 22.3). If an antibody binds to or near the DNA-binding domain it may disrupt complex formation, reducing the amount of complex formed (Fig. 22.3, track 11). Conversely, if the target epitope is distant from the DNA-binding domain, antibody binding will not be disrupted and formation of a ternary DNA–protein–antibody complex will produce a supershifted complex of lower mobility (Fig. 22.3, track 10).

In many instances the nature of the binding sequence provides a guide to possible binding proteins. Comparison can then be made of the migration of protein complexes formed on the sequence under study and known high affinity sites for the candidate protein. Protease clipping can be further used to confirm the equivalence of complexes formed on the two target DNA sequences (e.g., Watt and Molloy, 1993). For many proteins, cDNA clones can be used to express candidate proteins and binding and

complex migration compared between nuclear extracts and expressed proteins and between complexes formed by the expressed proteins on known binding sites and the sequence being analysed.

For complexes where the protein involved is unknown, gel retardation can be used to determine the molecular weight of a protein. This can be done indirectly through plotting mobility vs gel concentration (Orchard and May, 1993) or by cross-linking protein DNA complexes and subsequent electrophoresis on SDS–polyacrylamide gels (Miyamoto *et al.*, 1995). With the advent of increasingly sensitive proteomic technology, the direct identification of proteins using mass spectrometry has become feasible (see Woo *et al.*, 2002).

E. Kinetic Analysis

For purified proteins, gel retardation assays provide a convenient approach to the determination of association and dissociation rates and equilibrium-binding constants of protein–DNA complexes (Meisterernst *et al.*, 1988; Chodosh *et al.*, 1986). Accurate data can be obtained if the distribution between complex and free DNA and protein is not altered during gel loading and running. Bain and Ackers (1998) have described a cryogenic gel system where the rapid quenching of reactions by transfer to $-40°C$ and electrophoresis at this temperature results in almost complete stabilisation of species distribution. As well as its application in kinetic analysis, this gel system may allow visualisation of complexes that are present in solution but are not stable under conditions normally used in gel retardation experiments. With crude nuclear or cell extracts, the number of competing reactions for binding of proteins and DNA means that it is difficult to obtain good kinetic data (for discussion, see Cann, 1989), although measurement of dissociation rates is possible.

TABLE 22.3 Electrophoresis Buffers and Conditions

Buffer	Components	Running conditions
TNAE	6.7 mM Tris–HCl, pH 7.9 3.3 mM sodium acetate 1 mM Na$_2$EDTA Adjust the pH of a 50× stock buffer to 7.9	Gel 5% acrylamide, 40:1 acrylamide:bisacrylamide Prerun at 10–15V/cm for at least 30min. Gel run at 10–15V/cm for 2–3h
TG	50 mM Tris base 380 mM glycine 1.67 mM Na$_2$EDTA Prepare a ×5 stock solution	Gel 5% acrylamide, 30:1 or 80:1 acrylamide: bisacrylamide. Prerun at 10V/cm for at least 30min. Gel run at 15V/cm for 2–3h
TBE	89 mM Tris base 89 mM boric acid 2 mM EDTA, pH 8.3 Prepare a 5× stock solution	Gel 5% acrylamide, 40:1 or 30:1 acrylamide: bisacrylamide, prepared and run in either ½× or ¼× TBE Prerun at 10–15V/cm for at least 30min. Gel run at 10–15V/cm for 2–3h

F. Gel Preparation and Running

1. Gel Matrix

Acrylamide:bisacrylamide gels with acrylamide concentration 5% and ratios to crosslinker of 30:1 to 80:1 are used most commonly. Concentrations can be adjusted readily to provide the best resolution for individual complexes. Agarose gels can also be used in gel retardation assays; they can be useful for separation of large complexes such as nucleosomes or when using large DNA fragments as probes. With gel additives, agarose can also provide similar resolution to acrylamide gels (Chandrasekhar et al., 1998).

2. Electrophoresis Conditions

Gel retardation assays rely on the continued association of protein–DNA complexes during electrophoresis. To minimise dissociation of complexes, it is best to conduct electrophoresis at 4°C (in a cold room or refrigerated cabinet) and in low ionic strength buffers. Commonly used buffers and electrophoretic conditions are provided in Table 22.3. Because buffer conditions can differentially affect the migration and/or stability of complexes, it is valuable to test the different systems and choose the one that proves best for the complexes being studied.

Complex dissociation is minimised in the low salt TNAE buffer, but its low buffering capacity means that the running buffer needs to be recirculated. The higher salt buffers are more convenient in that they do not require recirculation and background nonspecific binding to the probe can be reduced. However, not all complexes are stable in the higher salt buffers. For convenience, electrophoresis at room temperature should also be tested, taking care to lower the voltage to about two-thirds that used in the cold in order to prevent the gel from overheating.

References

Anderson, R. D., Taplitz, S. J., Oberbauer, A. M., Calame, K. L., and Herschman, H. R. (1990). Metal-dependent binding of a nuclear factor to the rat metallothionein-I promoter. *Nucleic Acids Res.* **18**, 6049–6055.

Bain, D. L., and Ackers, G. K. (1998). A quantitative cryogenic gel-shift technique for analysis of protein-DNA binding. *Anal. Biochem.* **258**, 240–245.

Bendall, A. J., and Molloy, P. L., (1994). Base preferences for binding by the bHLH-Zip protein USF: Effects of MgCl$_2$ on specificity and comparison with binding of Myc family members. *Nucleic Acids Res.* **22**, 2801–2810.

Cann, J. R. (1989). Phenomenological theory of gel electrophoresis of protein-nucleic acid complexes. *J. Biol. Chem.* **264**, 17032–17040.

Chandrasekhar, S., Soubar, W. W., and Abcouwer, S. F. (1998). Use of modified agarose gel electrophoresis to resolve protein-DNA complexes for electrophoretic mobility shift assay. *Biotechniques* **24**, 217–218.

Chodosh, L. A., Carthew, R. W., and Sharp, P. A. (1986). A single polypeptide possesses the binding and transcription activities of the adenovirus major late transcription factor. *Mol. Cell. Biol.* **6**, 4723–4733.

Dignam, J. D., Lebowitz, R. M., and Roeder, R. G. (1983). Accurate transcription initiation by RNA polymerase II in a soluble extract from isolated mammalian nuclei. *Nucleic Acids Res.* **11**, 1475–1489.

Fei, Y., Matragoon, S., and Liou, G. I. (1995). Simple and efficient method for the preparation of nuclear extracts. *Biotechniques* **18**, 984–987.

Gorski, K., Carneiro, M., and Schibler, U. (1986). Tissue-specific *in vitro* transcription from the mouse albumin promoter. *Cell* **47**, 767–776.

Jing, D., Agnew, J., Patton, W. F., Hendrickson, J., and Beechem, J. M. (2003) A sensitive two-color electrophoretic mobility shift assay for detecting both nucleic acids and proteins in gels. *Proteomics* **3**, 1172–1180.

Laniel, M.-A., and Guerin, S. L. (1998). Improving sensitivity of the electrophoretic mobility shift assay by restricting tissue phosphatase activities. *Biotechniques* **24**, 964–969.

Luo, B., Perry, D. J., Zhang, L., Kharat, I., Basic, M., and Fagan, J. B. (1997). Mapping sequence specific DNA-protein interactions: A versatile quantitative method and its application to transcription factor XF1. *J Mol. Biol.* **266**, 479–492.

Meisterernst, M., Gander, I., Rogge, L., and Winnacker, E.-L. (1988). A quantitative analysis of nuclear factor I/DNA interactions. *Nucleic Acids Res.* **16**, 4419–4435.

Miyamoto, S., Cauley, K., and Verma, I. M. (1995). Ultraviolet cross-linking of DNA binding proteins. *Methods Enzymol.* **254**, 632–641.

Oliphant, A. R., Brandl, C. J., and Struhl, K. (1989). Defining the sequence-specificity of DNA-binding proteins by selecting binding sites from random-sequence oligonucleotides: Analysis of yeast GCN4 protein. *Mol. Cell. Biol.* **9**, 2944–2949.

Orchard, K., and May, G. E. (1993). An EMSA-based method for determining the molecular weight of a protein-DNA complex. *Nucleic Acids Res.* **21**, 3335–3336

Schreiber, E., Matthias, P., Muller, M. M., and Schaffner, W. (1989). Rapid detection of octamer binding proteins with 'mini-extracts' prepared from a small number of cells. *Nucleic Acids Res.* **17**, 6419

Slomiany, B. A., Kelly, M. M., and Kurtz, D. T. (2000). Extraction of nuclear proteins with increased DNA binding activity. *Biotechniques* **28**, 938–942

Watt, F., and Molloy, P. L. (1993). Specific cleavage of transcription factors by the thiol protease, m-calpain. *Nucleic Acids Res.* **21**, 5092–5100.

Woo, A. J., Dods, J. D., Susanto, E., Ulgiati, D., and Abraham, L. J. (2002). A proteomics approach for the identification of DNA binding activities observed in the electrophoretic mobility shift assay. *Mol. Cell. Proteomics* **1**, 472–478.

DNA Affinity Chromatography of Transcription Factors: The Oligonucleotide Trapping Approach

Suchareeta Mitra, Robert A. Moxley, and Harry W. Jarrett

I. INTRODUCTION

The purification of transcription factors is a complex topic. A brief search of genetic databases reveals over 10,000 transcription factor entries. Of these, several hundred have now been purified. Typically these purifications have involved some form of DNA affinity chromatography. Most recent purifications use specific, double-stranded oligonucleotide sequences attached covalently to a suitable chromatographic support. This topic has been reviewed elsewhere (Gadgil *et al.*, 2000).

We described a variant of this technique called oligonucleotide trapping (Gadgil and Jarrett, 2002), depicted in Fig. 23.1. A column ("the trap") is prepared with the single-stranded oligonucleotide ACACACACAC attached to CNBr-activated Sepharose through an aminoalkyl linker. Another DNA ("the probe") is prepared, which has a double-stranded region

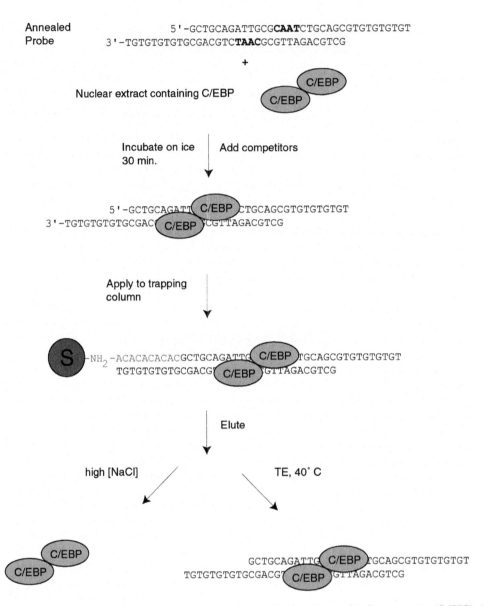

FIGURE 23.1 Oligonucleotide-trapping purification of the CAAT enhancer-binding protein (C/EBP) is shown schematically.

containing the element bound by a transcription factor and additionally has a GTGTGTGTGT single-stranded tail. The probe is mixed with a cell extract containing a transcription factor to be purified along with salt and various competitors and surfactants, which improve selectivity. The mixture is then applied to the trapping column and, after washing thoroughly, the column can be eluted either using high salt concentrations or using low salt and elevated temperatures.

The method has several advantages. The same "trap" column can be used with a variety of probes to purify different transcription factors and other DNA-binding proteins. The probe can be 5' end labeled to test new trap columns and, during protein purification, to follow the efficiency of column trapping of the protein–DNA complex. The labeled probe can be used directly in an electrophoretic mobility shift assay (EMSA) to follow transcription factor binding, to measure the amount of transcription factor present in various cell fractions, and to assess the effect of various competitors, detergents, and so on on this binding. The trapping method also has much higher capacity for transcription factors than traditional, covalent DNA affinity chromatography. This is because virtually all of the probe DNA is active and productively binds the transcription factor while much of covalently coupled DNA is usually inactive or inaccessible for binding (Gadgil and Jarrett, 2002; Massom and Jarrett, 1992). This article describes the current oligonucleotide trapping method practiced in our laboratory.

II. MATERIALS AND INSTRUMENTATION

CNBr-activated Sepharose (C-9142), heparin (H-3393), N,N,N',N'-tetramethylethylenediamine (T-9281), polydeoxyinosinic-deoxycytidylic acid [poly(dI), dC, P-4929], and igepal CA-630 (I-3021) are from Sigma. Acrylamide (161-0101) and N,N'-methylene-bisacrylamide (161-0201) are from Bio-Rad. Oligonucleotides are purchased from Integrated DNA Technologies (IDT), but other suppliers have also been used with good results. For oligonucleotides that will be coupled covalently to Sepharose (e.g., ACACACACAC), the "5' Amino Modifier C6" is added on the last (5') cycle. We have also used the Applied Biosystems Aminolink II reagent for this purpose and it also works well. We purchase unpurified

oligonucleotides at the 1-μmol scale which have "trityl off" and are deblocked.

Chromatography is either by simply using gravity flow with a fraction collector or using a Bio-Rad Biologic LP Chromatograph (Cat. No. 731-8300). Electrophoresis is with the Bio-Rad Mini-Protean II apparatus (Cat. No. 165-2944) and the PowerPac 200 power supply (Cat. No. 165-5050).

III. PROCEDURES

A. Preparation of Oligonucleotides

Solutions

1. *0.5 M EDTA*: 15 g Na₄EDTA, 3.2 g Na₂EDTA, 81 ml distilled water; autoclave for 45 min.
2. *1 M Tris, pH 7.5*: 60.6 g Tris base, 31.25 ml fresh 37% HCl, 441 ml distilled water, 0.5 ml 0.5 M EDTA. Check the pH of a 1:10 dilution. Autoclave for 1 h.
3. *TE*: 10 mM Tris, 1 mM EDTA, pH 7.5. To make 100 ml, add 1 ml 1 M Tris, pH 7.5, 0.2 ml 0.5 M EDTA, and sterile distilled water to complete 100 ml.
4. *3 M sodium acetate*: For 100 ml, add 40.8 g sodium acetate trihydrate, 5 ml glacial acetic acid, 0.2 ml 0.5 M EDTA, and 66 ml H₂O. Autoclave for 45 min.
5. *70% ethanol*: For 100 ml, mix 70 ml absolute ethanol with 30 ml distilled water. Keep in a −20°C freezer and use while cold.

Steps

1. Dissolve a dried, unpurified oligonucleotide (1 μmol) in 300 μl of TE, 30 μl 3 M sodium acetate, and add 1 ml of absolute ethanol for oligonucleotides longer than 12-mers and 1.32 ml ethanol for those between 6- and 11-mers.

2. Place the mixture in a −85°C freezer for 1h and then centrifuge at 4°C at 16,000 g for 12 min. Aspirate the supernatant carefully away.
3. Wash the pellet with 500 μl 70% ethanol (80% is used for shorter oligonucleotides), centrifuge again, and discard the supernatant.
4. Dry the oligonucleotide for 5 min in a Speed-Vac SC110 (Savant) at room temperature. Oligonucleotides purified by this simple ethanol precipitation are used for all our DNA-affinity chromatography experiments and we have found them to be of adequate purity. For long-term storage, the oligonucleotides are either left dry or dissolved in 500 μl TE, adjusted to 1 mM using their absorption at 260 nm (we use an Excel spreadsheet we configured to calculate molar absorptivities based upon a sequence; a similar calculator is available on the internet, and stored at −20°C. If stored in TE, oligonucleotides must be again precipitated and washed with ethanol prior to coupling, as Tris interferes with coupling.

B. Preparing the ACACACACAC- Sepharose Column

Solutions

1. *Coupling buffer*: 0.1 M NaHCO$_3$, pH 8.3, 0.5 M NaCl. To prepare 1 liter, dissolve 8.40 g NaHCO$_3$ in 800 ml distilled water, adjust the pH to 8.3 with HCl, add 29.22 g NaCl, and complete to 1 liter. Keep in a 4°C refrigerator and put on ice an hour before use.
2. *1 mM HCl*: 43 μl of concentrated (11.6 M) HCl in a total volume of 500 ml distilled water. Keep in a 4°C refrigerator and put on ice an hour before use.

3. *Blocking buffer*: 0.1 M Tris, pH 8.0, 0.5 M NaCl. For 1 liter, dissolve 12.1 g Tris base in 800 ml distilled water, adjust to pH 8 with HCl, add 29.22 g NaCl, and complete to 1 liter. Store at 4°C.
4. *1 M NaN$_3$*: Dissolve 6.50 g sodium azide to a total volume of 100 ml with distilled water.
5. *TE/azide*: 10 mM Tris, 1 mM EDTA, 10 mM NaN$_3$, pH 7.5. To make 100 ml, add 1 ml 1 M Tris, pH 7.5, 0.2 ml 0.5M EDTA, 1 ml 1 M NaN$_3$, and sterile distilled water to complete 100 ml.

Steps

1. Dissolve the dried, ethanol-precipitated 5′-[amino modifier C6]-ACACACACAC oligonucleotide, 200 nmol, in 2 ml coupling buffer, transfer to a 15-ml polypropylene screw-cap tube (Sarstedt # 62.554.002), and keep on ice.
2. Suspend 1.2 g CNBr-preactivated Sepharose in 100 ml ice-cold 1 mM HCl, cover with Parafilm, and invert occasionally while keeping in an ice-water slurry for 15 min.
3. Filter on a coarse sintered glass funnel (Pyrex brand, available from Fisher #36060-150C) and wash with an additional 100 ml ice-cold 1 mM HCl. Filter under vacuum until the filtered cake of resin just begins to form cracks and immediately scrape the Sepharose into the DNA-coupling buffer solution and mix with a vortex mixer.
4. The slurry should be watery enough to mix easily; if not, add an additional milliliters of coupling buffer. Mix the slurry in a cold room (4°C) overnight on a tube rotator (Cole Parmer Roto-Torque A-07637-00) at low speed just sufficient to keep the resin from settling.
5. The next day, filter the resin again, and wash five times with 5-ml portions of

blocking buffer. Determine the absorption at 260 nm of the combined washes and measure their volume (typically 25–30 ml) either gravimetrically or volumetrically. The amount of coupling is determined by difference using

$$\% \text{ coupling} = 100\% - \frac{A_{260} \times \text{volume (ml)}}{200 \times 10^{-9} \times 111,250 \times 10}$$

$$= 100 - \frac{A_{260} \times \text{volume (ml)}}{0.2225}$$

where $111,250 \, M^{-1} \text{cm}^{-1}$ is the calculated molar absorptivity of CACACACACA. For example, if the total washes were 30 ml, the absorbance would be 0.742 if none couples. If the washes actually have $A_{260} = 0.186$, the equation calculates 75% coupling. To convert this to nanomoles DNA per milliliter Sepharose, the 1.2-g preactivated Sepharose yields 3.5 ml/g or about 4.2 ml DNA–Sepharose. Thus 75% coupling would give $0.75 \times 200 \text{nmol}/4.2 \text{ml} = 36 \text{nmol DNA/ml Sepharose}$.

6. Leave the resin as a watery slurry in blocking buffer with mixing overnight in the cold room (4°C) and then put in TE/azide (10 mM Tris, pH 7.5, 1 mM EDTA, 10 mM NaN$_3$) and store at 4°C. We store resins in graduated 15-ml polypropylene tubes as 50% settled resin/50% liquid mixtures for a year or more with no loss of activity.

7. Then pack the (AC)$_5$-Sepharose into a 1-ml syringe column (Bio-Rad Poly-Prep empty columns, Cat. No. 731-1550 or Alltech Extract-Clean, Cat. No. 211101). This is done simply by adding 1 ml of the well-mixed 50% slurry to the column and allowing it to drain and overlaying with fresh buffer. We usually place an extra column frit on top of the 0.5-ml packed support bed to prevent disturbing it and keep the column in TE/azide when not in use. Assuming 75% coupling as described earlier, this column should bind 18 nmol of probe.

Pitfalls

There are essentially three kinds of failure we have observed as we help others do trapping.

1. Tris cannot be present during DNA coupling to CNBr-activated Sepharose. It can be removed effectively from DNA stored in TE by careful ethanol precipitation.

2. After hydration in 1 mM HCl and filtration, care must be taken not to introduce air into the Sepharose. Filter only until cracks begin to appear in the filtrate and stop, leaving a moist cake. Once air is introduced, the Sepharose particles will float, making column packing difficult or impossible.

3. Columns should usually only be used when EDTA is present. DNases all require divalent cations and, in the absence of EDTA, can destroy columns. In some cases, using divalent cations is unavoidable though and this will lessen column lifetime.

C. Preparing the Probe

The oligonucleotides, usually one for each strand, are obtained commercially and ethanol precipitated separately as described in Section III,A. In the case of C/EBP, an internally symmetric sequence [i.e., EP24(TG)$_5$] was used (Gadgil and Jarrett, 2002), requiring a single oligonucleotide. Probes are usually stored at −20°C as 1 mM solutions in TE. A portion of either strand can be labeled using the 5′ polynucleotide kinase reaction and [γ-^{32}P]ATP (Sambrook et al., 1989). In our laboratory, 5′ end labeling is performed with oligonucleotide in excess over ATP. Just prior to use, probe strands are annealed by diluting to 10–100 μM in TE0.1 (10 mM Tris, pH 7.5, 1 mM EDTA, 0.1M NaCl) and heating to 95°C for 5 min followed by a linear decrease to 4°C at 1 h in a thermocycler. They can also be annealed by boiling 50 ml water in a 100-ml beaker, removing it from the heat, and adding the sealed tube containing the DNA. After allowing the beaker to cool for

30 min, the tube is removed to an ice bath. In the case of EP24(TG)$_5$, after annealing we have

5'-GCTGCAGATTGCGC**AAT**CTGCA
GCGTGTGTGTGT-3'3'-TGTGTGTGTG
CGACGTC**TAA**CGCGTTAGACGTCG-5'

When first using a new probe or trap, we generally test the apparatus by 5' end labeling the probe (about 10,000 counts per minute of ^{32}P in 1–2 nmol DNA is sufficient) and applying it to the trap column in TE0.1 at 4°C. If all of the ^{32}P is attached to the DNA (i.e., proper care was taken to remove any excess [γ-^{32}P]ATP used for labeling) and if the trap column has sufficient capacity (e.g., 18 nmol in the example given for 75% coupling), greater than 95% of the counts should be retained by the trapping column. The column is then placed in a 37°C water bath and eluted with TE. All of the bound counts should elute. This test can also be included during purification itself, i.e., the 5' end-labeled probe can be used during purification to monitor column performance.

D. Preparing the Cell Extract

Most transcription factors are in the nucleus and are prepared from nuclear extracts. However, for a new transcription factor, this is not at all certain and should be ascertained. C/EBP is found in the nucleus of rat liver. In this case, a nuclear extract prepared by the method of Gorski *et al.* (1986) was used as the starting point. Other methods for preparing nuclear extracts can also be used. Other fractions should also be checked for activity, however, to be certain of localization and to follow the success of the preparation. In working with other transcription factors, we have found that extracting the nucleus with $0.36\,M$ (NH$_4$)SO$_4$ as in the usual method (Gorski *et al.*, 1986) did not effectively release the transcription factor and higher salt, i.e., $0.6\,M$ (NH$_4$)SO$_4$, was required. Therefore, monitoring each fraction as cells are fractionated and perhaps altering the fractionation

procedure to accommodate the protein of interest are usually worthwhile.

Monitoring is accomplished by preparing serial dilutions of each cell fraction and adding these to an electrophoretic gel shift assay (EMSA). This is usually a qualitative analysis. For example, we typically use three-fold serial dilutions and note for each fraction (cytosol, nuclear extract) the dilution that gave an approximate 50% of the maximal shift obtained. This 50% shift is thus a rough unit of activity. Where the bulk of activity resides is determined from the dilution required for 50% shift and accounting for the volume of each fraction.

Alternatively, several kits are available for preparing nuclear extracts (i.e., the ones from Panomics, Inc., Cat. No. AY2002 or Active Motif, Cat. No. 40010) and nuclear extracts already prepared from a variety of cell lines are available commercially (see Active Motif's catalogue). While these may not be suitable in all cases, they can certainly save time when applicable.

For C/EBP, the rat liver nuclear extracts prepared (Gadgil and Jarrett, 2002) are dialyzed into 25 mM HEPES, pH 7.6, 40 mM KCl, 0.1 mM EDTA, 1 mM dithiothreitol, and 10% glycerol, adjusted to 5 mg/ml protein, and stored in 0.2-ml aliquots at −85°C.

E. Electrophoretic Mobility Shift Assay

Here we describe the complete assay in use in our laboratory, including directions for preparing the gel. Alternatively, we have also used 4–12% Bio-Rad Tris/glycine ready gels (with 0.25× TBE as the running buffer) with good results.

Solutions

1. 5× *TBE*: To make 1 liter, dissolve 30.03 g Tris base, 15.25 g boric acid, 10 ml *0.5M* EDTA, and add distilled water to 1000 ml and store at room temperature

2. *5M NaCl*: Combine 29.2 g NaCl, 89 ml distilled water, and 0.1 ml 0.5*M* EDTA. Autoclave for 45 min and store at room temperature.

3. *5× incubation buffer (for 8 ml)*: Combine 400 μl 1*M* Tris–Cl (pH 7.5), 80 μl 0.5*M* EDTA (pH 8.0), 320 μl 5*M* NaCl, 1.6 ml glycerol, and 3.2 μl β-mercaptoethanol. Add distilled water to 8 ml and store at 4°C for up to 1 week.

4. *Acrylamide*: Combine 29 g acrylamide, 1 g N,N'-methylene-bisacrylamide, and distilled water to a total of 100 ml. Store at 4°C for up to 1 month.

5. *10% ammonium persulfate*: Weigh 0.1 g ammonium persulfate and add 1 ml distilled water. Discard after 1 day.

6. *N,N,N',N'-Tetramethylethylenediamine (TEMED)*

7. *1 μg/μl dI.dC*: Dissolve 1 mg polydeoxyinosinic–deoxycytidylic acid in 1 ml TE

8. *10× TE0.1*: For 1 liter, combine 12.1 g Tris base, 20 ml 0.5*M* EDTA, and 800 ml distilled water. Titrate to pH 7.5 with HCl, dissolve 58.44 g NaCl, and complete to 1000 ml with distilled water.

9. *TE0.4*: For 100 ml, mix 10 ml 10× TE.1, 6 ml 5*M* NaCl, and complete to 100 ml

10. *TE1.2*: For 100 ml, mix 10 ml 10× TE.1, 22 ml 5*M* NaCl, and complete to 100 ml. Store solutions 8–10 at room temperature.

1. Gel Preparation

Steps

1. Prepare the gel (5% acrylamide) in an 16 × 100-mm test tube by combining 7.7 ml distilled water, 0.5 ml 5× TBE, 1.7 ml acrylamide, 0.1 ml 10% ammonium persulfate, and 20 μl TEMED.

2. Mix well and quickly pour into the assembled plates and place the comb of a Bio-Rad protein II minigel apparatus.

3. Let the gel polymerize.

4. Fill the gel tank with 0.25× TBE (15 ml 5× TBE completed to 300 ml with distilled water). Remove the comb, wash, and fill the wells with 0.25× TBE.

5. For some transcription factors, preelectrophorese the gel for 30 min at 100 V.

2. Mobility Assay

Steps

1. In a microtube, combine 5 μl 5× incubation buffer, 1 μl 1 μg/μl poly (dI-dC), 1–5 μg transcription factor or nuclear extract, 20,000 cpm ^{32}P-labeled annealed oligonucleotides, and distilled water for 25 μl total.

2. Mix gently and incubate at room temperature for 20 min.

3. Add 2 μl of bromphenol blue (0.1% in distilled water) to each sample and mix gently.

4. Load 20–25 μl onto gel.

5. Run gel at 150 V for 1 h until the dye front is ~1 mm from bottom.

6. Dry gel overnight or place in a water-tight Zip-Lock bag.

7. Expose film overnight at −85°C using an intensifying screen or use a Phosphorimager.

F. Oligonucleotide Trapping

By adding different dilutions of nuclear extract to a gel shift, performed using a known amount of DNA, usually a total shift of the DNA can be obtained. By observing what dilution was just sufficient to give a total shift (or, better, use densitometry to determine what gives a 50% shift), the approximate concentration of the DNA-binding activity can be obtained. This method should be applied to a new transcription factor to estimate its concentration. We have experience with only a small number of transcription factors, but the highest concentration we have found of a transcription factor in a nuclear extract is about 40 fmol/μl (40 nM). This number may not be meaningful to all transcription factor/nuclear extract combinations but does provide a starting point for estimation. If a nuclear extract is diluted 10-fold into TE0.1 and an annealed probe is added to give a concentration of 40 nM or greater, the DNA should be in 10-fold or greater excess over the transcription factor. At this very low concentration of DNA, very little nonspecific binding should occur and yet all of the transcription factor should be bound. To test if the conditions are correct, EMSA can be performed under the conditions of chromatography to test whether binding is complete. Thus, if the nuclear extract is diluted 10-fold into TE0.1 and increasing amounts of 5'-end-labeled DNA probe are added (from 0.1 to 1000 nM) and the mixture is applied to EMSA, all of the DNA should be shifted at the lower concentrations but eventually a point will be reached where all of the DNA is not shifted because it is now in excess. An excess of about 10-fold is appropriate.

We also use a similar approach to determine which competitors and detergents we can use during trapping. We have found that heparin [an inhibitor of double-stranded DNA binding (Gadgil and Jarrett, 1999)], in concentrations as high as 4 mg/ml, and $(dT)_{18}$ (a competitor of single-stranded DNA binding), as high as 20 μM, have little effect on the DNA binding of many transcription factors but can diminish nonspecific binding greatly. However, we have found some transcription factors, notably some members of the E2F family, do not bind DNA even when very small amounts of heparin are used (unpublished data). Poly(dI-dC) at 0.05 mg/ml can greatly diminish nonspecific binding greatly. Similarly, we find that 0.1% igepal, Triton, or Tween detergents do not adversely affect most gel shifts but we have found exceptions. We usually test various concentrations of each of these reagents in an EMSA experiment before using them in chromatography.

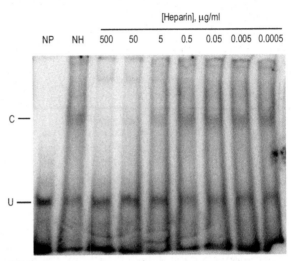

FIGURE 23.2 Electrophoretic mobility shift assay (EMSA) of C/EBP in the presence of heparin. EMSA was performed as in Section III,E in the presence of the heparin concentrations shown. NP, no protein added showing the position of the unshifted (U) oligonucleotide. NH, no heparin added to the complete assay mixture showing the position of the shifted complex (C).

Such an experiment to determine the concentration of heparin to use in the purification of rat liver C/EBP is shown in Fig. 23.2. Concentrations of heparin of 0.5 μg/ml or less have no effect on the gel shift observed and 0.5 μg/ml would then be chosen tentatively for use in trapping. The same experiment is then performed with each component to be tested (detergents, dIdC, etc.) to find the highest concentration that can be used safely. As a last test, a mixture containing all the components is then tested to ascertain that there is no synergism between the reagents that could affect DNA binding adversely.

Trapping is then performed using this tested combination. What follows is a set of conditions that work well for rat liver C/EBP.

Steps

1. Combine 1 ml 10X TE0.1, 1 μl 5 mg/ml heparin (0.5 μg/ml final), 90 μl 1 mM $(dT)_{18}$ (9 μM final), 0.5 ml 1 mg/ml dI-dC (50 μg/ml

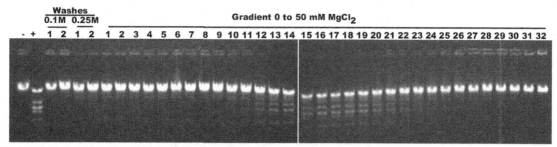

FIGURE 23.3 Catalytic chromatography of *Eco*RI using the trapping approach on a 4.6 × 50-mm (AC)5-silica column. Fractions were assayed as described previously (Jurado *et al.*, 2000) by digestion of λ phage. A minus sign shows control reaction without *Eco*RI, and a plus sign shows digestion with pure *Eco*RI. After loading 0.4 ml of the RY13 extract, the column was washed with 10 ml of a buffer containing 1 m*M* EDTA and 0.1*M* KCl and then with 0.25*M*. Following washing, the column was eluted with a linear gradient to 50 m*M* MgCl$_2$. The chromatography conditions are described in more detail elsewhere (Jurado *et al.*, 2000).

final), and distilled water to complete 8 ml. Cool on ice 30 min.

2. Add 1 ml rat liver nuclear extract and mix gently. Add 20 μl of 100 μ*M* annealed EP24GT trapping oligonucleotide (final is 200 n*M* strand, 100 n*M* duplex) and mix gently. Incubate on ice for 30 min.

3. Chromatography. All operations take place in a 4°C cold room. All solutions are kept on ice. Apply the trapping mixture to the 0.5 ml (AC)$_5$-Sepharose column at about 0.3 ml/min.

4. Wash the column with 20 ml TE0.4. Elute the column by either of two procedures (see Fig. 23.1)

5a. High salt: Elute the column with 10 ml TE1.2 (10 m*M* Tris, pH 7.5, 1 m*M* EDTA, 1.2*M* NaCl).

5b. Temperature: Place the column in a 37°C water bath and elute with 10 ml 37°C TE (no salt).

6. Column fractions are then assayed by EMSA. In either case, the conditions for EMSA are not ideal. For high salt elution, the salt can interfere with electrophoresis and DNA binding; therefore, only 1–2 μl can be added safely to the 25-μl gel-shift mixture. For temperature, the protein–DNA complex elutes and the DNA acts as a "cold

competitor" in EMSA and diminishes the amount of shift observed. However, in either case, a gel shift is observed sufficiently to identify the active fractions. Our experience has been that temperature elution gives somewhat higher purity than salt elution (Gadgil and Jarrett, 2002).

G. Other Applications

We have extended the trapping method in two directions. We have developed a CNBr-activated HPLC silica (Jurado *et al.*, 2002) and produced (AC)$_5$-silica. This allows trapping to be performed using the much higher mass transfer characteristics of 7-μm, 300-Å pore silica. Second, we have extended the method to restriction enzymes and used catalytic means for elution. An example of this technique is shown in Fig. 23.3. In this case, the trapping oligonucleotide is

Ecogt   5′-GCATGC**GAATTC**GCATG
         TGTGTGTGT
         3′-CGTAC**GCTTAAG**CGTA

which of course binds to the *Eco*RI restriction endonuclease. This oligonucleotide was used to trap *Eco*RI from a crude bacterial

(*E. coli* strain RY13) extract (400 µl) in the presence of EDTA. *Eco*RI binds DNA in the absence of Mg^{2+} (presence of EDTA) but is catalytically inactive (Jurado *et al.*, 2000). Once bound to the column and washed thoroughly, the enzyme was eluted catalytically by adding Mg^{2+} (50 mM) to the column buffer. The enzyme binds Mg^{2+}, digests the column DNA, and elutes. It is detected in Fig. 23.3 by its characteristic digestion of λ phage DNA (in fractions 11–23).

Once columns are eluted catalytically, the DNA is converted to product (in this case, digested DNA) and are no longer useful for affinity chromatography. However, by using the trapping approach, the digested DNA is simply removed (by washing with 37°C TE or 70°C water) to return to $(AC)_5$-silica and then the EcoGT DNA is replaced (i.e., trapped) for the next purification. This strategy allows DNA, rather than columns, to become a consumable reagent in the catalytic chromatography of enzymes such as endonucleases.

IV. CONCLUSIONS

Trapping has proven to be a useful technique for the purification of transcription factors. It allows the protein–DNA complex to form in solution at low concentrations and under conditions that can be tested beforehand using the EMSA assay with the same DNA.

It has now been extended to DNases and can probably also be applied to DNA or RNA polymerases and to DNA repair enzymes, although we have not yet done so.

Acknowledgment

This work was supported by the National Institutes of Health (GM43609).

References

Gadgil, H., and Jarrett, H. W. (1999). Heparin elution of transcription factors from DNA-Sepharose columns. *J. Chromatogr. A* **848**, 131–138.

Gadgil, H., and Jarrett, H. W. (2002). An oligonucleotide trapping method for purification of transcription factors. *J. Chromatogr. A* **966**, 99–110.

Gadgil, H., Jurado, L. A., and Jarrett, H. W. (2000). DNA affinity chromatography of transcription factors. *Anal. Biochem.* **290**, 147–178.

Gorski, K., Carneiro, M., and Schibler, U. (1986). Tissue-specific *in vitro* transcription from the mouse albumin promoter. *Cell* **47**, 767–776.

Jurado, L. A., Drummond, J. T., and Jarrett, H. W. (2000). Catalytic chromatography. *Anal. Biochem.* **282**, 39–45.

Jurado, L. A., Mosley, J., and Jarrett, H. W. (2002). Cyanogen bromide activation and coupling of ligands to diol-containing silica for high performance affinity chromatography: Optimization of conditions. *J. Chromatogr. A* **971**, 95–104.

Massom, L. R., and Jarrett, H. W. (1992). High-performance affinity chromatography of DNA. II. Porosity effects. *J. Chromatogr. A* **600**, 221–228.

Sambrook, J., Fritsch, E. F., and Maniatis, T. (1989). "Molecular Cloning, a Laboratory Manual," 2nd Ed. Cold Spring Harbor Laboratory Press, Cold Spring Harbor, NY.

MASS SPECTROMETRY METHODS FOR DETERMINATION OF PROTEIN IDENTITY AND PROTEIN MODIFICATIONS

24

Protein Identification and Sequencing by Mass Spectrometry

Leonard J. Foster and Matthias Mann

I. INTRODUCTION

In the postgenomic world, research priorities are shifting toward understanding the function of gene products, and proteomics is the term given to large-scale determination of gene product function, starting with where and when the products are expressed. The underlying technology required for mass spectrometry-based proteomic experiments is very young and still undergoing rapid development in both hardware and software areas (Aebersold and Goodlett, 2001; Aebersold and Mann, 2003;

Mann *et al.*, 2001; Washburn *et al.*, 2002). The most powerful and most popular method for elucidating the protein composition of highly complex samples is proteolytic digestion of the proteins to peptides followed by single- or multidimensional high-pressure liquid chromatography (HPLC or LC) with on-line coupling to electrospray ionization tandem mass spectrometry (MS/MS) to generate peptide fragmentation spectra . These fragments are then compared to theoretical fragments predicted from amino acid sequence databases to arrive and protein identifications. We prefer quadrupole time-of-flight hybrid mass spectrometers, sacrificing the ease of use of ion trap-type spectrometers for higher resolution data. This article describes the general procedures used in our laboratory for sequencing and identifying proteins from highly complex samples. Because this is not a literature review, the reference list is not comprehensive and does not necessarily refer to the original description of a given technique.

II. MATERIALS AND INSTRUMENTATION

Urea (Cat. No. U5128), thiourea (Cat. No. T8656), dithiothreitol (DTT, Cat. No. D9163), absolute ethanol (EtOH, Cat. No. E7023), iodoacetamide (Cat. No. I1149), and heptafluorobutyric acid (HFBA, Cat. No. H7133) are from Sigma. LysC is from Wako (Osaka, Japan, Cat. No. 12502543). Sequencing-grade porcine trypsin (Cat. No. V511C) is from Promega, acetonitrile (Cat. No. 34881) is from Riedel-da Haën, methanol (Cat. No. M/4056/17) is from Fisher, and acetic acid (Cat. No. 6052) is from J. T. Baker. All water used here is "MilliQ"-quality distilled, deionized water. The following consumables are all obtained from the indicated sources: C18 Empore extraction disks (3M, Cat. No. 2215), P200 pipette tips (Gilson but any laboratory plastics supplier will suffice), 22-gauge flat-tip syringes (Hamilton, Cat. No. 90134), LC columns (New Objective, FS 360-100-8-N-20-C15), and 50- and 20-μm-inner-diameter fused silica capillary tubing (Polymicro Technologies LLC, Cat. Nos. 020375 and 050375). Vydac prototype 3-μm C18 beads (Cat. No. 218MSB3) are a kind gift from Grace Vydac (Hesperia, CA). The helium pressure cells used in our laboratory are custom made by a local metal workshop, but similar instruments can be purchased from Brechbühler AG. The HPLC system used in these protocols is the Agilent 1100 Series with 0.2- to 20-μl/min flow rate. The hybrid quadrupole TOF mass spectrometer is from MDS Sciex

and Applied Biosystems. All peptide fragmentation data are searched against the appropriate databases using a dual processor LinuxOS Mascot search engine (Matrix Science).

III. PROCEDURES

A. Sample Preparation

1. In-Solution Digestion

In-solution digestion can be used where the protein sample contains no detergent and is relatively simple (i.e., <300 proteins).

Solutions

1. *8 M urea*: $8 M$ urea in $10 mM$ HEPES, pH 8.0. To make 10 ml, dissolve 4.80 g urea and 23.8 mg HEPES in 10 ml water. Adjust pH with NaOH. Store at room temperature.
2. *6 M urea/2 M thiourea*: $6 M$ urea, $2 M$ thiourea in $10 mM$ HEPES, pH 8.0. To make 10 ml, dissolve 3.60 g urea, 23.8 mg HEPES, and 1.52 g thiourea in water. Adjust pH as necessary with NaOH. Solutions of thiourea often contain insoluble particles so it is a good idea to centrifuge this sample at $5000 g$ for 10 min to clarify it. Store at 4°C.
3. *8 M guanidine*: $8 M$ guanidine HCl, pH 1.5. To make 10 ml, dissolve 7.65 g guanidine HCl in water. Adjust pH with HCl.
4. *Digestion buffer*: $50 mM$ NH_4HCO_3. To make 10 ml, dissolve 40 mg NH_4HCO_3 in 10 ml water. Store at room temperature.
5. *Iodoacetamide stock solution*: 0.5 μg/μl iodoacetamide in digestion buffer. To make 10 ml, dissolve 5 mg iodoacetamide in 10 ml digestion buffer. Separate into 100-μl aliquots and store at −20°C.
6. *DTT stock solution*: 0.5 μg/μl DTT in water. To make 10 ml, dissolve 5 mg DTT in 10 ml water and store at −20°C.

7. *LysC stock solution*: 0.5 μg/μl. To make 1 ml, dilute 0.5 μg in 1 ml digestion buffer. Separate into small aliquots and store at −20°C.

Steps

1. Solubilize the protein pellet (either precipitated from solution or a centrifuge pellet from a subcellular fractionation) in $8M$ urea, $6M$ urea/$2M$ thiourea or $8M$ guanidine. Selection of the particular denaturant will be sample specific, but urea/thiourea works best for membrane proteins. Keep the volume as small as possible.
2. Add 1 μg DTT/50 μg sample protein and incubate for 30 min at room temperature. If material is limiting, then the protein mass needs only to be roughly estimated.
3. Add 5 μg iodoacetamide/50 μg sample protein and incubate for 20 min at room temperature.
4. Add 1 μg LysC/50 μg sample protein and incubate for 3 h or overnight, at room temperature.
5. Dilute sample with 4 volumes digestion buffer, add 1 μg trypsin/50 μg sample protein, and incubate overnight at room temperature. Digested peptides may be stored at −20°C indefinitely.

2. In Gel Digestion

This procedure is as described by Shevchenko *et al.*, (1996). Single-dimension SDS–PAGE can be used to reduce sample complexity but sample recovery is likely much less than in-solution digestions and so should only be used for complex (>300 proteins) samples where sample amount is not limited.

Solutions

1. *Digestion buffer*: 50 mM NH_4HCO_3. To make 10 ml, dissolve 40 mg NH_4HCO_3 in 10 ml water. Store at room temperature.

2. NH_4HCO_3/*EtOH*: 50% digestion buffer/50% EtOH. Combine equal volumes of digestion buffer and neat ethanol. Store at room temperature.
3. Trypsin solution: Dilute trypsin stock solution to 12.5 ng/μl with digestion buffer. Prepare *immediately* prior to use to minimize autocatalysis.
4. *Iodoacetamide*: 55 mM iodoacetamide in digestion buffer. To make 1 ml, dissolve 10.2 mg iodoacetamide in 1 ml digestion buffer. Prepare fresh.
5. *DTT*: 10 mM DTT in water. To make 1 ml, dilute 10 μl of a $1M$ DTT solution in 990 μl water and store at −20°C.
6. *Extraction solution*: 3% trifluoroacetic acid (TFA), 30% acetonitrile. To make 1 ml, dilute 300 μl acetonitrile and 30 μl TFA in 670 μl water.

Steps

1. Excise individual stained gel bands and/or molecular weight ranges and place gel pieces in clean 1.5-ml microfuge tubes.
2. Chop each gel piece into smaller pieces approximately 1 mm per side using a scalpel.
3. Wash the gel pieces twice with NH_4HCO_3/EtOH for 20 min each at room temperature. Discard the supernatant each time. For each of the steps described here, enough solution should be used to cover the gel pieces.
4. Dehydrate the gel pieces by incubating for 10 min in absolute EtOH. Discard solution afterward.
5. Reduce the proteins by incubating for 45 min in DTT at 56°C. Discard solution afterward.
6. Block free sulphydryl groups by incubating for 30 min in iodoacetamide at room temperature. Discard solution afterward.
7. Wash gel pieces once with digestion buffer for 20 min at room temperature. Discard supernatant afterward.
8. Dehydrate pieces as in step 4.

9. Wash gel pieces once as in step 7.
10. Dehydrate gel pieces <u>twice</u> as in step 4.
11. Remove remaining ethanol from gel pieces by vacuum centrifugation.
12. Digest proteins with trypsin overnight at 37°C. Add enough trypsin solution to cover the dehydrated gel pieces. When the gel has swelled as much as possible (~20 min), remove excess trypsin solution and cover gels in digestion buffer.
13. Add 2 μl TFA to the digestion, quickly finger vortex the solution, and separate the liquid from the gel pieces, storing the liquid in a clean microfuge tube.
14. Extract the gel pieces by adding extraction solution to cover the gel. Shake the mixture vigorously for 5 min at room temperature. Remove the liquid and combine with that from step 13.
15. Dehydrate gel pieces in acetonitrile for 10 min at room temperature. Combine supernatant from this step with that from steps 13 and 14.

3. Desalting, Filtering, and Concentration

This procedure is identical to that described by Rappsilber et al. (2003).

Solutions

1. *Sample buffer*: 1% TFA, 5% acetonitrile. To prepare 1 ml, dilute 10 μl TFA and 50 μl acetonitrile in 940 μl water.
2. *Buffer B*: 0.02% HFBA, 0.5% acetic acid, 80% acetonitrile in water. To make 500 ml, combine 400 ml acetonitrile, 2.5 ml acetic acid, and 50 μl HFBA and top to 500 ml with water.

Steps

1. Prepare as many desalting columns as necessary by punching out small disks of C18 Empore filter using a 22-gauge flat-tipped syringe and ejecting the disks into

P200 pipette tips. Ensure that the disk is securely wedged in the bottom of the tip.
2. Condition a column by forcing 5 μl of methanol through the Empore disk with a syringe fitted to the end of the pipette tip.
3. Remove any remaining organic solvent in the column by forcing 5 μl of sample buffer through.
4. Prepare the peptide sample for binding to reverse-phase material. For in-solution digestions, dilute the sample 3× with buffer A and pH the resulting solution to <2.5 with acetic acid. For in-gel digestions, dry down the extracted peptides in a vacuum centrifuge, resuspend in 100 μl sample buffer, and pH the resulting solution to <2.5 with acetic acid.
5. Force the acidified peptide sample through the C18 column.
6. Wash the column with 10 μl of sample buffer.
7. Elute the peptides from the C18 material using 5 μl buffer B. Elute directly into a microfuge tube.

B. LC/MS/MS

1. Column Preparation

This procedure is identical to that described by Rappsilber et al. (2003).

Solutions

1. *Matrix slurry*: Place a few cubic millimeters of dry Vydac matrix, a small magnetic stirbar, and 300 μl methanol in a 1.5-ml flat-bottomed sample vial.
2. *Buffer B as described earlier.*
3. *Buffer A*: 0.5% acetic acid, 0.02% HFBA. To make 500 ml, dilute 2.5 ml acetic acid and 50 μl HFBA in 500 ml water.

Steps

1. Insert the vial of matrix slurry in the helium pressure cell and position of the

cell on a magnetic stir plate. Ensure that the plate is capable of rotating the bar inside the vial.

2. Insert an empty fused silica column into the pressure cell and tighten the seal. The back end of the column should rest in the matrix slurry but high enough that it is not bumped by the rotating stirbar.

3. Apply 30 to 50 bar helium to the system. The beads should be visible collecting in the column tip.

4. Allow the packing to proceed until the desired column length is achieved. Usually we use columns of between 7 and 10 cm in length.

5. Slowly release the pressure from the system. The column should remain tightly packed. If the packing separates, then it usually indicates that the tip of the column is blocked in some manner. Columns can be emptied by reversing the column orientation in the pressure cell and forcing methanol through the column backward.

6. Before loading any peptides onto the column, wash any methanol from the matrix by forcing buffer A through the column for 5 min using the pressure cell with 50 bar helium. At this pressure the flow rate should be 0.3 to 0.5 μl/min.

7. Column resolution can be checked using the mass spectrometer as a readout by analyzing a known amount (we use 100 fmol) of trypsin-digested bovine serum albumin. In a 20-min linear gradient from 5 to 80% buffer B, peptides should elute in approximately 20 s (width at half-maximum).

2. Liquid Chromatography

This procedure is similar to that described by Rappsilber *et al.* (2002) and Foster *et al.* (2003) with some modifications. For a comprehensive review of more advanced LC plumping, see Rappsilber *et al.* (2003).

Solutions

Buffer B and sample buffer as described earlier.

Steps

1. Remove all liquid from the desalted peptide sample by vacuum centrifugation.

2. Resuspend the peptide sample in 3 μl sample buffer. Alternatively, if only a small fraction of the desalted sample is to be analyzed, then instead of evaporating and resuspending, it can be directly diluted sufficiently with sample buffer to reduce the acetonitrile content to <5%. The high organic solvent content here will prevent peptides from binding to the reverse-phase matrix.

3. Measure the output rate of the HPLC pump required to achieve a flow rate through the column of 200 nl/min. Volumes this low can be measured by collecting the flow through in a graduated glass capillary.

4. Load the peptide sample onto the analytical column prepared earlier using the pressure cell again. Place the microfuge tube containing the sample in the cell, insert the analytical column, tighten the seal, and apply 50 bar pressure. Semiaccurate volumes can be loaded in this way by simply measuring the volume of liquid collecting at the column tip.

5. During sample loading, prepare the HPLC plumping (Fig. 24.1) and gradient program (Fig. 24.2). The HPLC system should be situated as close to the mass spectrometer orifice as possible (50 cm separating the two allows enough working space).

6. Flush the HPLC plumbing for 5 min with 7% buffer B in buffer A to remove air bubbles from the system.

7. Reset the flow to 0.00 μl/min and insert the analytical column containing the sample peptides into the plumbing system (Fig. 24.1).

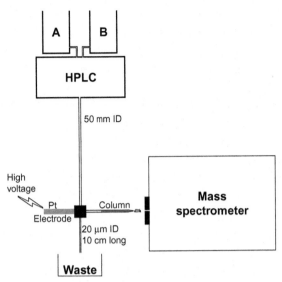

FIGURE 24.1 Simplest HPLC plumbing arrangement.
The HPLC pump forces a mixture of buffers A and B out of the pump at a high flow rate. This high flow rate is then split, with most of the volume going through 20-μm capillary tubing (to provide back pressure) to a waste collection vessel and the remaining flow (200nl/min) being forced through the analytical column. Electrospray voltage (1800V) is applied to a platinum (Pt) electrode in the fourth port of the splitter, spraying the sample into the spectrometer for analysis.

3. Acquiring Mass Spectra

This procedure is similar to that described previously (Blagoev *et al.*, 2003; Foster *et al.*, 2003; Rappsilber *et al.*, 2002).

Steps

1. Calibrate the mass spectrometer using a calibration standard containing at least two ions of known masses spanning the mass region to be measured.
2. Set up an acquisition method for the LC/MS experiment. Typically the method should be approximately 30min longer than the HPLC elution program to allow time for the gradient delay and setup (see Fig. 24.2A for a typical elution gradient profile). Current mass spectrometers work by performing a few short experiments that are repeated over

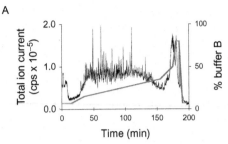

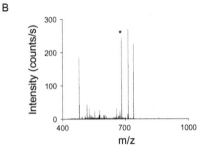

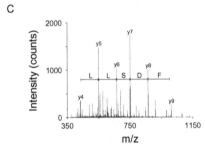

FIGURE 24.2 Sample LC, MS, and MS/MS traces.
(A) Typical elution profile of a reversed-phase HPLC program and the associated total ion chromatogram registered by the mass spectrometer detector. Most of the peptides will elute from the analytical column between 15 and 35% B and so the majority of analysis time is spent in this range. (B) A typical survey or MS scan. Multiply charged ions such as the one indicated by an asterisk (*) detected in this spectra would then be selected for subsequent fragmentation. (C) A typical fragmentation or MS-MS scan. The Y-ions used to identify this peptide are labeled.

and over for the entire length of the LC run. Each cycle typically involves one survey scan (MS, see Fig. 24.2B for an example) from which the software chooses between two and four of the most intense peptides for product ion scans (called an MS–MS scan or fragmentation scan, see Fig. 24.2C) before starting the cycle again. The cycle length and scan times are defined by the user and should be adjusted to get sufficient sampling across an LC peak. For instance, if peptides elute from the LC column in 10 s and a cycle time of 15 s was in use, then it would be possible for a peptide to elute from the column and never be recorded by a survey scan (and thus never be chosen for sequencing). The other major factor in deciding cycle times is sensitivity. The scan times should be long enough such that sufficient ion statistics are obtained for confident identifications. Cycle times are largely dependent on the duty cycle and the sensitivity of the mass spectrometer. For the QSTAR Pulsars in use in our laboratory, the following settings are typical: one 1.0-s MS time-of-flight (TOF) experiment followed by three or four 1.5-s information-dependent MS/MS TOF experiments, only multiply charged ions selected for sequencing, 3 min and 0.9 Da exclusion windows, enhancement around 700 Da, MS and MS/MS scan windows of 400 to 1200 Da, 35 counts/s IDA threshold, 1800 to 2000 V spray voltage.

3. Position the analytical column in front of the orifice plate and use the tuning mode of the mass spectrometer software to adjust the column position for maximum sensitivity.

4. Start the HPLC pump to the predetermined flow rate (see Section III,B,2) using a mix of 5% B in A.

5. Add a sample to the acquisition queue that calls the method described in step 2.

6. Start this sample and open the file to check that the detector is receiving a stable signal.

7. Start the elution program on the HPLC described in Section III,B,2.

C. Protein Identification

1. Mascot Searching

There are a number of commercial products for searching measured fragmentation spectra against theoretical spectra predicted from sequence databases. We use the Mascot search engine in our laboratory.

Steps

1. Mascot cannot interpret raw mass spectra itself and must by given a peak list to work with. Different mass spectrometer manufacturers use different software to do this but the output is all generally similar. For spectra acquired with ABI Analyst, fragmentation spectra are processed from the LC/MS file using the IDA Processor supplied with Analyst. If the mass spectrometer is tuned properly, then the default settings in IDA Processor are a good starting point for optimal extraction.

2. Use Mascot to search the peak list file against the protein sequence database of interest with. Standard settings for Mascot v1.8 are shown in Fig. 24.3. We typically search any human samples against the IPI database (to download, go to http://srs.ebi.ac.uk/srs6bin/cgi-bin/wgetz?-page+LibInfo+-id+1hEHl1KZ36s+-lib+IPI) rather than NCBI due to the high redundancy of the latter. The UniProt database from the creators of IPI and with financial backing from NIH will likely replace all general databases as the library of choice for such searches.

2. Understanding Mascot Output, Confirming Identifications, and Internal Calibration

Steps

1. Peptides are coded in Mascot according to the rules listed in Table 24.1. Only checked queries should be further considered for positive identifications. Thus, any protein with no checked queries can and should be immediately discarded from the

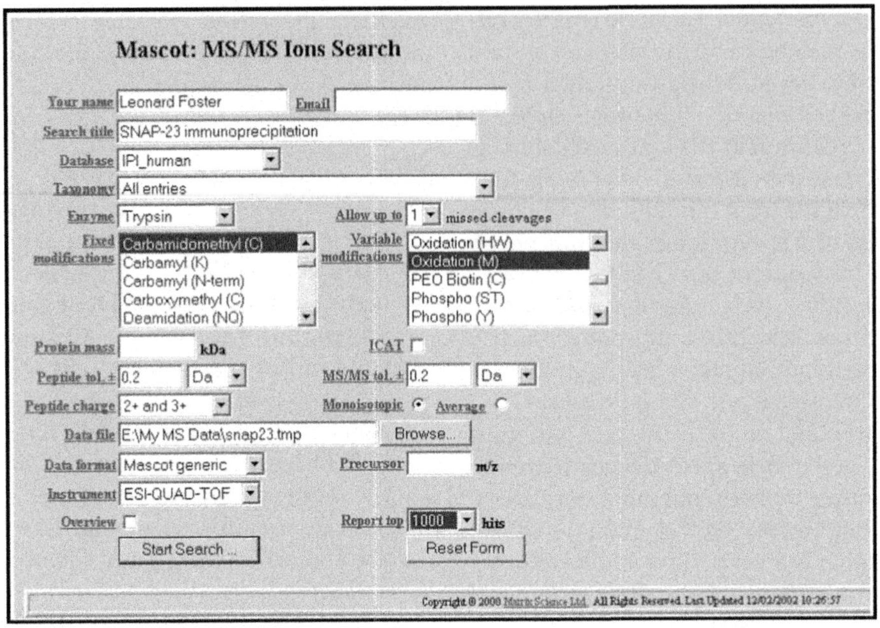

FIGURE 24.3   Typical parameters for a Mascot search.

"hit list", as their identifications were based on spectra who fit different peptides more closely.

2. With the remaining queries some guidelines can be established to minimize time-consuming manual verification. Depending on the database searched and the conditions used for the search, the significance level in Mascot scores will shift. With the conditions specified in Fig. 24.3 we typically accept any protein with a single IonsScore <45 or two IonsScores <30 with no further verification.

3. Any remaining proteins in the list should be verified manually by direct examination of the fragmentation spectra of the peptides used for the identification. For a quadrupole-TOF hybrid instrument, we look for a Y-ion series of at least three fragments and a Mascot score that is distinct from all other matches to that spectra. Many additional sequence-specific factors can be considered but are beyond the scope of this article (Rappsilber et al., 2003).

4. While certain types of mass spectrometers have inherently high mass accuracy, spectra obtained from these instruments can often be improved even further. By simply calculating the mass errors for each identified peptide (the difference between the measured mass and the calculated mass for the amino acid sequence), a regression correction can be calculated that, when applied back against the whole data set, can greatly improve the mass accuracy. Using the highest-scoring peptides from a Mascot search, we have obtained mass accuracies <10 ppm, representing up to 10-fold improvements over the raw data. While there will always be outliers, peptides with a mass deviation larger than twice the average mass deviation should be inspected carefully.

TABLE 24.1 Coding Scheme for Queries (Peptides) in Mascot

Coding method	Meaning
Bold type	First use of spectra in the output list
Light type	Subsequent use of spectra
Red type	Highest-scoring match of spectra
Black type	Less than the highest-scoring match of spectra
Checked query	Highest-scoring use of spectra

IV. COMMENTS

A. Keratin

There are, of course, many biological studies where the identification of keratins by mass spectrometry would be interesting (i.e., studies of intermediate filaments), but the vast majority of keratin identifications are unwanted. Fortunately, keratin contamination does not have to be accepted as a *fait d'accompli*, as careful sample handling can severely reduce and sometimes eliminate keratins altogether.

1. Check all solutions before use—with any sign of precipitates, or "floaties," new solutions should be prepared.
2. Use Milli-Q-quality water for everything.
3. Work in a laminar flow hood where possible. Remember that flow hoods are *not* fume hoods, however, so do not use large volumes of volatile or hazardous chemicals in them.
4. *Never* bring the sample into contact with anything that may have been touched by fingers or had dust fall on it. Implicit in this is to keep unused sample tubes and pipette tips in covered containers.
5. When working with solutions, take care not to pass anything over the top of the open container, as particles may fall off. Beyond this, always enter a container with a pipettor or forceps with the container tilted so that any dust that might fall does not end up in the container.

B. Gloves

We have not yet resolved the gloves/no gloves issue in our laboratory. While they certainly help prevent flakes of skin from falling off our hands, they may increase static charges on sample tubes, thereby attracting air-borne particles. Regardless, by following the general rules described earlier, any additional benefits of gloves seem to be minimal, as those who use them in our laboratory generally see no more or less keratin contamination than those who do not.

V. PITFALLS

While much of the sample handling and data acquisition can be automated fairly easily, there is currently no commercially available software designed to handle the enormous volumes of data that this procedure is capable of generating. For single experiments the manual verification of spectra assignments and data compilation is a relatively simple task but one that grows exponentially more complex with each additional set of results added to the data set. Interpretation and analysis are additionally hampered by the lack of standards in the field for what constitutes an "identified protein." It is hoped that open source or public domain packages will become available soon to make these processes more efficient, transparent, and comparable between groups.

An additional pitfall is the observation that two analyses of the same sample will often not give identical lists of protein components. This is not due to the database search software but rather to the data the software has available. For unknown reasons the mass spectrometer does not choose the same set of peptides for sequencing in duplicate runs, even under identical conditions using the very simplest samples (i.e., purified bovine serum albumin). Proteins that are abundant enough to have three or four (or more) peptides selected for sequencing will likely be found in parallel analyses, but in complexity-limited samples where a large fraction of identifications are based on single peptides, an investigator should not expect to identify the same proteins in a second analysis. To get around this problem, we typically analyze a single sample multiple times and/or attempt to fractionate the sample prior to reversed-phase HPLC (either at protein or peptide levels).

References

Aebersold, R., and Goodlett, D. R. (2001). Mass spectrometry in proteomics. *Chem. Rev.* **101**, 269–295.

Aebersold, R., and Mann, M. (2003). Mass spectrometry-based proteomics. *Neture* **422**(6928), 198–207.

Blagoev, B., Kratchmarova, I., Ong, S. E., Nielsen, M., Foster, L. J., and Mann, M. (2003). A proteomics strategy to elucidate functional protein-protein interactions applied to EGF signaling. *Nature Biotechnol.* **21**.

Foster, L. J., de Hoog, C. L., and Mann, M. (2003). Unbiased quantitative proteomic analysis of lipid rafts reveals high specificity for signalling factors. *Proc Natl Acad Sci USA* **100**, 5813–5818.

Mann, M., Hendrickson, R. C., and Pandey, A. (2001). Analysis of proteins and proteomes by mass spectrometry. *Annu. Rev. Biochem.* **70**, 437–473.

Rappsilber, J., Andersen, J. S., Ishihama, Y., Ong, S. E., Foster, L. J., and Mann, M. (2003). A recipe collection for the identification of peptides in complex mixtures. *Sci. STKE.*

Rappsilber, J., Ishihama, Y., and Mann, M. (2003). Stage (STop And Go Extraction) tips for MALDO, nanoelectrospray, and LC/MS sample pre-treatment in proteomics. *Anal. Chem.* **175**, 663–670.

Rappsilber, J., Ryder, U., Lamond, A. I., and Mann, M. (2002). Large-scale proteomic analysis of the human spliceosome. *Genome Res.* **12**, 1231–1245.

Shevchenko, A., Wilm, M., Vorm, O., and Mann, M. (1996). Mass spectrometric sequencing of proteins silver-stained polyacrylamide gels. *Anal. Chem.* **68**, 850–858.

Washburn, M. P., Ulaszek, R., Deciu, C., Schieltz, D. M., and Yates, J. R., 3rd (2002). Analysis of quantitative proteomic data generated via multidimensional protein identification technology. *Anal. Chem.* **74**, 1650–1657.

Proteome Specific Sample Preparation Methods for Matrix-Assisted Laser Desorption/Ionization Mass Spectrometry

Martin R. Larsen, Sabrina Laugesen, and Peter Roepstorff

I. INTRODUCTION

Early in the 1990s a number of groups demonstrated that it was possible to identify proteins in databases based on peptide mass maps obtained by mass spectrometric (MS) analysis of peptides derived by specific enzymatic proteolysis of a given protein (Henzel *et al.*, 1993; Mann *et al.*, 1993; Pappin *et al.*, 1993; James *et al.*, 1993). Shortly after that, the field of proteomics started to expand concurrent with the development of more sensitive mass spectrometers so that even very low amounts of gel-separated

proteins could be identified. Presently, most mass spectrometers routinely allow analysis of peptides or proteins in the low femtomole level.

With the development of sensitive mass spectrometers, the real limitation for the analysis of complex and contaminated peptide and protein mixtures is sample preparation prior to mass spectrometric analysis. Therefore, major efforts have been invested into developing new methods for sample preparation prior to MS.

In the traditional sample preparation method for matrix-assisted laser desorption/ionization

(MALDI) MS, the analyte and matrix solutions are mixed directly on the MALDI target, referred to as the dried droplet method (Karas and Hillenkamp, 1988; Hillenkamp *et al.* 1991). Alternatively, the matrix and analyte solutions can be mixed in a test tube prior to application onto the target. The tolerance of this method toward nonvolatile contaminants is limited to the efficiency of the crystallization process and to sample washing postmatrix crystallization. Additionally, it has been reported that the matrix solution conditions (especially the solvent) have a large influence on the quality of the MALDI-MS analysis (Cohen and Chait, 1996). With the introduction of the fast evaporation or thin-layer method (Vorm *et al.*, 1994), an increase in sensitivity and resolution was obtained. Here, the matrix is dissolved in a highly volatile organic solvent, which is applied onto the target to create a thin homogeneous layer of matrix crystals. The analyte solution is placed on top of the film and dried at ambient temperature followed by rinsing, resulting in decreased alkali metal adduct formation and a concomitant increase in signal intensity and sensitivity. The fast evaporation method was further developed with the inclusion of nitrocellulose (NC) in the matrix (Jensen *et al.*, 1996; Kussmann *et al.*, 1997). The presence of NC reduced the intensity of the alkali metal ion adducts, presumably because it binds the alkali metal ions very strongly, thereby preventing them from entering the gas phase. A noteworthy improvement in sample preparation for both MALDI-MS and electrospray ionization (ESI) MS came with the introduction of custom-made disposable microcolumns as a fast cleanup step prior to MS, first introduced by Wilm and Mann (1996). They used a pulled capillary needle in which they packed a small volume of reversed-phase chromatographic material. The analyte molecules were eluted directly into a nano-ESI capillary needle by centrifugation prior to mass analysis. Later, this sample preparation was simplified by using Eppendorf GELoader

tips packed with chromatographic material or prepacked microcolumns (e.g., ZipTips, Millipore) (Gobom *et al.*, 1997, Kussmann *et al.*, 1997; Erdjument-Bromage *et al.*, 1998). Elution of the analyte from the microcolumns with matrix solution was later demonstrated to be a very efficient sample preparation method for MALDI (Gobom *et al.*, 1999). The use of microcolumns allows upconcentration and desalting of highly diluted samples prior to MS analysis, resulting in an increase in the sensitivity and of the overall quality of the mass spectra. In addition, significantly higher sequence coverage from peptide mass maps is observed, which is important not only for unambiguous protein identification, but especially for complete characterization of posttranslational modifications in proteomics.

This article reports a number of protocols currently used in our laboratory for MALDI-MS sample preparation optimized for proteomic research, i.e., high sensitivity, high tolerance towards low molecular weight contaminants, and high sequence coverage. We focus mainly on sample preparation methods involving sample cleanup using microcolumn technology, as traditional methods such as the dried droplet and the fast evaporation method have been described extensively in Roepstorff *et al.* (1998).

II. MATERIALS AND INSTRUMENTATION

The GELoader pipette tips are from Eppendorf (Hamburg, Germany). The chromatographic column materials, Poros R2 and Poros Oligo R3, are from PerSeptive Biosystems (Framingham, MA). α-Cyano-4-hydroxycinnamic acid (4HCCA) and activated charcoal (C-5510) are from Sigma (St. Louis, MO). 2,5-Dihydroxybenzoic acid (DHB) and 2-hydroxy-5-methoxybenzoic acid (HMB) are from Aldrich. Acetonitrile, formic acid (FA), and trifluoroacetic acid (TFA) are all analytical grade and obtained from different manufacturers. Disposable syringe (1 ml) are from BD Plastipak. The water is from a Milli-Q system (Millipore, Bedford, MA).

MALDI mass spectra are recorded on a Bruker Reflex III mass spectrometer (384 sample plate inlet), a Perseptive Voyager STR mass spectrometer, or a TOF-Spec 2E (Micromass, Manchester, UK). All are equipped with delayed extraction.

III. PROCEDURES

A. Preparation of Matrix Solutions

1. Matrix Solutions Containing 4HCCA

a. *4HCCA (I)*: Dissolve 10–20 mg 4HCCA in 1 ml 70% acetonitrile/0.1% TFA
b. *4HCCA (II)*: Dissolve 10–20 mg 4HCCA in 1 ml acetone/water (99/liter, v/v)

2. Matrix Solutions Containing 2,4,6-THAP

THAP: Dissolve 10 mg 2,4,6-THAP in 0.5 ml acetonitrile/water (70/30, v/v)

3. Matrix Solutions Containing DHB

DHB: Dissolve 20 mg DHB in 0.5 ml acetonitrile/water (50/50, v/v)

4. Mixed Matrix Solutions

a. *4HCCA/DHB*: Dissolve 20 mg 4HCCA in 1 ml acetonitrile/5% FA (70:30, v/v). Dissolve 20 mg DHB in 1 ml acetonitrile/0.1% TFA (70:30, v/v). The two solutions are then combined in a 1:1 volume ratio
b. *DHB/HMB (super DHB)*: Dissolve 9 mg DHB and 1 mg HMB in 0.5 ml acetonitrile/water (20/80, v/v)

B. Sample Preparation for MALDI-MS Analysis

1. Traditional Dried Droplet Method

The dried droplet method originally introduced by Karas and Hillenkamp (1988) is the oldest sample preparation method and is, in many applications, still the preferred one for MALDI-MS. It is surprisingly simple, it provides good results for different types of samples, and it can tolerate moderate concentrations of low molecular weight contaminants. This matrix preparation method is used preferentially in proteomics for high throughput automated peptide mass mapping of proteins present in relative high abundance. The following matrices are traditionally used with this method: 4HCCA (I), THAP, and DHB.

Steps

1. Mix equal volumes of analyte and matrix solutions and, if appropriate, 0.1% TFA or 2% FA on the sample support.
2. Let the sample dry at ambient temperature. Alternatively, evaporation of the solvent can be assisted by a stream of inert gas.
3. If the matrix solution is 4HCCA or THAP, then the matrix crystals can be washed by depositing 10 μl of 0.1% TFA on top of the dried preparation and then removing it carefully with a piece of paper tissue.

Note: A common problem with the dried droplet method is the accumulation of analyte/ matrix crystals in the periphery of the sample surface, especially in the presence of low molecular weight contaminants. For DHB, the analytes tend to associate with the big crystals that form at the periphery of the sample surface, whereas salts are found predominantly in the smaller crystals formed in the centre of the sample surface. This is the reason why it is often necessary to search for "sweet" spots on the sample surface. Inclusion of acid in the first step increases tolerance towards low molecular weight contaminants and helps in the crystallization process.

2. Mixed Matrix Dried Droplet Method

The use of matrix mixtures was reported in the 1990s by Karas and coworkers (Tsarbopoulos et al., 1994). They described a matrix mixture consisting of DHB and HMB (super DHB), which was used predominantly for the analysis of glycosylated proteins. Mixed matrices presumably combine the different properties of the individual matrices to a property superior to the individual matrices. Our group developed a mixed matrix solution consisting of 4HCCA and DHB, which has proven to be very useful in analysis of low amounts of peptides and proteins (Laugesen and Roepstorff, 2003). Here the hydrophilic matrix (DHB) forms crystals in the outer edge of the spot, whereas the hydrophobic 4HCCA matrix forms crystals in the centre of the spot. An increased tolerance towards salts is observed, presumably because the peptides cocrystalize with matrix in the centre, whereas the salt molecules are upconcentrated in the hydrophilic DHB edge. An example of the performance of the 4HCCA/DHB matrix compared to traditional dried droplet using only one of the matrices is shown in Fig. 25.1.

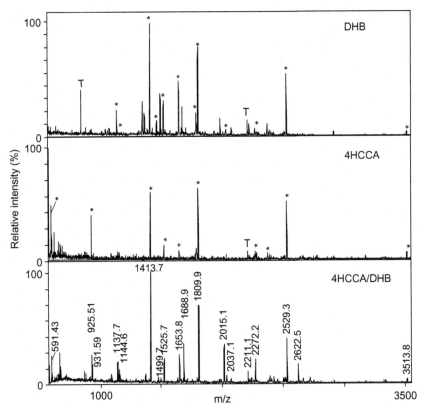

FIGURE 25.1   MALDI-MS peptide mass maps of peptides derived by tryptic digestion of a silver-stained spot obtained from 2D-PAGE of proteins extracted from barley grain. Spectra were obtained with DHB, CHCA, and the matrix mixture preparations. Asterisks indicate signals matching protein z-type serpin from barley. Commonly observed trypsin autodigest signals are indicated by T.

Steps

1. Mix the two matrix solutions in a small Eppendorf tube in a proper ratio (typically 1:1) and vortex intensively.
2. Mix an equal volume of analyte, matrix mix solution, and, if appropriate, 0.1% TFA or 2% formic acid on the sample support.
3. Let the solution dry at ambient temperature.

Note: These matrix preparations are not washable, as DHB is soluble in water/0.1% TFA solution. Desorption with the laser is performed in the centre of the preparation.

3. Fast Evaporation Thin-Layer/Sandwich Method

The fast evaporation method is used less frequently in proteome analysis nowadays. However, several laboratories prepare a layer of matrix prior to spotting sample and matrix on top. The method is especially useful to ensure even crystallization when anchor chip targets are used.

Steps

1. Place a small droplet (0.5 µl) of the matrix solution [4HCCA (II)] onto the MALDI target so that it spreads out. Evaporation of the solvent results in a homogeneous layer of matrix crystals.
2. Apply 0.5 µl 2% FA or 0.1% TFA, 0.5 µl analyte solution, and 0.5 µl 4HCCA (I) on top of the thin matrix layer. The addition of FA or TFA solution is not necessary if the analyte solution is already acidified.
3. Let the solution dry at ambient temperature.
4. The matrix crystals can be washed as described earlier.

Note: The fast evaporation methods are used exclusively with 4HCCA and THAP as matrices.

4. Micropurification Method

The dominating sample preparation method used for proteomics in our laboratory and many other laboratories involves the use of microcolumns, either homemade GELoader tip (Eppendorf) columns or prepacked microcolumns (e.g., ZipTips, Millipore). This sample preparation method is compatible with 4HCCA, DHB, HMB, and the DHB/4HCCA mixture.

This section describes the use of GELoader tip microcolumns and refers to the manufacturer's protocol for the ZipTips.

Steps

1. Make a partially constricted GELoader pipette tip by squeezing the narrow end. The two most common ways to generate a partially constricted GELoader pipette tip are illustrated in Fig. 25.2. Put the narrow end of a GELoader tip flat on a hard surface. Then roll a 1.5-ml microfuge tube over the final 1 mm of the tip. Alternatively, fix the narrow end of the GELoader tip using a flat-surface forceps and twist the tip to close the end.
2. Prepare a slurry of 100–200 µl chromatographic material, e.g., Poros R2, Oligo R3, or Graphite powder, in 70% acetonitrile (approximately 1.5 mg/100 µl).
3. Load 20 µl 70% acetonitrile in the top of the constricted GELoader tip followed by approximately 0.5 µl of the chromatographic

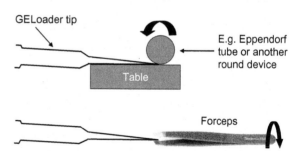

FIGURE 25.2 Preparation of a constricted GELoader tip can be performed in a number of different ways. One way is to roll a 1.5-ml Eppendorf tube over the narrow end of the tip with gentle pressure to squeeze the approximately last 0.5 mm of the tip flat (top). The other strategy is to use a forceps to twist the last 1 mm of the tip (bottom).

material slurry. Use a 1-ml syringe fitted to the diameter of the GELoader tip via a disposable pipette tip to press the liquid down gently, thereby generating a small column at the end of the constricted tip. Push tall liquid out of the column before performing the next step.

4. Apply 20 μl of 2% FA or 0.1% TFA to the top of the column. Press approximately 10 μl through the column for equilibration. Leave the remaining 10 μl on top of the column bed.

5. Mix the analyte with the remaining acid solution and use gentle air pressure to press the liquid through the column. Leave approximately 3 μl of solvent on top of the column. Do not dry the column in this step!

6. Apple 10–20 μl of 2% FA or 0.1% TFA on top of the remaining solution and use air pressure to wash the column. In this step, all liquid should be pushed out of the column.

7. Elute the peptides with 0.5 μl matrix solution directly onto the MALDI MS target. The eluent should be spotted in several small droplets on the target for a further concentration. Most peptides are only present in the two first droplets, but some may come later depending on the size and hydrophobicity. As an alternative the peptides can be eluted with increasing concentrations of organic solvent and applied by the dried droplet method.

8. Depending on the size of the column and the abundance/concentration of the analyte molecules that have been loaded, the column can be reused from 2 to 10 times after extensive washing with 70–100% acetonitrile.

Notes: Using microcolumns, the low molecular weight contaminants can be removed easily from the analyte molecules and diluted samples are concentrated readily with a resulting increase in sensitivity and sequence coverage. An example is shown in Fig. 25.3. Several types of reversed-phase material can be used for column packing. Poros R2 and Poros Oligo R3

have proven excellent in our hands. Graphite powder has been used as an alternative to reversed-phase material, giving superior recovery of hydrophilic peptides and higher sequence coverage (Larsen *et al.*, 2002). Examples of the use of Poros R2 and graphite column materials in proteome analysis are given in Fig. 25.4.

The same procedure as described for GELoader tip microcolumns can be used with commercially available ZipTip microcolumns, resulting in increased sensitivity compared to the protocol described by the manufacturer.

The performance of the homemade microcolumn is superior to that of the commercially available ZipTips, especially for analysis of very low abundant proteins from 2D gels. An example of the performance of Poros R2 versus ZipTip C_{18} is shown in Fig. 25.5.

C. Column Material for GELoader Tip Microcolumns Used in Proteomics

The most common chromatographic material used for microcolumns in proteomics is reversed-phase material. However, other kind of materials can be applied with success.

i. Poros R2 and Oligo R3 are the main reversed-phase material used in proteomics to desalt and concentrate peptides prior to mass spectrometry. Poros R2 is designed for the general separation of proteins, peptides, and nucleic acids. The binding strength is similar to low carbon-loading C8 or C18 supports. The Oligo R3 medium is designed for hydrophilic peptides and nucleic acids and is similar to high carbon-loading C18 supports.

ii. Graphite powder can be used as an alternative to RP material as a peptide cleanup medium, especially for small hydrophilic peptides or phosphopeptides (Chin and Papac, 1999; Larsen *et al.*, 2002). However, it also works very well for all other peptides if they are eluted with the 4HCCA matrix (Larsen *et al.*, 2002).

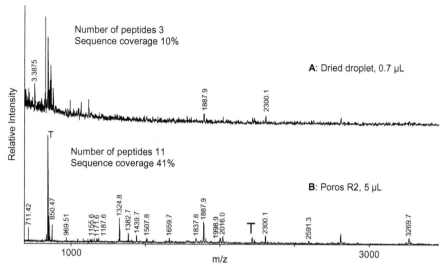

FIGURE 25.3 MALDI-MS peptide mass maps of peptides derived by tryptic digestion of a protein from a 2D gel of a membrane preparation from *Pseudomonas aeruginosa* (hypothetical protein, NC_002516). (A) Spectrum obtained using the dried droplet method with 0.7 μl peptide solution. Three peptides were detected covering 10% of the protein. The protein could not be identified using this peptide mass map. (B) Spectrum obtained after desalting and concentration of 5 μl peptide solution using a Poros R2 GELoader tip microcolumn. Eleven peptides were detected, and the protein could be identified easily with 41% sequence coverage. T indicates peptides originating from tryptic autodigestion.

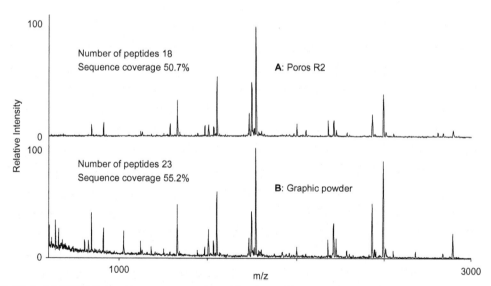

FIGURE 25.4 MALDI-MS peptide mass maps of peptides derived by tryptic digestion of a protein from a 2D gel of a membrane preparation from *Pseudomonas aeruginosa* (probable porin, NC_002516). An aliquot of the peptide solution was desalted and concentrated using a Poros R2 and a graphite powder GELoader tip microcolumn, respectively. (A) Spectrum obtained from the Poros R2 column. A total of 18 peptides could be assigned to the sequence of the probable porin covering 50.7% of the sequence. (B) Spectrum obtained from the graphite powder column. Twenty-three peptides covering 55.2% of the sequence could be assigned.

V. MASS SPEC METHODS FOR DETERMINATION OF PROTEIN IDENTITY AND PROTEIN MODIFICATIONS

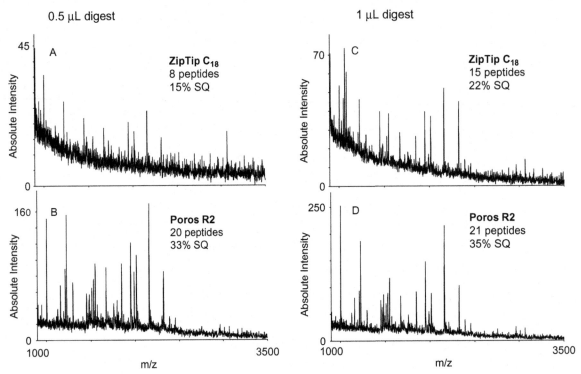

FIGURE 25.5 Comparison of GELoader tip microcolumns with the commercially available ZipTip C_{18}. A weak Coomassie-stained spot on a 2D gel of a membrane preparation from *Pseudomonas aerugionosa* [Fe(III)-pyochelin receptor precursor, NP_252911] was submitted to in-gel digestion using trypsin in a total volume of $20\,\mu l$ buffer. Aliquots (0.5 and $1\,\mu l$) were desalted and concentrated on the two different microcolumns. Spectra obtained from the ZipTip C_{18} microcolumn purification of 0.5- and 1-μl aliquots are shown in A and C, respectively. Spectra obtained from the GELoader tip microcolumn purification of 0.5- and 1-μl aliquots are shown in B and D, respectively. As judged from the intensity of the peptide signals (absolute intensities), GELoader tip microcolumns are between 5 and 10 times more sensitive than the commercially available ZipTip columns.

Alternatively, it can be used to desalt and concentrate carbohydrates prior to mass spectrometric analysis.

iii. Anion (e.g., Poros HQ) and cation (e.g., Poros S) exchange columns are used less frequently for protein identification purposes in proteomics but can be advantageous for specific purposes, such as the isolation of specific peptides when the goal is full characterization of proteins. Protocols for the use of these types of column materials are available from the resin manufacturer's home page (www.appliedbiosystems.com).

Note: Many other types of column material can be packed into the small GELoader tip microcolumns, including immobilized metal affinity material, immobilized enzymes, and immobilized DNA/RNA.

D. Selection of Sample Preparation Method in Proteomics

The choice of matrix and sample preparation method for MALDI-MS analysis of peptides in proteome analysis is very dependent on the amount and complexity of the sample to be analysed. In general, peptides derived by proteolytic digestion of proteins visible by Coomassie blue require less sensitive sample preparation methods than proteins visible by more sensitive staining procedures, such as silver staining or fluorescent staining (e.g., Sypro Ruby). For the first category of peptides, a standard dried droplet or sandwich method is fast and straightforward and results in detection of sufficient peptides for ambiguous protein identification. In addition, these methods are automated easily and therefore suitable for high-throughput proteomics. For a lower amount of starting material, the dried droplet or sandwich methods will not provide enough peptide signals due to suppression effects by low molecular weight contaminants or simply because the sample is too diluted. In such cases, desalting and concentration using microcolumns will be needed. As judged from the intensity of the peptides in the example shown in Fig. 25.5, the homemade GELoader tip microcolumns are 5–10 times more sensitive than commercially available tips. However, commercially available microcolumns are easy to handle and are compatible with several different liquid-handling robots, making them suitable for high-throughput proteomics. The combined use of different matrices and several types of microcolumns may be needed if the goal is to obtain the highest possible sequence coverage. This is relevant for observation of posttranslationally modified peptides and for determination of the processing sites when the proteins are truncated.

E. Final Comments

The sample preparation procedures are under constant development and may also vary according to the individual experimentalist. Therefore, this article should merely serve as a source of inspiration. It must, however, be kept in mind that the quality of all solvents and chemicals is of paramount importance for success. Therefore, all new batches of solvents and chemicals must be checked with a standard protein digest before being used. In addition, we always aliquot the solvents immediately after positive testing, as any solvent will be contaminated after opening the bottle a few times in the laboratory.

References

Chin, E. T., and Papac, D. I. (1999). The use of a porous graphitic carbon column for desalting hydrophilic peptides prior to matrix-assisted laser desorption/ionization time-of-flight mass spectrometry. *Anal. Biochem.* **273**(2), 179–185.

Cohen, S. L., and Chait, B. T. (1996). Influence of matrix solution conditions on the MALDI-MS analysis of peptides and proteins. *Anal. Chem.* **68**(1), 31–37.

Erdjument-Bromage, H., Lui, M., Lacomis, L., Grewal, A., Annan, R. S., McNulty, D. E., Carr, S. A., and Tempst, P. (1998). Examination of micro-tip reversed-phase liquid chromatographic extraction of peptide pools for mass spectrometric analysis. *J. Chromatogr. A* **826**(2), 167–181.

Gobom, J., Nordhoff, E., Ekman, R., and Roepstorff, P. (1997). Rapid micro-scale proteolysis of proteins for MALDI-MS peptide mapping using immobilized trypsin. *Int. J. Mass Spectrom. Ion Proc.* **169/170**, 153–163.

Gobom, J., Nordhoff, E., Mirgorodskaya, E., Ekman, R., and Roepstorff, P. (1999). Sample purification and preparation technique based on nano-scale reversed-phase columns for the sensitive analysis of complex peptide mixtures by matrix-assisted laser desorption/ionization mass spectrometry. *J. Mass Spectrom.* **34**, 105–116.

Henzel, W. J., Billeci, T. M., Stults, J. T., and Wong, S. C. (1993). Identifying proteins from two-dimensional gels by molecular mass searching of peptide fragments in protein sequence databases. *Proc. Natl. Acad. Sci. USA* **90**, 5011–5015

Hillenkamp, F., Karas, M., Beavis, R. C., and Chait, B. T. (1991). Matrix-assisted laser desorption/ionization mass spectrometry of biopolymers. *Anal. Chem.* **63**(24), 1193A–1203A.

James, P., Quadroni, M., Carafoli, E., and Gonnet, G. (1993). Protein identification by mass profile fingerprinting. *Biochem. Biophys. Res. Commun.* **195**, 58–64

Jensen, O. N., Podtelejnikov, A., and Mann, M. (1996). Delayed extraction improves specificity in database searches by matrix-assisted laser desorption/ionization peptide maps. *Rapid Commun. Mass Spectrom.* **10**(11), 1371–1378.

Karas, M., and Hillenkamp, F. (1988). Laser desorption ionization of proteins with molecular masses exceeding 10,000 daltons. *Anal. Chem.* **60**, 2299–2301.

Kussmann, M., Nordhoff, E., Nielsen, H. R., Haebel, S., Larsen, M. R., Jacobsen, L., Jensen, C., Gobom, J., Mirgorodskaya, E., Kristensen, A. K., Palm, L., and Roepstorff, P. (1997). MALDI-MS sample preparation techniques designed for various peptide and protein analytes. *J. Mass Spectrom.* **32**, 593–601.

Larsen, M. R., Cordwell, S. J., and Roepstorff, P. (2002). Graphite powder as an alternative or supplement to reversed-phase material for desalting and concentration of peptide mixtures prior to matrix-assisted laser desorption/ionization mass spectrometry. *Proteomics* **2**, 1277–1287.

Laugesen, S., and Roepstorff, P. (2003). Combination of Two Matrices Results in Improved Performance of MALDI MS for Peptide Mass Mapping and Protein Analysis. *J. Am. Soc. Mass. Spectrom.* **14**: 992–1002.

Mann, M., Højrup, P., and Roepstorff, P. (1993). Use of mass spectrometric molecular weight information to identify proteins in sequence databases. *Biol. Mass Spectrom.* **22**, 338–345.

Pappin, D. J. C., Højrup, P., and Bleasby, A. J. (1993). Rapid identification of proteins by peptide-mass finger printing. *Curr. Biol.* **3**, 327–332.

Roepstorff, P., Larsen, M. R., Rahbek-Nielsen, H., and Nordhoff, E. (1998). Sample preparation methods for matrix assisted laser desorption/ionization mass spectrometry of peptides, proteins and nucleic acids. *In* "Cell Biology: A Laboratory Handbook," 2nd Ed. Vol. 4, pp. 556–565.

Tsarbopoulos, A., Karas, M., Strupat, K., Pramanik, B. N., Nagabhushan, T. L., and Hillenkamp, F. (1994). Comparative mapping of recombinant proteins and glycoproteins by plasma desorption and matrix-assisted laser desorption/ionization mass spectrometry. *Anal. Chem.* **66**(13), 2062–2070.

Vorm, O., Roepstorff, P., and Mann, M. (1994). Matrix, surfaces made by fast evaporation yield improved resolution and very high sensitivity in MALDI TOF. *Anal. Chem.* **66**, 3281–3287.

Wilm, M., and Mann, M. (1996). Analytical properties of the nanoelectrospray ion source. *Anal. Chem.* **68**(1), 1–8.s

In-Gel Digestion of Protein Spots for Mass Spectrometry

Kris Gevaert and Joël Vandekerckhove

I. INTRODUCTION

Modern techniques for studying complex protein mixtures utilize two main analytical techniques: (a) one- or two-dimensional polyacrylamide gel electrophoresis (1D or 2D PAGE) for dividing the protein mixture into its individual components and (b) accurate mass spectrometry (MS) for identifying these proteins. Different gel-staining procedures are available for visualizing gel-separated proteins, of which most are compatible with further MS analysis.

Different types of Coomassie brilliant blue staining (Bennett and Scott, 1971; Wilson, 1983) and silver-staining procedures (Merril *et al.*, 1979) are by far the most used visualization protocols prior to mass spectrometric analysis. Proteins spots of interest, e.g., protein spots of which the staining intensity (thus concentration) and/or the position (modification status) differs between two compared samples, are excised from the gel and cleaved into peptides after protease addition. The generated peptide mixture elutes out of the gel passively and is analyzed by a variety of chromatographic and mass spectrometric methods (e.g., reviewed by Aebersold and Mann, 2003).

Frequently, trypsin is used to generate peptides, cleaving at the carboxy-terminal side of lysine and arginine, generating peptides of an average size of about 10 amino acids. For most proteins, this leads to a sufficient number of analyzable peptides and thus to unambiguous identification. This article describes a standard protocol for *in-gel* tryptic protein digestion leading to MS-based protein identification and characterization.

II. MATERIALS AND INSTRUMENTATION

A Sentry ionizing air blower (product number 4003143) is from Simco, Deerlijk, Belgium. Singly wrapped sterile stainless steel scalpels (product number 10.295.10MN) are from BCB Ltd., Cardiff, UK. Singly wrapped Eppendorf Biopur Safe-Lock micro test tubes (product number 0030 121.589) are from VWR International Belgium, Leuven, Belgium. Water (product number 4218) and acetonitrile (product number 9017) used to prepare the different buffers are of the best quality available and are both from Malinckrodt Baker B.V., Deventer, The Netherlands. Sequencing grade modified trypsin (product number V5111) is from Promega Benelux BV, Leiden, The Netherlands. Ammonium hydrogen carbonate (product number A-6141) is from Sigma-Aldrich Corp., St. Louis, MO. Recombinant, proteomics grade trypsin from *Pichia pastoris* (product number 3 357 228) can be obtained from Roche Diagnostics GmbH, Penzberg, Germany. Peptide synthesizer graded trifluoroacetic acid (product number PTS6045) is from Rathburn Chemicals Ltd., Walkerburn, Scotland, UK.

III. PROCEDURE

In-Gel Digestion of Proteins Separated in Polyacrylamide Gels

The following procedure is derived from a general procedure that has been described previously (Rosenfeld *et al.*, 1992), but to which important modifications are added. Generally, this approach can be used to digest proteins *in gel* independent from the procedure that was used to visualize them, although care must be taken, as some silver-staining protocols use amino acid cross-linkers that render most of the proteins inaccessible for the employed protease. In the following procedure, trypsin is used to digest the protein into peptides. In principle, every available protease can be used as long as the final buffer conditions (pH, additives, etc.) are met.

Solutions

1. *Wash solution*: Prior to *in-gel* digestion, excised gel pieces containing the proteins are washed extensively with an aqueous acetonitrile solution (1/1, v/v) of the highest purity available. This solution is preferentially freshly prepared in small volumes (e.g., 10 ml).
2. *Digestion buffer*: This buffer must always be freshly prepared and consists of 50 mM ammonium bicarbonate dissolved in water of the highest purity available. The pH of this buffer does not need any adjustment and will be around 7.8.

Steps

1. Following Coomassie or silver staining, wash the polyacrylamide gels in water to remove excess substances, derived from the electrophoresis and staining procedure, from the gel matrix. Repeat this washing step at

least three times (for 30 min each), each time using fresh volumes of water.

2. Then transfer the gel to a clean container and keep moist during the excision of protein spots. Immediately prior to gel spot excision, open the container and blow a stream of ionized air over it. This air stream almost completely eliminates charging of plastic material by static electricity by neutralizing the surface of the working area. When a neutralizing air stream is not used, dust particles will be attracted to the gel container, the polyacrylamide gel, gloves, scalpel, and microtest tubes.

3. Excise protein spots using a sterile scalpel or a hollow stainless steel needle, and excise only the heart of the protein spot (i.e., the most intensely coloured gel area). Using the scalpel blade or a clean forceps, transfer each individual protein spot to a Biopur microtest tube. Prior to excising the next protein spot, wash the scalpel (and the forceps) extensively with methanol to remove small gel particles that may stick to it and thus cross-contaminate other protein spots.

4. Cover each protein spot with wash solution and incubate for 15 min at room temperature. During this wash step, gel pieces will shrink and buffer components (and Coomassie) are efficiently extracted from the gel.

5. Centrifuge the tubes briefly, remove the wash solution, and submerge the gel pieces for a second time in this solution for 15 min at room temperature.

6. Following a second centrifugation step, remove the wash solution and transfer the tubes to a centrifugal vacuum concentrator in which the gel pieces are dried to complete dryness (this takes about 10 min).

7. Rehydrate the dried gel pieces with 10 μl (corresponding to 0.1 μg) of a freshly prepared 0.001% (w/v) trypsin solution in 50 mM ammonium bicarbonate (pH 7.8) for 10 min at room temperature, after which excess trypsin solution is removed.

8. Subsequently submerge the rehydrated gel pieces in 50 mM ammonium bicarbonate buffer (pH 7.8); depending up the size of the excised gel pieces, between 50 and 100 μl of buffer must be added.

9. Close the tubes and place into a thermostatically controlled incubator, and tryptic digestion proceeds overnight at 37°C.

10. Terminate protein digestion by adding 10 μl of 10% (v/v) trifluoroacetic acid (TFA).

11. Following a brief centrifugation step, remove and transfer the supernatant containing the peptide mixture to a new tube, which is frozen at −20°C until further analysis (e.g., by mass spectrometric techniques).

IV. COMMENTS

1. It has been shown that the removal of metallic silver from silver-stained proteins by oxidation with hydrogen peroxide prior to *in-gel* digestion not only increases the overall sensitivity of MS-based protein identification, but also augments the coverage of the sequence of the analyzed protein (Sumner *et al.*, 2002). One of the side effects is that methionines and carbamidomethylated cysteines will be (at least partly) oxidized to their sulfoxide derivatives (Gevaert *et al.*, 2002). This will increase the complexity of the generated peptide mixtures in two ways: (a) all peptides containing methionine and/or cysteine will be present as couples (oxidized and nonoxidized forms) and (b) neutral losses of the side chains of the oxidized amino acids give rise to satellite peaks (Steen and Mann, 2001), which may cause problems during MALDI-MS analysis.

In contrast, this added complexity might help in identifying proteins, as peptides containing rare amino acids (methionine and cysteine) can be recognized in the peptide mass maps and verify the database findings of search algorithms.

2. When analyzing Coomassie-stained protein spots by MALDI-MS-based peptide mass fingerprinting (reviewed by Cottrell, 1994), it is important to remove as much of the noncovalently bound Coomassie molecules as possible during the washing steps (steps 4, 5, and 6 of the procedure). Coomassie molecules ionize rather efficiently in MALDI mode and thereby suppress ionization of peptide molecules. Even more severe, in some cases, Coomassie ions will saturate the detector of the mass spectrometer, making it less sensitive for peptide ions striking it. Furthermore, these Coomassie ions disturb the lower mass section of the peptide mass maps, possibly masking some important small peptide ions.

3. The accuracy of protein identification largely depends on the total number of peptides that are available for analysis. Proteins that have many hydrophobic patches are sometimes difficult to digest, as either the protease has limited access to the chain of the protein or the liberated peptides are too hydrophobic and thus do not elute well out of the gel pieces. For such proteins, the efficiency of protein digestion is improved by adding detergents or chaotropes that partly unfold the protein in the gel, thereby exposing more cleavage sites. Unfortunately, the choice of such additional components is rather limited, as many of them block the action of the protease or interfere with further reversed-phase high-performance liquid chromatographic (RP-HPLC) analysis or with mass spectrometric analyses. One detergent that has proven to be particularly useful for the digestion of membrane proteins is octyl-β-glucopyranoside

(van Montfort et al., 2002). This detergent increases the recovery of hydrophobic peptides following digestion and does not interfere with subsequent MS analysis. From our experience, we have found that, especially when working with trypsin, fairly high concentrations of urea (up to $4M$) in the protein digestion buffer can be tolerated without any severe effect on the activity of the protease. Urea notably increases the total coverage of the analyzed protein and thus simplifies protein identification by mass spectrometry. The main drawback when using urea is carbamylation of free amino groups by ammonium cyanate present in urea solution, which, in addition to making less sites available for tryptic cleavage, makes the peptide mixtures more complex.

4. Particularly when employing MALDI-MS for protein identification, it is important to remove any substances that might interfere with the matrix crystallization and the ionization processes. Such substances include salts, buffering substances, chaotropes, and detergents. Several procedures have been described (Gevaert et al., 1997) and are compatible with the in-gel digestion procedure described here.

5. Recombinant trypsin made in the yeast Pichia pastoris has been made available commercially. The main advantage of this protease over other commercially available proteases is the complete absence of nontryptic proteolytic activity. Other types of proteases, although treated with specific inhibitors of commonly copurifying proteases, still lead to nontryptic peptides, which complicate peptide mass maps and make unambiguous protein identification more complex. For example, the most prominent contamination found in trypsin isolated from mammalian organs is chymotrypsin. Because chymotrypsin cleaves at the C-terminal side of five different amino acids, upon prolonged

incubation of protein substrates with "chymotrypsin-contaminated" trypsin, this leads to very complex peptide mass maps.

6. Transient or permanent modifications of amino acid side chain are molecular switches to control the activity of a great number of proteins/enzymes. Mass spectrometric analysis is the sole technology available for the in-depth analysis of the vastness of modifications that occur *in vivo*. However, when digesting a gel-separated protein with only one enzyme, too few ionizable peptides will be generated so as to cover the complete sequence of the analyzed protein and thus many possible modifications may escape analysis. Therefore, it is advantageous to use multiple enzymes either in sequence or together in order to generate many protein fragments and thus cover as much of the protein sequence as possible. Combinations of site-specific and unspecific proteases have been used for that purpose (e.g., MacCoss *et al.*, 2002).

V. PITFALLS

1. One of the major pitfalls encountered frequently is contamination of protein spots with human epidermal keratins. During *in-gel* protein digestion, keratins that are present digest as well and will give a characteristic and easily recognizable pattern of peptide masses in the mass maps obtained (Parker *et al.*, 1998). To our experience, it is very important to wear clean gloves during all steps of the procedure, i.e., when mounting and handling polyacrylamide gels, when excising stained proteins out of the gel, and when handling the test tubes containing these spots prior to digestion. Next to this, the staining procedures are preferentially kept as short as possible and in the smallest

volume possible, as it has been shown that a long exposure of gels to air increases the risk for keratin contamination (Sinha *et al.*, 2001). In our laboratory, we minimize this problem by working in a laminar air flow hood during the excision of protein spots and during preparation of the protein digests. Furthermore, the gels and the test tubes are constantly in a flow of neutralizing air by which electrostatic charging of plastic surface, which attracts dust particles (and thus keratins), is reduced.

2. Because it is very difficult to estimate the amount of protein present in a gel, in some cases too much protease will be added. Although many of the proteases that are available commercially are so-called "proteomics grade", during typical incubation times, they give rise to autodigestion products. These peptides can, in some cases, be used as internal standards, by which the peptide mass will be determined more accurately (e.g., Li *et al.*, 1997) or, when overrepresented, suppress ionization of peptides from the analyzed proteins, thus hampering protein identification.

Acknowledgment

K.G. is a Postdoctoral Fellow of the Fund for Scientific Research—Flanders (Belgium) (F.W.O.-Vlaanderen).

References

Aebersold, R., and Mann, M. (2003). Mass spectrometry-based proteomics. *Nature* **422**, 198–207.

Bennett, J., and Scott, K. J. (1971). Quantitative staining of fraction I protein in polyacrylamide gels using Coomassie brillant blue. *Anal. Biochem.* **43**, 173–182.

Cottrell, J. S. (1994). Protein identification by peptide mass fingerprinting. *Pept. Res.* **7**, 115–124.

Gevaert, K., Demol, H., Puype, M., Broekaert, D., De Boeck, S., Houthaeve, T., and Vandekerckhove, J. (1997). Peptides adsorbed on reverse-phase chromatographic beads as targets for femtomole sequencing by post-source decay matrix assisted laser desorption ionization-reflectron time of flight mass spectrometry (MALDI-RETOF-MS). *Electrophoresis* **18**, 2950–2960.

Gevaert, K., Van Damme, J., Goethals, M., Thomas, G. R., Hoorelbeke, B., Demol, H., Martens, L., Puype, M., Staes, A., and Vandekerckhove, J. (2002). Chromatographic isolation of methionine-containing peptides for gel-free proteome analysis: Identification of more than 800 *Escherichia coli* proteins. *Mol. Cell. Proteomics* **1**, 896–903.

Li, G., Waltham, M., Anderson, N. L., Unsworth, E., Treston, A., and Weinstein, J. N. (1997). Rapid mass spectrometric identification of proteins from two-dimensional polyacrylamide gels after in gel proteolytic digestion. *Electrophoresis* **18**, 391–402.

MacCoss, M. J., McDonald, W. H., Saraf, A., Sadygov, R., Clark, J. M., Tasto, J. J., Gould, K. L., Wolters, D., Washburn, M., Weiss, A., Clark, J. I., and Yates, J. R., 3rd (2002). Shotgun identification of protein modifications from protein complexes and lens tissue. *Proc. Natl. Acad. Sci. USA* **99**, 7900–7905.

Merril, C. R., Switzer, R. C., and Van Keuren, M. L. (1979). Trace polypeptides in cellular extracts and human body fluids detected by two-dimensional electrophoresis and a highly sensitive silver stain. *Proc. Natl. Acad. Sci. USA* **76**, 4335–4339.

Parker, K. C., Garrels, J. I., Hines, W., Butler, E. M., McKee, A. H., Patterson, D., and Martin, S. (1998). Identification of yeast proteins from two-dimensional gels: Working out spot cross-contamination. *Electrophoresis* **19**, 1920–1932.

Rosenfeld, J., Capdevielle, J., Guillemot, J. C., and Ferrara, P. (1992). In-gel digestion of proteins for internal sequence analysis after one- or two-dimensional gel electrophoresis. *Anal. Biochem.* **203**, 173–179.

Sinha, P., Poland, J., Schnölzer, M., and Rabilloud, T. (2001). A new silver staining apparatus and procedure for matrix-assisted laser desorption/ionization-time of flight analysis of proteins after two-dimensional electrophoresis. *Proteomics* **1**, 835–840.

Steen, H., and Mann, M. (2001). Similarity between condensed phase and gas phase chemistry: Fragmentation of peptides containing oxidized cysteine residues and its implications for proteomics. *J. Am. Soc. Mass Spectrom.* **12**, 228–232.

Sumner, L. W., Wolf-Sumner, B., White, S. P., and Asirvatham, V. S. (2002). Silver stain removal using H_2O_2 for enhanced peptide mass mapping by matrix-assisted laser desorption/ionization time-of-flight mass spectrometry. *Rapid Commun. Mass Spectrom.* **16**, 160–168.

van Montfort, B. A., Canas, B., Duurkens, R., Godovac-Zimmermann, J., and Robillard, G. T. (2002). Improved in-gel approaches to generate peptide maps of integral membrane proteins with matrix-assisted laser desorption/ionization time-of-flight mass spectrometry. *J. Mass Spectrom.* **37**, 322–330.

Wilson, C. M. (1983). Staining of proteins on gels: Comparisons of dyes and procedures. *Methods Enzymol.* **91**, 236–247.

27

Peptide Sequencing by Tandem Mass Spectrometry

John R. Yates, III, David Schieltz, Antonius Koller, and John Venable

I. INTRODUCTION

Microcolumn reversed-phase HPLC electrospray ionization tandem mass spectrometry (ESI-MS/MS) is a rapid and sensitive technique for the analysis of complex mixtures of peptides (Hunt *et al.*, 1986; Griffin *et al.*, 1991). This technique can be used to determine the amino acid sequence of unknown peptides, to verify the structure of proteins, and to determine posttranslational modifications. In particular, the strength of this approach is the analysis of peptides in complicated mixtures. Two approaches for sequence analysis of peptides are described: low flow rate infusion (microelectrospray ionization) and reversed-phase microcapillary liquid chromatography tandem mass spectrometry (Gale and Smith, 1993; Emmett and Caprioli, 1994; Andren *et al.*, 1994; Wilm and Mann, 1994; Davis *et al.*, 1995). Algorithms have been developed to help interpret tandem mass spectra to derive a sequence *de novo* and an example of their use is illustrated.

II. MATERIALS AND INSTRUMENTATION

Solvents are from Fisher Scientific (Springfield, NJ): J. T. Baker, 2-propanol 9095-02, acetonitrile OPTIMA A996-4, J. T. Baker, acetic acid 6903-05. Fused silica capillaries are from Polymicro Technologies (Tucson, AZ): 375 μm o.d. × 20 μm i.d., TSP375020, 365 μm o.d. × 100 μm i.d., TSP200100. Reversed-phase chromatographic supports are from Zorbax Eclipse XDB. The column packing device is homemade and is shown in Fig. 27.1. The microelectrospray platform was built in the The Scripps Research Institutes instrument shop (Fig. 27.2). The XYZ manipulator is from Newport Corp. (Irvine, CA). MT-XYZ and the high voltage connector suitable for the ThermoFinnigan electrospray voltage are from Lemo USA (www.lemo.com) (San Jose, CA; P/N 4-89626). A P-2000 laser puller is from Sutter Instruments (Novato, CA).

The electrospray ionization tandem mass spectrometers, LCQ and Q-TOF mass spectrometers, are from ThermoFinnigan (San Jose, CA) and Waters (Milford, MA), respectively. Pumps for HPLC and HP1100 are from Agilent (Palo Alto, CA). DeNovoX is provided with the BioWorks software from ThermoFinnigan.

III. PROCEDURES

A. Preparation of Microcolumns for Reversed-Phase HPLC

Solutions

1. *HPLC solvent A*: To make 1 liter, add 5 ml of formic acid to 995 ml of distilled and deionized water.
2. *HPLC solvent B*: To make 1 liter, add 5 ml of formic acid to a solution of 200 ml of distilled and deionized water and 795 ml of acetonitrile.

Steps

1. Construct microcapillary columns according to Gatlin *et al*. Rinsea 50-cm piece of fused silica capillary (365 μm o.d. × 100 μm i.d.) with 2-propanol and dry. Remove the polyimide coating with a low temperature flame in the middle of the capillary. Position the capillary on a laser puller and pull to create two columns each with tips containing 5-μm openings. The settings on the laser puller for 365 × 100-μm i.d. capillary are heat 270, filament 0, velocity 30, delay 128, and pull 0. Repeat this program three times before the capillary is pulled. This program produces a 5-μm tip, although the exact program may depend on mirror alignment and capillary positioning. *Caution*: Conditions to pull a capillary may vary from puller to puller.
2. The packing device is depicted in Fig. 27.1A. Insert the blunt end of the column into the swagelock fitting in the top of the packing device. To pack the column, fill a polypropylene Eppendorf centrifuge tube (1.5 ml) with ~100 μg of packing material and 1 ml of methanol and sonicate briefly to suspend the material and minimize aggregation. Place the solution in a high-pressure packing device (Fig. 27.1B), insert the column, and place the end of the column in the solution (Fig. 27.1C). Use helium gas at a pressure of ~500 psi to drive the packing material into the column. Continue packing until the material fills a length of the capillary corresponding to 10–20 cm. Allow the pressure to slowly drop to zero. Condition the column by rinsing with 100% solvent B and slowly reducing the percentage of solvent B until it reaches initial HPLC conditions (100% solvent A). Use a linear gradient of 100% solvent A to 20% solvent A over 30 min to finish conditioning the column.
3. Configuration for microcolumn HPLC is shown in Fig. 27.2 and 27.3.

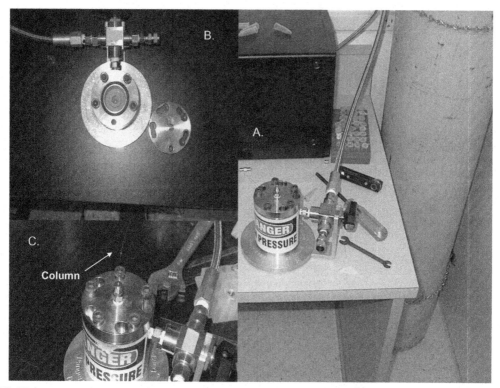

FIGURE 27.1 Pneumatic device for packing capillary columns and loading samples. A: Device is shown connected to gas line. B: Device is shown with the top removed showing the inside. A slot can be seen where an Eppendorf tube can be inserted for either column packing or sample loading. C: Pneumatic device with a fused silica capillary column.

Gradient: linear, 30–60 min, 0 to 100% (80:20 acetonitrile: 0.5% acetic acid)

Flow: 100 μl/min, flow split to final flow rate 200–300 nl/min

Columns: 100 μm i.d. × 20–25 cm, Phenomonex AQUA, 5-μm particles

B. Low Flow Rate Infusion Tandem Mass Spectrometry of Peptides

Media and Solutions

1. *1:1 0.5% acetic acid/ methanol*: To make 100 ml, add 2.5 ml of glacial acetic acid to 47.5 ml of distilled and deionized water and add 50 ml of HPLC grade methanol. Store in closed container at room temperature.

Steps

1. To accomplish low flow rate infusion, create a liquid metal junction. Valco fitting (P/N MU1XCTI), titanium is preferred because it is a biologically inert material. The fused silica capillary is 375 μm outer diameter (o.d) ×20 μm inner diameter (i.d).

2. To infuse a sample, connect an entrance line of fused silica capillary to a syringe pump or to a device to pneumatically drive the liquid at a flow rate of 10–200 nl/min. Strip a fused silica microelectrospray needle (4 cm of 375 μm o.d. × 20 μm i.d.) of its polyimide coating near the exit and then connect to the Valco fitting. Place a voltage (900–2000 V) on the fitting to transfer the potential to the liquid and form the electrospray. Place the

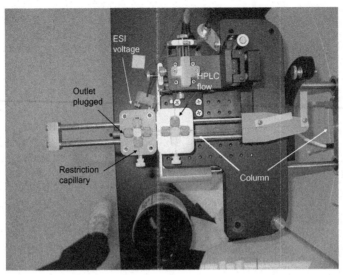

FIGURE 27.2   Microelectrospray stage for capillary liquid chromatography. HPLC solvent is initially split using a peak microTee to another microTee. This microTee contains a gold electrode for application of the ESI voltage and a restriction capillary to create back pressure and direct flow to waste.

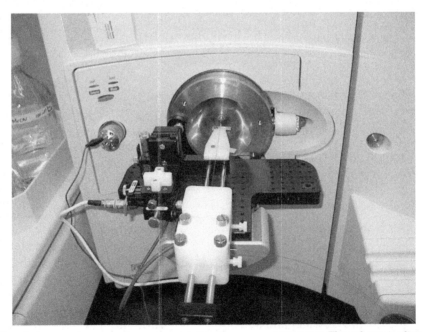

FIGURE 27.3   Microelectrospray source is shown connected to the mass spectrometer. The column is aligned with the opening in the heated capillary using the XYZ manipulator.

V. MASS SPEC METHODS FOR DETERMINATION OF PROTEIN IDENTITY AND PROTEIN MODIFICATIONS

exit end of the electrospray needle close (0.1–0.5 mm) to the entrance of the heated capillary.

3. The advantage of low flow rate infusion is that a small volume of sample can be infused over a long time period (5–30 min). This allows time to record a mass spectrum and then to begin acquiring tandem mass spectra of ions observed in the mass spectrum. Low flow rate infusion works quite well for reasonably uncomplicated peptide mixtures. Complicated peptide mixtures, such as those derived from proteolytic digestion of large proteins or mixtures of proteins, can suffer from ion suppression. The tandem mass spectrum shown in Fig. 27.4 was acquired through infusion and subjected to *de novo* sequence analysis as discussed later.

C. Analysis of Peptides by Microcolumn Liquid Chromatography Microelectrospray Ion Trap Mass Spectrometry (LCQ)

Steps

1. Insert the packed 10- to 20-cm column into an Upchurch PEEK tee. Insert the blunt end of the column into the tee and direct

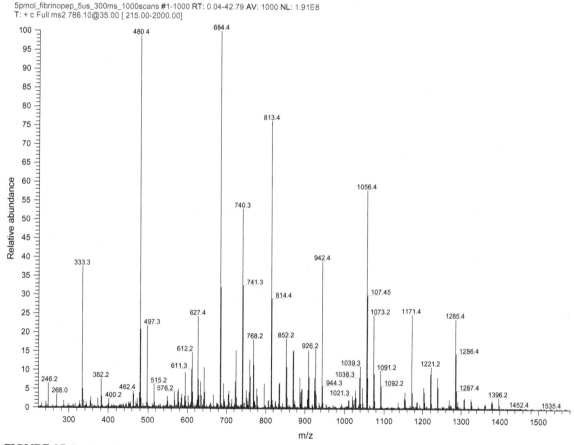

FIGURE 27.4   A tandem mass spectrum of the doubly charged ion, (M+2H) + 2 786.10, of Glu-Fibrinopeptide.

the solvent from the HPLC into one of the other arms of the tee. Insert a length of fused silica into the remaining arm of the tee and connect to a second Upchurch PEEK tee. Insert into one of the arms a gold wire so that a length of the wire protrudes approximately 1–2 cm (Fig. 27.2). Attach the electrospray voltage to the gold wire. In the remaining arm of the tee, insert a length of fused silica 365 × 5 μm. The fused silica tubing is used to create a restriction to the HPLC flow to force 100–300 nl/min through the microcolumn and this solvent is directed to waste. The microelectrospray source connected to an ion trap mass spectrometer is depicted in Fig. 27.2.
Electrospray voltage: −1800–2000 V

2. An aliquot of sample is injected to record the molecular weight of the peptides.
Scan mass range: 400–1500 amu
Electron multiplier voltage: 1000 V

3. Sequence analysis of peptides is performed during a second HPLC analysis by selecting the precursor ion with a 3u Full Width at Half Height (FWHH) wide notch and exciting the ions within the ion trap to induce fragmentation. Collision energies are on the order of 35 eV. The fragment ions are then scanned out of the ion trap to an electron multiplier set at 1000 V.

4. Alternatively, the LCQ mass spectrometer can automatically acquire tandem mass spectra over the course of an LC analysis. The analysis incorporates a unit mass resolution scan to find the most abundant ion above a preset ion abundance threshold. This ion is isolated and a high mass resolution scan is acquired. By resolving the isotopic peaks of the isolated ion, the charge state can be determined. For example, if the m/z distance between the isotope peaks is 1, then the charge state is +1. If the m/z distance is 0.5, then the charge state is +2. The LCQ can resolve a +4 charge state up to an m/z value of 1500. The tandem mass

spectrum for Glu-fibrinopeptide is shown in Fig. 27.4.

IV. COMMENTS

Under low-energy, multiple collision conditions, peptides fragment primarily at the amide bonds, producing sequence-specific fragmentation patterns. When fragment ions are produced the charge can be retained on the N terminus of the fragment of the ion to form ions of type b. If the charge is retained on the C-terminal fragment, the ion is of type y. There is no easy method to distinguish a b-ion from a y-ion without derivatization of the peptide. The mass spectrum shown in Fig. 27.6 was obtained by a ThermoFinnigan LCQ Deca ion trap mass spectrometer. A hallmark of this instrument is the ability to automatically acquire tandem mass spectra. After a mass spectrum is obtained, the data system identifies the most abundant ion and then acquires a tandem mass spectra in the next scan.

To demonstrate the sequencing process, the MS/MS spectrum for the peptide shown in Fig. 27.4 is interpreted. The process is demonstrated on a tandem mass spectrum acquired on an LCQ ion trap mass spectrometer. The process is identical to the one followed for triple quadrupole mass spectrometers as described in Yates et al. (1994) and McCormack et al. (1994). The mass spectrum is produced by collisional activation of a doubly charged ion, $(M + 2H)^{2+}$, at m/z 786.10. The protonated molecular weight of this peptide is calculated by multiplying the m/z value by 2 and subtracting the weight of one proton to give a value of 1571.2 (average mass). Subtraction of the highest mass fragment ion, m/z 1396.2 (1571.2 − 1396.2 = 175.0), from the (M + H) + value yields a mass of 175. The mass fits C-terminal Arg. If this peptide was created through trypsin digestion, then this is a good assignment. Subtraction of the ion at

1396.2 from the next highest ion, 1285.4, yields a mass difference of 110.8, which does not correspond to a mass for one of the common amino acids. This situation strongly suggests the ion at 1285.4 is of a different type. Assuming the ion at 1285.4 is of type y and that y-ions tend to predominate in tandem mass spectra of doubly charged peptide ions created through trypsin digestion, then it would be worthwhile to start from this point in the spectrum. A window can be set for where the next ion should reside in the spectrum using a range between a mass of 57 (Gly) and 186 (Trp). This window assumes no modifications to amino acids that would be larger then Trp. The next ion that fits the aforementioned criteria is 1171.4, which corresponds to a mass of 114 or Asn. The difference between the next ion at 1056.4 and 1171.4 is 115 for Asp and then the difference between this ion and 942.4 is 114. Thus far a sequence of Asn–Asp–Asn has been determined. The next abundant ion in the spectrum is 813.4, which nets a difference of 129 or Glu. An abundant ion is observed at 740.3, which yields a difference of 73. This value does not correspond to the weight of a common amino acid and thus can be temporarily discounted. Proceeding to the next ion to determine if a mass difference exists with the ion at 813.4 that provides a better fit, the ion at 684.4 is tried. This ion produces a mass difference of 129 and we now have a sequence of Asn–Asp–Asn–Glu–Glu. The next set of ions are less abundant and the largest of this group 627.4 gives a mass difference of 57 corresponding to Gly. If the other ions are tried, none produce as close a fit with an amino acid residue mass. The next two ions are rather obvious at 480.4 and 333.3. These ions yield a sequence of Phe–Phe. Two ions remain in the low end of the mass spectrum but only 246.2 produces a good fit and the amino acid Ser is obtained. The sequence now corresponds to Asn–Asp–Asn–Glu–Glu–Gly–Phe–Phe–Ser. By using the presumption the ion series used to interpret the sequence was of type y, this sequence corresponds to the N-terminal sequence of the peptide. We know the C-terminal residue of the peptide is Arg so a calculation of the difference between 175 and 246.2 produces a mass of 71 and thus Ala. Completion of the sequence will require closer scrutiny of ions at the ends of the spectrum. At the high mass end of the spectrum between ions at 1285.4 and 1396.2 there are several ions. The ion at 1384.6 is 99 Da from the ion at 1285.4 and could be a Val residue. The corresponding ion at 286 is present at the low m/z end of the spectrum, confirming this residue as the likely sequence. The difference between 1384.6 and 1571.2 is 186.6. This corresponds to a Trp residue or a combination of two or more amino acids. Several combinations of amino acids will fit this mass: Glu–Gly, Asp–Ala, or Ser–Val. There is insufficient information in the spectrum to differentiate these possible sequences. The Ser–Val sequence can be eliminated by converting the peptide to its methyl ester as neither Ser nor Val should convert to a methyl ester with a *resulting* mass increase of 14Da. The final sequence for the peptide derived from this spectrum is [EG,DA,SV]NDNEEGFFSAR.

De Novo Sequencing Algorithm

Results from two different *de novo* sequencing algorithms are shown in Fig. 27.5 and 27.6. The DeNovoX program automatically determines a complete or partial amino acid sequence of an unknown peptide by interpreting its MS/MS spectra. It is capable of identifying peptides with posttranslational modifications, mutations, or single amino acid deletion in the peptide. The probability-based program, which is optimized for data collected with the ThermoFinnigan LCQ ion trap instrument, takes data files that contain the m/z values and intensities of ions in MS/MS spectra together with the information on the respective parent ion mass and charge state as input. The data files are processed by the DeNovoX program in a fully automated mode. The program takes into account over 25 different factors relevant to *de novo* peptide sequencing. Amino acid tables that specify the amino acids considered for the search can be selected by the user and can include all major posttranslational modifications. All other parameters are set by default in the program, including the mass tolerance of 0.8Da for fragment ions. A list of suggested sequence tags and completed sequences with respective probability values are included in the output of the program, in addition to assigned spectra. The sequence interpretation with the highest probability is WVNDNEEGFFSAR and is shown in Fig. 27.5. The correct sequence is EGNDNEEGFFSAR. Clearly the first two amino acid residues are incorrect for reasons similar to those encountered in the manual interpretation. Ion trap tandem mass spectra are often lacking the necessary sequences to definitively identify the N-terminal amino acid residues. These sequences can then be used for further database searches (e.g., BLAST search).

Peptide tandem mass spectrometry data obtained by Q-TOF mass spectrometers can be sequenced using the MassSeq program from MicroMass. *De novo* sequencing is performed by using the MaxEnt 3/MassSeq programs in the Waters-Micromass Masslynx 3.5 software package. The raw tandem mass spectrum is first processed with the Massive Inference algorithm (MaxEnt 3) (Jaynes, 1957; Skilling,

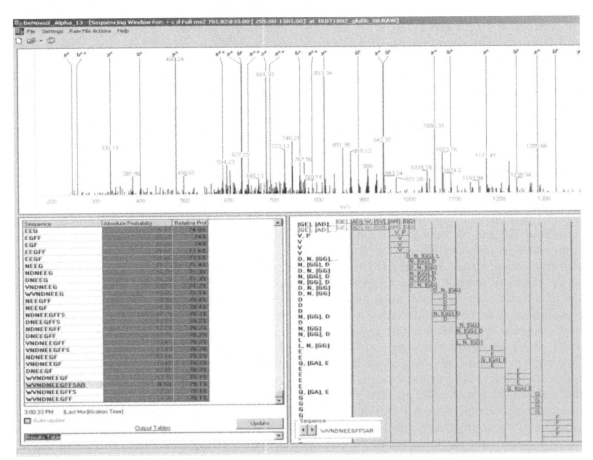

FIGURE 27.5 Interpretation output from the DeNovoX program on the tandem mass spectrum shown in Figure 27.4.

1989), which produces a spectrum where any higher charge state ion is collapsed into a single charge state and the isotopic distribution of each peak is folded into the lowest mass member (^{12}C isotope). Interpretation of the processed spectrum is performed by the MassSeq program, which generated candidate sequences based on the molecular mass of the peptide. Because the total number of possible sequences is very large, a terminated Markov Chain Monte Carlo algorithm is employed to generate a smaller, yet representative set of sequences. Predicted spectra are generated from each of the candidate sequences using probabilistic fragmentation and then each are compared to the MaxEnt 3 processed spectrum. The

final result is a list of sequences, which are scored based on their probability of a correct match to the MaxEnt 3 spectrum. Within each resulting peptide sequence, each amino acid is assigned a probability as to how likely the residue is correct assigned to that position in the sequence. Therefore, this method provides the ability to sequencing peptides *de novo* and have information on the degree of likelihood that the sequence is correct. Q-TOF spectra often contain more information about the sequence and the complete sequence can often be inferred. Interpretation of a tandem mass spectrum of Glu-fibrinopeptide is shown in Fig. 27.6. Both algorithms produce a nearly complete sequence.

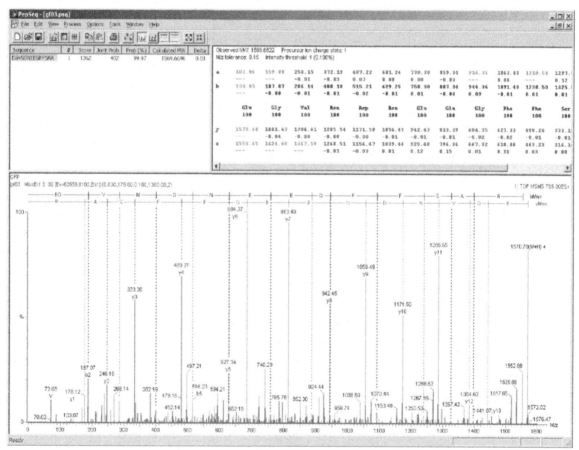

FIGURE 27.6 Interpretation output of the MassSeq program using a tandem mass spectrum acquired on a Q-TOF mass spectrometer (data not shown).

V. MASS SPEC METHODS FOR DETERMINATION OF PROTEIN IDENTITY AND PROTEIN MODIFICATIONS

V. PITFALLS

1. Results from sequencing should be checked by derivatizing the peptide to ensure that the appropriate mass shift is observed. One derivatization method consists of forming the peptide methyl ester by adding 2N methanolic HCl to the dry peptide. A second tandem mass spectrum can be acquired for the derivatized peptide to determine if the fragment ions also shift by the correct mass increment.

2. Distilled and deionized water must be used in the HPLC solvents to avoid the production of sodium or potassium adducts.

References

Andren, P. E., Emmett, M. R., and Caprioli, R. M. (1994). Micro-electrospray—zeptomole-attomole per microliter sensitivity for peptides. *J. Am. Soc. Mass Spectrom.* **5**, 867–869.

Davis, M. T., Stahl, D. C., Hefta, S. A., and Lee T. D. (1995). A microscale electrospray interface for online, capillary liquid-chromatography tandem mass-spectrometry of complex peptide mixtures. *Anal. Chem.* **67**, 4549–4556.

Emmett, M. R., and Caprioli, R. M. (1994). Micro-electrospray mass spectrometry: Ultra-high-sensivity analysis of peptides and proteins. *J. Am. Soc. Mass Spectrom.* **5**, 605–613.

Gale, D. C., and Smith, R. D. (1993). Small volume and low flow rate electrospray ionization mass spectrometry of aqueous samples. *Rapid Commun. Mass Spectrom.* **7**, 1017–1021.

Gatlin, C. L., Kleemann, G. R., Hays, L. G., Link, A. J. & Yates, J. R. (1998). Protein identification at the low femtomole level from silver-stained gels using a new fritless electrospray interface for liquid chromatography-microspray and nanospray masss spectrometry. *Anal Biochem* **263**, 93–101.

Griffin, P. R., Coffman, J. A., Hood, L. E., and Yates, J. R. III (1991). Structural analysis of proteins by HPLC-MS and HPLC-MS/MS using electrospray ionization on a triple quadrupole mass spectrometer. *Int. J. Mass Spectrom. Ion Proc.* **111**, 131–149.

Hunt, D. F., Yates, J. R. III, Shabanowitz, J., Winston, S., and Hauer, C. R. (1986). Protein sequencing by tandem mass spectrometry. *Proc. Natl. Acad. Sci. USA* **84**, 620–623.

Jaynes, E. T. (1957). Information theory and statistical mechanics, I. *Phys. Rev.* **106**, 620–630.

Kennedy, R. T., and Jorgenson, J. W. (1989). Preparation and evaluation of packed capillary liquid chromatography columns with inner diameters from 20 to 50 μm. *Anal. Chem.* **56**, 1128–1135.

McCormack, A. L., Eng, J. K., and Yates, J. R. III (1994). Peptide sequence analysis on quadrupole mass spectrometers. *Methods* **6**, 274–283.

Skilling, J. (1989). *In* "Maximum Entropy and Bayesian Methods" (J. Skilling, ed.), p. 45. Kluwer Academic, Dordrecht.

Wilm, M. S., and Mann, M. (1994). Electrospray and taylor-cone theory, doles beam of macromolecules at last. *In. J. Mass Spectrom. Ion Proc.* **136**, 167–180.

Yates, J. R. III, McCormack, A. L., Hayden, J., and Davey, M. (1994). Sequencing peptides derived from the class II major histocompatibility complex by tandem mass spectrometry. *In* "Cell Biology: A Laboratory Handbook" (J. E. Celis, ed.), pp. 380–388. Academic Press, San Diego.

Direct Database Searching Using Tandem Mass Spectra of Peptides

John R. Yates III and William Hayes McDonald

I. INTRODUCTION

Whole genome sequencing has provided a sequence infrastructure to the enormous benefit of mass spectrometry and protein biochemistry. The ability to match tandem mass spectra to sequences in protein or nucleotide databases allows accurate and high throughput identification of peptides and proteins (Eng *et al.*, 1994; Yates *et al.*, 1995a,b, 1996). Furthermore, this data analysis capability, when combined with automated acquisition of tandem mass spectra, allows direct analysis and identification of proteins in mixtures.

Applications of this procedure to rapidly survey the identities of proteins in protein complexes, subcellular compartments, cells, or tissues are made possible because of the combination of tandem mass spectrometry and database searching (McCormack *et al.*, 1995; Link *et al.*, 1997, 1999; Washburn *et al.*, 2001; Florens *et al.*, 2002). The method allows both automated and accurate searches of protein and nucleotide databases, as well as the use of spectra of posttranslationally modified peptides. Modifications such as phosphorylation, methylation, and acetylation can be identified through the analysis of digested protein complexes (Cheeseman *et al.*, 2002; MacCoss *et al.*, 2002). A detailed description of the manner in which tandem mass spectra are processed and then used to search databases is described.

II. MATERIALS AND INSTRUMENTATION

Tandem mass spectra of peptides were obtained by electrospray ionization tandem mass spectrometry (Finnigan MAT, San Jose, CA) as described previously (Gatlin *et al.*, 1998; Verma *et al.*, 2000; Washburn *et al.*, 2001).

Several computer algorithms now exist to search tandem mass spectra through sequence databases. Information about these programs can be accessed through the following URLs: http://www.matrixscience.com, http://prospector.ucsf.edu, http://fields.scripps.edu/sequest, and http://www.proteometrics.com. This article discusses the use of SEQUEST, as we have the most experience with this program. SEQUEST searches were performed on a PC Linux computing cluster described elsewhere using a parallelized version of the SEQUEST algorithm (Sadygov, 2002). Methods for data reduction, preliminary scoring, and cross-correlation analysis used in the computer algorithm have been described in detail elsewhere (Eng *et al.*, 1994; Yates *et al.*, 1995a,b). Search results are analyzed with the program DTAselect (Tabb *et al.*, 2002). Information on how to obtain the DTAselect program can be found at http://www.sequest.org. This program can accommodate output from the SEQUEST and Mascot search programs. Databases can be obtained over the Internet using anonymous ftp to the National Center for Biotechnology Information (NCBI) (ftp://ncbi.nlm.nih.gov). Databases such as the GenBank database of nucleotide sequences, the NRP (nonredundant protein) database, and dbEST—the collection of expressed sequence tag sequences—can be obtained from the NCBI site. The complete *Saccharomyces cerevisiae*

sequence can be obtained from Stanford Genomic Resources (http://genome-www.stanford.edu).

III. PROCEDURES

A. Method for Database Search with Tandem Mass Spectra

Steps

1. Convert raw, binary data from an MS/MS file to ASCII in the following format:

   1734.9   2
   110.3   49422.8
   112.3   32433.3
   112.9   65452.1
   :   :

   The first line in the file contains (M + H)+ and charge state information. Conversion of binary data is accomplished automatically by SEQUEST with data files from ThermoFinnigan mass spectrometers. Most mass spectrometers can convert proprietary data file formats to a form that can be used by the database search programs or the search programs can read the formats directly. The next task calculates the molecular weight of the peptide based on its precursor *m/z* value. This calculation is accomplished with the program *2 to 3*, which identifies the charge state of the precursor ion by identifying related fragment ion pairs (Sadygov *et al.*, 2002). Fragment ion pairs sum to the molecular weight of the peptide and by initially assuming that the precursor ion is doubly charged and checking that related fragment ion pairs exist in the spectrum will verify this assumption. A molecular weight based on a +3 charge state is then calculated and the existence of fragment ion pairs. The correct molecular weight will produce a much larger number of related ion pairs. This program is designed to use nominal resolution spectra where the charge state based on isotope spread cannot be calculated. If the spectrum does not contain a minimum amount of fragmentation information to identify the charge state, then that spectrum is not used for database searching.

2. For SEQUEST searches, search parameters can be set in the param.h header file. This includes the database to be searched, whether it is a nucleotide or protein database, and whether to perform calculations using average or monoisotopic masses. The set of sequence ions (types b and y) to be considered can also be selected as well as the relative abundance values to be used during the theoretical reconstruction of tandem mass spectra. The masses of amino acids to consider for modified peptides can be input. The mass tolerance for the peptide $(M + H)+$ can be set as well. Enzyme or chemical cleavage specificity can be entered from the following list: no enzyme, trypsin, chymotrypsin, clostripain, cyanogen bromide, iodoso benzoate, proline endopeptidase, *Staphylococcus aureus* V8 protease, lysine endoproteinase, arginine endoproteinase, AspN-endoproteinase, and elastase.

3. The tandem mass spectrometry data file is then processed in two ways. First the precursor ion is removed and the remaining ions are normalized. This process is not required when analyzing ion trap mass spectra. All but the top 200 most abundant ions are then removed from the search file. This processed spectrum is used to search the database. To perform the cross-correlation analysis described later, a second file from the tandem mass spectrum is created. The spectrum is divided into 10 equal regions and within each region the ions are normalized to the most abundant ion. The molecular weight of the peptide in the tandem mass spectrum is calculated directly from the precursor ion.

4. A search of the database involves scanning each entry to find linear combinations of amino acids, proceeding from the N to the C terminus, that are within some tolerance of the mass of the peptide represented by the tandem mass spectrum. Sequence selection can also be guided by the cleavage specificity of the protease used to create the peptide, including consideration of incompletely digested peptides from either side of the primary sites, or it can be performed with no assumptions about how the peptide was created. If a nucleotide database is searched, the nucleotide sequences are translated "on the fly" to protein sequences in 6-reading frames (Yates *et al.*, 1995b). Chemical modifications can be considered by changing the amino acid mass used to calculate the masses of the peptides. The modified amino acid is then considered at every occurrence in the sequence (Yates *et al.*, 1995a,b).

5. Once an amino acid sequence is within the defined mass tolerance, a preliminary evaluation is performed (Eng *et al.*, 1994). First, the number (n_i) of predicted fragment ions that match ions observed in the spectrum within the fragment ion mass tolerance and their abundances (i_m) are summed. If an ion series is continuous, i.e., if consecutive sequence ions are present, then a component of the score, β, is incremented. A sequence that matches a continuous set of ions is weighted more heavily than one that matches a few sequence ions randomly. If an immonium ion is present in the spectrum (not usually the case with Ion Trap MS/MS data), then the associated amino acid must be present in the sequence under consideration or an additional component of the score, ρ, is increased or decreased correspondingly. The total number of predicted sequence ions is also noted (n_t). A score is calculated for each amino acid sequence by using the following relationship in Eq. (1),

$$Sp = (\Sigma im) * ni * (1 + \beta) * (1 + \rho)/n_x. \quad (1)$$

6. Each of the top 500 scoring sequences are subjected to a cross-correlation analysis. This is performed by reconstructing a model tandem mass spectrum for each of the amino acid sequences in the list of 500 and comparing each one to the processed experimental tandem mass spectrum (step 2) by using a cross-correlation function. The cross-correlation function is a very sensitive signal processing method used to compare the coherence of two signals (Owens, 1992). This is performed, in effect, by translating one signal across another. If two signals are the same or very similar, the correlation function should maximize when there is no offset between the signals. A cross correlation score is computed for each of the 500 amino acid sequences.

7. Cross-correlation values are normalized by dividing the XCorr value by the autocorrelation of the experimental tandem mass spectrum. The formula used to calculate a normalized cross-correlation value is shown by Eq. (2). By normalizing correlation scores for tandem mass spectra of +1, +2, and +3 ions, the scores are roughly equivalent and a statistical confidence for each sequence assignment can be assessed (MacCoss *et al.*, 2002).

$$XCorr = \frac{Corr(Exp, Theo)}{\sqrt{Corr(Exp, Exp) * Corr(Theo, Theo)}}$$
$$(2)$$

B. Data Assembly and Filtering

Steps

1. For very large analyses, such as multidimensional liquid chromatography experiments (MudPIT), extremely large data sets can be produced. A program,

DTASelect, is used to assemble, filter, and display the results from search (Tabb *et al.*, 2002). The program is especially suited for the analysis of data produced in multidimensional liquid chromatography separations together. DTASelect can be used with output from SEQUEST or Mascot searches. Figure 28.1 shows output from DTASelect from a multidimensional analysis of the *S. cerevisiae* proteasome. In the first column, the probability the match is correct is shown. The closer the value is to 1 the higher the probability the match is correct. This entry in the table is linked to the tandem mass spectrum and a representation of the quality of the sequences is fit to the spectrum. How well other sequences fit to the spectrum can be viewed from this page as well. Each of the predicted b- and y-ions are aligned with matching fragment ions in the spectrum. Mass differences between predicted fragment ions and fragments ions observed in the spectrum are shown in the upper right hand side of the page. At the bottom of the figures are check boxes to indicate the validation status of the fit of the spectrum to the sequence (Fig. 28.2). Access to individual search data allows the inspection of search results to look for sequence similarity between sequences.

2. In the second and third columns marked XCorr and DeltaCn, respectively, are scores for the tandem mass spectrum's match to the sequence. The values under XCorr are obtained from the cross-correlation analysis and are normalized as described earlier. The larger the value, the closer the fit between the experimental tandem mass spectrum and the model tandem mass spectrum constructed from the sequence. The DeltaCn value is the difference between the normalized XCorr values for the first and second search results. The larger DeltaCn is the more dissimilar the first and second answers will be.

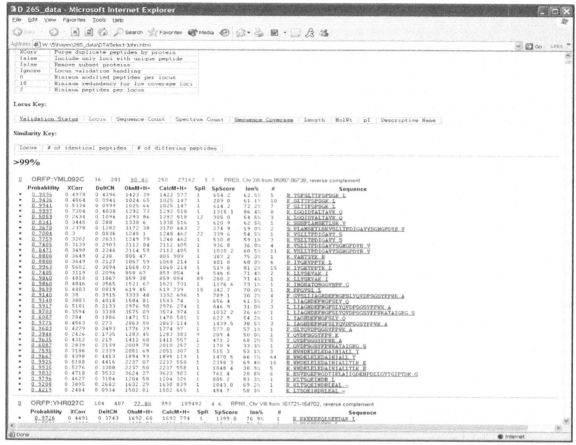

FIGURE 28.1   DTASelect output for analysis of the *S. cerevisiae* proteasome.

3. The fourth and fifth columns show the (M + H)+ value calculated from the precursor ion chosen for MS/MS and the (M + H)+ value calculated from the sequence, respectively.

4. Columns SpR, SpScore, and Ion% show the rank of the peptide match based on the preliminary score, the preliminary score and the percentage of ions matched between the experimental spectrum, and sequence ions predicted from the sequence. Generally, short peptides produce an SpR that is less than one. The correct result usually matches a large percentage of ions, Ion%, between those predicted and those observed. Larger peptides can result in a low percentage of

ion matches because of the limited mass range of the ion trap mass spectrometer.

5. In the last column the matched sequence is shown. The actual matching sequence is preceded and terminated by a period. Amino acid residues before the preceding period and after the terminating period are shown to illustrate the enzymatic cleavage sites.

6. Peptide matches are grouped by gene locus, and information pertaining to the protein is shown in the header. From left to right the information shown in the header describes the validation status of peptides matched to that protein, the locus name for the gene, the number of peptides identified, the spectrum

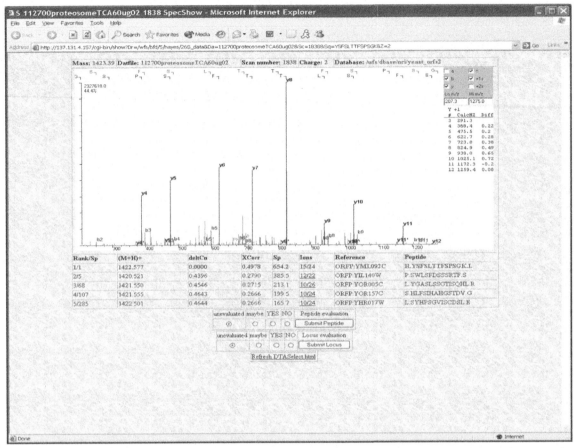

FIGURE 28.2 Following the link from the probability value displayed in the first column (Fig. 28.1), information for individual searches of tandem mass spectra can be found. This is useful to determine if a close score is the result of identifying two closely related sequences. The tandem mass spectrum is shown as well as the match to the sequence.

count indicating the total number of spectra matched (redundant spectra are included), percentage sequence coverage, which is linked to a visual display of coverage and overlap (Fig. 28.3), protein length, molecular weight, pI, and the descriptive name for the protein.

IV. COMMENTS

SEQUEST software can provide highly accurate protein identifications. The highly specific information represented in a tandem mass spectrum allows proteins present in mixtures to be identified, as each tandem mass spectrum is a specific address to a protein in the same manner that an amino acid sequence can be highly specific for a protein. Generally, several or more spectra are obtained for each protein, but a protein can be identified on the basis of one tandem mass spectrum if the following criteria are met: the amino acid sequence represented by the tandem mass spectrum is at least seven amino acids in length, the tandem mass spectrum contains a sufficient number of sequence ions

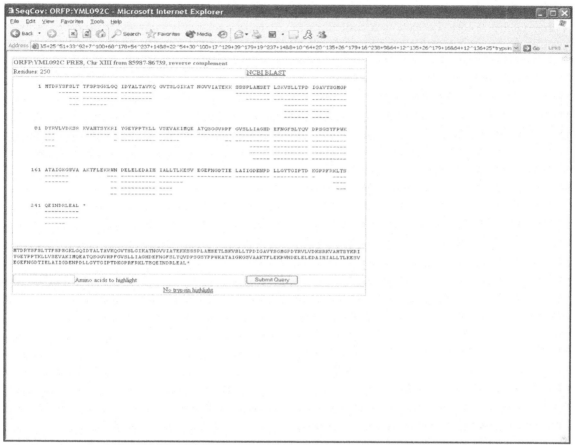

FIGURE 28.3 Following the link from the percentage in the locus header, sequence coverage of the protein can be observed. This figure shows regions of the protein where peptides were identified. This information is useful when a search of modification is desired. Regions lacking sequence coverage could represent sites of modification, areas that were difficult to digest, or areas where digestion yields too many short peptides. By using the input box at the bottom of the page, specific amino acid residues can be highlighted. Sites of K and R in the sequence are shown.

to allow validation of the identified sequence, and the amino acid sequence is unique to a single protein within the organism from which the protein was derived.

When evaluating a match between a spectrum and a sequence, several scores should be considered. The greater the probability value, the more statistically significant the match will be. These probability scores are based on empirical measures. The DTASelect program does retain search results with lower probabilities as

their significance may increase if many higher scoring spectra match to the same protein. The normalized cross-correlation value provides a mathematical evaluation of how close the experimental MS/MS fits to a model derived from the sequence. The limitations of this method relate to the ability to model a sequence as a spectrum and the predictability of peptide fragmentation. The DeltaCn value reflects how well the spectrum matches the model spectrum in comparison to background matches from the database. The XCorr value is an inherent property of the spectrum and changes as a function of spectral quality, whereas DeltaCn changes as a function of spectral quality and database size (e.g., background). Thus these two values provide a good quantitative measure of the match quality.

SEQUEST is ideal for creating a high-throughput, automated system for the identification of proteins from two-dimensional gels. By combining an autosampler with an HPLC protein, digests can be automatically injected and separated and tandem mass spectra automatically acquired. Tandem mass spectra can be acquired through an instrument control program on the ThermoFinnigan LCQ series mass spectrometer allowing unattended acquisition. Proteins present in mixtures can also be identified through the acquisition of tandem mass spectra and matching spectra to their respective proteins. Database sequence errors can be tolerated by the adjustment of parameters such as the ion series and mass tolerance to be used in a search. Database searching using tandem mass spectra enables "shotgun proteomics" or the analysis of digested protein mixtures. Unlike a 2 Dimensional Gel Electrophoresis analysis, it is only after a database search relationships among spectra that can be determined. For relatively simple protein mixtures, combining the "shotgun proteomics" method with multi-dimensional liquid chromatography can yield reasonably good protein sequence coverage.

V. PITFALLS

1. All tandem mass spectra should be validated against the sequence matched in the database.
2. SEQUEST will match a tandem mass spectrum to a similar, mass conserved sequence if the correct sequence is not present in the database.

3. SEQUEST can match tandem mass spectra with a relatively poor signal-to-noise ratio to the correct sequence, but the spectrum should be of sufficient quality for validation or the match should be considered tentative.

References

Cheeseman, I. M., Anderson, S., *et al.* (2002). Phospho-regulation of kinetochore-microtubule attachments by the *Aurora kinase* Ipl1p. *Cell* **111**(2), 163–172.

Eng, J. K., McCormack, A. L., *et al.* (1994). An approach to correlate tandem mass spectral data of peptides with amino acid sequences in a protein database. *J. Am. Soc. Mass Spectrom.* **5**, 976–989.

Florens, L., Washburn, M. P., *et al.* (2002). A proteomic view of the *Plasmodium falciparum* life cycle. *Nature* **419**(6906), 520–526.

Gatlin, C. L., Kleemann, G. R., *et al.* (1998). Protein identification at the low femtomole level from silver-stained gels using a new fritless electrospray interface for liquid chromatography-microspray and nanospray mass spectrometry. *Anal Biochem* **263**(1), 93–101.

Link, A. J., Carmack, E., *et al.* (1997). A strategy for the identification of proteins localized to subcellular spaces: Application to *E. coli* periplasmic proteins. *Int. J. Mass Spectrom. Ion Proc.* **160**, 303–316.

Link, A. J., Eng, J., *et al.* (1999). Direct analysis of protein complexes using mass spectrometry. *Nature Biotechnol.* **17**(7), 676–682.

MacCoss, M. J., McDonald, W. H., *et al.* (2002). Shotgun identification of protein modifications from protein complexes and lens tissue. *Proc. Natl. Acad. Sci. USA* **99**(12), 7900–7905.

MacCoss, M. J., Wu, C. C., *et al.* (2002). Probability-based validation of protein identifications using a modified SEQUEST algorithm. *Anal. Chem.* **74**(21), 5593–5599.

McCormack, A. L., Eng, J. K., *et al.* (1995). Microcolumn liquid chromatography-electrospray ionization tandem mass spectrometry. *Biochem. Biotech. Appli. Electrospray Ionization Mass Spectrom.* **619**, 207–225.

Owens, K. G. (1992). Application of correlation analysis techniques to mass spectral data. *Appl. Spectrosc. Rev.* **27**(1), 1–49.

Sadygov, R. G., Eng, J., Durr, E., Saraf, A., McDonald, H., MacCoss, M. J., and Yates, J. R. III (2002). Code developments to improve the efficiency of automated MS/MS spectra interpretation. *J. Proteome Res.* **2**, 211–215.

Tabb, D. L., McDonald, H. W., Yates, J. R., III (2002). DTASelect and Contrast: Tools for aseembling and comparing protein identifications from shotgun proteomics. *J. Proteome Res.* **1**(1), 21–36.

Verma, R., Chen, S., *et al.* (2000). Proteasomal proteomics: Identification of nucleotide-sensitive proteasome-interacting proteins by mass spectrometric analysis of affinity-purified proteasomes. *Mol. Biol. Cell* **11**(10), 3425–3439.

Washburn, M. P., Wolters, D., *et al.* (2001). Large-scale analysis of the yeast proteome by multidimensional protein identification technology. *Nature Biotechnol.* **19**(3), 242–247.

Yates, J. R., 3rd, Eng, J. K., *et al.* (1995a). Mining genomes: Correlating tandem mass spectra of modified and unmodified peptides to sequences in nucleotide databases. *Anal. Chem.* **67**(18), 3202–3210.

Yates, J. R., 3rd, Eng, J. K., *et al.* (1995b). Method to correlate tandem mass spectra of modified peptides to amino acid sequences in the protein database. *Anal. Chem.* **67**(8), 1426–1436.

Yates, J. R., 3rd, McCormack, A. L., *et al.* (1996). Mining genomes with MS. *Anal. Chem.* **68**(17), 534A–540A.

Identification of Proteins from Organisms with Unsequenced Genomes by Tandem Mass Spectrometry and Sequence-Similarity Database Searching Tools

Adam J. Liska and Andrej Shevchenko

I. INTRODUCTION

The analysis of proteomes by mass spectrometric methods that correlate peptide fragments from proteins with database entries *in silico* has been dependent on the sequencing of genomes. Mass spectrometry and database sequences have enabled the analysis of the human, mouse, and *Arabidopsis* proteomes, among others. Due to the high homology between living organisms at the molecular level, it is possible to use the available protein sequences accumulated in databases from a range of organisms as a reference for the identification of proteins from organisms with unsequenced genomes by sequence-similarity database searching. As research continues in organisms such as *Xenopus*, maize, cow, and others with limited database sequence resources, sequence-similarity searching is a powerful method for protein identification. This article focuses on MS BLAST (Shevchenko *et al.*, 2001) and MultiTag (Sunyaev *et al.*, 2003) as bioinformatic methods for the identification of proteins by the interpretation of tandem mass spectra of peptides and sequence-similarity searching.

II. MATERIALS AND INSTRUMENTATION

In analyses where MS BLAST is utilized for protein identification, tandem mass spectra of peptides can be acquired with any ionization source and mass spectrometer that enables *de novo* sequence prediction: nanoelectrospray, LC/MS/MS, o-MALDI (MALDI quadrupole TOF), MALDI TOF-TOF, QqTOF, triple quad, PSD MALDI-TOF, and ion trap. Alternatively, in analyses where MultiTag is applied, tandem mass spectra of peptides can be acquired with any ionization source and tandem mass spectrometer that enables the creation of peptide sequence tags (Mann and Wilm, 1994): nanoelectrospray, LC/MS/MS, QqTOF, triple quad, or other novel system.

Software for *de novo* sequence prediction from tandem mass spectra is often included in the software packages associated with mass spectrometers: BioMultiview, BioAnalyst (both are from MDS Sciex, Canada), BioMassLynx (Micromass Ltd, UK), BioTools (Bruker Daltonics, Germany), and DeNovoX (ThermoFinnigan). The Lutefisk program (Johnson and Taylor, 2000) can be acquired from http://www.hairyfatguy.com/Sherpa/. BioAnalyst with the ProBlast processing script can generate a complete MS BLAST query automatically from multiple-spectra files acquired by nanoelectrospray or LC/MS/MS (Nimkar and Loo, 2002). A web browser such as Internet Explorer or Netscape is also required to gain access to the MS BLAST web interface located at http://dove.embl-heidelberg.de/Blast2/msblast.html. An independent BLAST computer (Paracel BlastMachine system) may also be purchased and installed for rapid and private MS BLAST operation.

Sequence tags that contain a few confidently designated amino acid residues and mass values that lock the short sequence stretch into the length of the peptide can often be generated from tandem mass spectra with the software packages associated with mass spectrometers. Microsoft Excel or an alternative spreadsheet program is required for compiling of search results before submission to MultiTag. BioAnalyst software has an associated processing script that produces a list of database search results from a generated list of sequence tags to accelerate spectra processing with MultiTag.

III. PROCEDURES

A. Identification of Proteins by MS BLAST Database Searching

1. MS BLAST is a specialized BLAST-based tool for the identification of proteins by sequence-similarity searching that utilizes peptide sequences produced by the interpretation of tandem mass spectra (Shevchenko *et al.*, 2001). The algorithm and principles of BLAST sequence-similarity searching are reported in detail elsewhere (Altschul *et al.*, 1997). A useful list of BLAST servers accessible on the web is provided in Gaeta (2000).

2. Peptide sequences are generated from the interpretation of tandem mass spectra from the analysis of a single in-gel or in-solution digest of an unknown protein, edited and assembled into a query list for the MS BLAST search. If tandem mass spectra were interpreted by *de novo* sequencing software, disregard relative scores and use the entire list of candidate sequences (or some 50–100 top scoring sequence proposals per fragmented

peptide precursor) (Fig. 29.1). Automated interpretation of tandem mass spectra often requires adjustment of parameters that affect the quality of predicted sequences. It is therefore advisable to test the settings in advance using digests of standard proteins and to adjust them if necessary. Note that the settings may depend on a charge state of the fragmented precursor ion. Use only the standard single-letter symbols for amino acid residues. If the software introduces special symbols for modified amino acid residues, replace them with standard symbols.

3. When interpreting MS/MS spectra manually, try making the longest possible sequence stretches, although their accuracy may be compromised. For example, it is usually

difficult to interpret unambiguously fragment ion series at the low m/z range because of abundant peaks of chemical noise and numerous fragment ions from other series. In this case, it is better to include many complete (albeit low confidence) sequence proposals into the query rather than using a single (although accurate) three or four amino acid sequence stretch deduced from a noise-free high m/z segment of the spectrum.

4. Gaps and ambiguities in peptide sequences can occur due to the fragmentary nature of tandem mass spectra of peptides. Some *de novo* sequencing programs may suggest a gap in the peptide sequence that can be filled with various isobaric combinations of amino acid residues. For example,

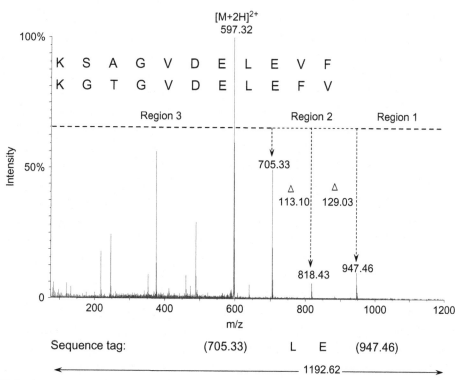

FIGURE 29.1 A spectrum with *de novo*-predicted amino acid sequences and a manually constructed sequence tag. Multiple candidate peptide sequences can be generated from a single spectrum for MS BLAST analysis, whereas MultiTag requires one sequence tag per spectrum.

DTPS[…]HYNAR, […] = [S, V] or [D, A]

If one or two combinations were suggested, include all variants into a searching string:

-DTPSSVHYNAR-DTPSVSHYNAR-
DTPSDAHYNAR-DTPSADHYNAR-

If more combinations were possible, the symbol X can be used instead to fill the gap. Zero score is assigned to X symbol in PAM30MS scoring matrix and therefore it matches weakly any amino acid residue:

-DTPSXXHYNAR-

Note that MS BLAST is sensitive to the number of amino acid residues that are filling the gap. If the gap could be filled by a combination of two and three amino acid residues, consider both options in the query

-DTPSXXHYNAR-DTPSXXXHYNAR-

5. Isobaric amino acids need to be altered in the MS BLAST query. L stands for Leu (L) and Ile (I). Z stands for Gln (Q) and Lys (K), if undistinguishable in the spectrum. Use Q or K if the amino acid residue can be determined. The query string needs to be further altered for cleavage site specificity. If the proposed sequence is complete, a putative trypsin cleavage site symbol B is added prior to the peptide sequence:

… -BDTPSVDHYNAR-

It is often difficult to determine two amino acid residues located at the N terminus of the peptide. In this case, present them as

… -BXXPSVDHYNAR-…

MS BLAST will then consider BXX residues in possible sequence alignments.

6. The regular BLAST search must be altered in options and settings for an MS BLAST query:

NOGAP is absolutely essential, it turns off gapped alignment method so that only high-scoring pairs (HSPs) with no internal gaps are reported.

SPAN1 is absolutely essential, it identifies and fetches the best matching peptide sequence among similar peptide sequences in the query. Therefore the query may contain multiple partially redundant variants of the same peptide sequence without affecting the total score of the protein hit.

HSPMAX 100 limits the total number of reported HSPs to 100. Set it to a higher number (e.g., 200) if a large query is submitted and a complete list of protein hits (including low confidence hits) is required in the output.

SORT_BY_TOTALSCORE places the hits with multiple high scoring pairs to the top of the list. Note that the total score is not displayed, but can be calculated, if necessary, by adding up scores of individual HSPs.

EXPECT: It is usually sufficient to set EXPECT at 100. Searching with higher EXPECT (as, 1000) will report many short low-scoring HSPs, thus increasing the sequence coverage by matching more fragmented peptides to the protein sequence. Note that low scoring HSPs do not increase statistical confidence of protein identification. The EXPECT setting also does not affect the scores of retrieved HSPs.

MATRIX: PAM30MS is a specifically modified scoring matrix. It is not used for conventional BLAST searching.

PROGRAM: blast2p.

DATABASE: nrdb95 are default settings of the MS BLAST interface.

FILTER: Filtering is set to "none" default. However, if the sequence query contains many repeating stretches (as… EQEQEQ…), filtering should be set to "default."

At the EMBL web interface, all parameters are preset and only the number of fragmented peptides and query sequences need to be input.

7. Space all candidate sequence proposals obtained from MS/MS spectra with a "−" (minus) symbol and merge them into a single text string that can be pasted directly into the query window at the MS BLAST web interface (Fig. 29.2). The query may contain space symbols, hard returns, numbers, and so on, as the server ignores them. For example, it is convenient to keep masses of precursor ions in the query, as it makes retrospective analysis of data much easier. Statistical evaluation is a very important element of MS BLAST protocol, as the query typically comprises

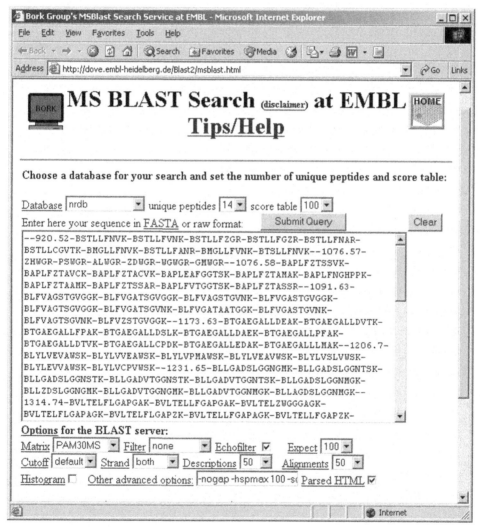

FIGURE 29.2 The MS BLAST web interface. A generated query is pasted in the input window, the number of tandem mass spectra from which sequences were derived is input in "unique peptides," and all other settings are set automatically according to MS BLAST parameters.

many incorrect and partially redundant peptide sequences. Note that the statistics of conventional BLAST searching are not applicable and therefore ignore reported E-values and P-values. Thresholds of statistical significance of MS BLAST hits were estimated in a computational experiment and scoring thresholds were set conditionally on the number of reported HSPs and the size of the database searched. Experimental MS BLAST hits are evaluated based on the number of fragmented precursors (this value is entered in the search parameters), and confident hits appear in red, borderline hits in green, and random matches in black at the web interface cited earlier (Fig. 29.3).

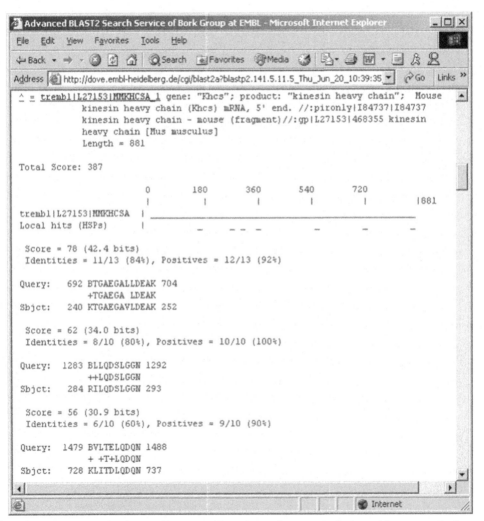

FIGURE 29.3 A section of the MS BLAST output where three of seven matched peptides to one database entry are shown. A list of matching database entries with the highest significance match at the top of the list is generated. Significant hits are color coded for easy data interpretation.

8. MS BLAST can, in principle, be used to search protein, EST, and genomic databases. The EMBL site only supports protein BLAST searching due to available computational capacity. A script can be written to retrieve specific genomic sequences that lie within the aligned peptides from a *tblastn* search against specific genomic databases. This search enables the use of unannotated genomic sequences and makes it possible to identify novel genes in large genomes. However, in both EST and genomic searches, different scoring schemes would need to be developed and installed locally. To set up a local BLAST searching engine, WU-BLAST 2.0 can be acquired from http://blast.wustl.edu/.

B. Identification of Proteins by MultiTag Database Searching

1. MultiTag is a software program that sorts compiled results from database searches with partial and complete sequence tags and calculates the significance of matches that align multiple sequence tags (for a complete description, see Sunyaev *et al.*, 2003). MultiTag is based on error-tolerant searching with multiple partial sequence tags. This technique enables the correlation of search results from multiple searches with sequence tags representative of numerous spectra and gauges the significance of those matches.

2. Sequence tags should be generated manually from tandem mass spectra acquired in the analysis of a single in-gel or in-solution digest of an unknown protein or proteins. Some mass spectrometer software enables the automatic prediction of sequence tags; however, for best results it is advisable to make sequence tags by manual interpretation or gauge the accuracy of the automatic prediction software with a standard protein prior to the analysis of unknown samples. One sequence tag per spectrum should be made using prominent Y-ions, usually larger than the multiply charged precursor. Sequence tags made with two to four amino acids each from multiple MS/MS spectra should be compiled in a text file list that includes the tag followed by the parent mass.

(360.20)FLL(733.44)918.64
(561.27)LA(745.40)935.48
(866.41)DEA(1181.52)1422.59

3. Each sequence tag is used to search a protein database, and the results from four searches are compiled in a spreadsheet. For the most specific results, mass tolerances should be narrow, taking into consideration the best accuracy of the mass spectrometer employed. The database is first searched using the complete sequence tag:

(866.41)DEA(1181.52)

This search will only find proteins that contain peptides with exactly these amino acids, spaced with exact amino acids residues that give mass combinations to make up the gaps to the peptide's termini. The second search allows for one error within the amino acid representation itself:

(866.41)D?A(1181.52)

The third search allows for errors between the analyzed peptide and the database at the C terminus of the peptide (searching with regions 1 and 2 only):

DEA(1181.52)

The fourth search allows for errors between the analyzed peptide and the database at the N terminus of the peptide (searching with regions 2 and 3 only):

(866.41)DEA

In the first column of the results table, the parent mass should be followed by an "NC",

"E", "N", or "C" for sequence tag search results from complete, one-error, regions 1 and 2 (N-terminal match), and regions 2 and 3 (C-terminal match), respectively. The second, third, fourth, fifth, and six columns should include the amino acid sequence of the peptide matched, the molecular weight of the protein the tag matched, the accession number of the entry, the name of the protein, and the species name, respectively (Fig. 29.4). Compile the search results in a spreadsheet from all searches with the sequence tags generated, and save results as a text file.

4. Submit the search results table to MultiTag. Designate the mass tolerance used for searching (in daltons), input the approximate number of entries in the database searched, input the list of tags used to generate search results, and compute significance. Results will be sorted, with the database entry containing the most unlikely correlation event at the top of the list, and probabilities are calculated (Fig. 29.5). Results can be evaluated based on the number of sequence tags matched, E-values and P-values [count (predicted)]. E-values lower than

Nr	Tag Mass	Sequence	Mass(kDa)	DB Accession	Protein name	Species
0	1173.628E	SAAKKVKNAEK	47.465796	gi|14210646	(AY033620) putative RNA-binding pro	Unknown
1	1173.628E	TGAEHLWLTR	27.823135	gi|7019377	(NM_013393) cell division protein Ft	Human
2	1173.628E	TGAEHLWLTR	27.487108	gi|13386002	(NM_026510) RIKEN cDNA 2310037B18 [	Mouse
3	1173.628E	SAANKALNDKK	15.032396	gi|12744797	AF323725_1 (AF323725) PsaN precurso	Unknown
4	1173.628E	ASAEILSVDRV	47.31585	gi|15611313	(NC_000921) EXODEOXYRIBONUCLEASE LA	Helicobacter Pylori
5	1173.628E	ASAEILSVDRV	47.455069	gi|15644887	(NC_000915) exonuclease VII, large	Helicobacter Pylori
6	1173.628E	TGAETLWEEAK	12.263558	gi|401181	THGF_TOBAC FLOWER-SPECIFIC GAMMA-THIO	Unknown
7	1173.628E	SAAEKRQEK	79.528375	gi|18375979	(AL356173) conserved hypothetical p	Unknown
8	1173.628E	SAAERKRQEK	82.102377	gi|11359450	T49456 hypothetical protein B14D6.8	Unknown
9	1173.628E	SANEKKSINVK	143.453267	gi|17224297	AF218388_1 (AF218388) apoptotic pro	Rat
10	1173.628E	SANEKKSINVK	143.435205	gi|13027436	(NM_023979) apoptotic protease acti	Rat
11	1173.628E	SAAEAQATRGR	21.850893	gi|13162112	(AL512667) putative tetR-family tra	some Streptomyces
12	1173.628N	GTAEQPRLFVG	32.418744	gi|7294725	(AE003544) CG7547 gene product [Dros	Fruit Fly
13	1173.628N	SAAEQWKQDL	74.867942	gi|17227422	(NC_003267) ORF_ID:all8048~unknown	Unknown
14	1173.628N	ASAEQRATQTI	36.747253	gi|17481280	(AB062896) vomeronasal receptor 1 A	Mouse
15	1173.628N	ASAEQRATQTI	35.05824	gi|3892596	(Y12724) pheromone receptor 2 [Mus m	Mouse
16	1173.628N	ASAEQRATQTI	35.84985	gi|17481276	(AB062895) vomeronasal receptor 1 A	Mouse
17	1173.628N	ASAEQRATQTI	36.402747	gi|18558569	(AY065464) vomeronasal receptor V1R	Mouse
18	1173.628N	ASAEQRATQTI	34.605305	gi|16716523	(NM_053218) vomeronasal 1 receptor,	Mouse
19	1173.628N	ASAEKGIASVRS	13.48812	gi|15802561	(NC_002655) orf, hypothetical prote	Escherichia Coli
20	1173.628N	SAAEQSGLDKNG	35.440047	gi|12620486	AF322012_67 (AF322013) ID142 [Brady	Unknown
21	1173.628N	ASAEKKRQATS	56.223343	gi|8570440	AC020622_1 (AC020622) Contains simil	Human
22	1173.628N	ASAEKKRQATS	64.846084	gi|15223502	(NM_100069) hypothetical protein [A	Mouse-Ear Cress
23	1173.628N	SAAEKLSEETL	272.279457	gi|4874311	AC006053_15 (AC006053) unknown prote	Mouse-Ear Cress
24	1173.628N	SAAEKLSEETL	60.547133	gi|15081785	(AY048285) At2g25730/F3N11.18 [Arab	Mouse-Ear Cress
25	1173.628N	SAAEKLSEETL	277.4555	gi|18400918	(NM_128132) unknown protein [Arabid	Mouse-Ear Cress
26	1173.628N	ASAEKKAEKSE	105.810124	gi|15900468	(NC_003028) translation initiation	Streptococcus Pyogenes
27	1173.628N	ASAEKYPHEF	31.188912	gi|17988633	(NC_003318) DIPEPTIDE TRANSPORT ATP	Unknown
28	1173.628N	GTAEKMPTTSR	13.777599	gi|3204328	(AJ008500) gag protein [Human immuno	Human
29	1173.628N	GTAEKMPSTTR	13.71449	gi|3204368	(AJ008521) gag protein [Human immuno	Human
30	1173.628N	TGAEKRSFVAD	80.784195	gi|7302767	(AE003803) CG4878 gene product [alt	Fruit Fly
31	1173.628N	SAAEKIVVYSGG	50.711733	gi|17231331	(NC_003272) unknown protein [Nostoc	Unknown
32	1173.628N	SAAEKAVSAPPR	55.045677	gi|13471797	(NC_002678) ATP-binding protein of	Unknown
33	1173.628N	SAAEKFDVSMT	24.872427	gi|10956719	(NC_002490) conjugal transfer prote	Unknown
34	1173.628N	ASAEKEQIAQI	144.720391	gi|16555336	(AY056833) chitin synthase [Anophel	Unknown
35	1173.628N	GTAEQHGRNVK	46.230479	gi|16759094	(NC_003198) putative IS element tra	Unknown
36	1173.628N	GTAEQHIKEGK	51.501952	gi|695769	(X84038) transposase [Xanthobacter au	Unknown
37	1173.628N	GTAEKGGLAIGDT	86.792704	gi|8894820	(AL360055) putative ABC transport sy	some Streptomyces
38	1173.628N	SAAEKDKGKQE	10.64305	gi|18550306	(XM_103535) hypothetical protein XP	Human
39	1173.628N	TGAEKAPKSPSK	13.977534	gi|6009909	(AB018242) histone H2A-like protein	Unknown
40	1173.628N	ASAEQCGRQAGG	33.741525	gi|7798662	AF135145_1 (AF135145) class I chitin	Unknown
41	1173.628N	GTAEKMPNTSR	13.580314	gi|3204322	(AJ008497) gag protein [Human immuno	Human
42	1173.628N	GTAEKMPNTSR	13.697865	gi|3355417	(AJ011213) gag protein [Human immuno	Human
43	1173.628N	GTAEKMPNTSR	13.807781	gi|3204271	(AJ008470) gag protein [Human immuno	Human
44	1173.628N	GTAEKMPNTSR	13.637407	gi|3204303	(AJ008487) gag protein [Human immuno	Human
45	1173.628N	GTAEKMPNTSR	13.552304	gi|3204338	(AJ008505) gag protein [Human immuno	Human

FIGURE 29.4 Compiled and formatted search results from sequence tag searching are input in the MultiTag software via opening a text-formatted results file.

1×10^{-3} and P-values lower than 1×10^{-4} can be considered significant matches. Final E-values are highly dependent on the number of tags submitted for database searching; more tags will tend to diminish the significance of the alignment of multiple tags to one database entry. Reported P-values are less affected by the number sequence tags in the query. P-values reflect an approximation of the probability that the tags that are aligned will match randomly to an entry in a database of a specific size and at a specific mass accuracy, while neglecting the query size (number of sequence tags). Low P-values (but with higher E-values) are good indicators of alignments having borderline significance that need further manual evaluation to conclude a confident identification. Three partial sequence tags are normally specific enough to identify one entry in a database of 1,000,000 entries, at a mass accuracy of 0.1Da.

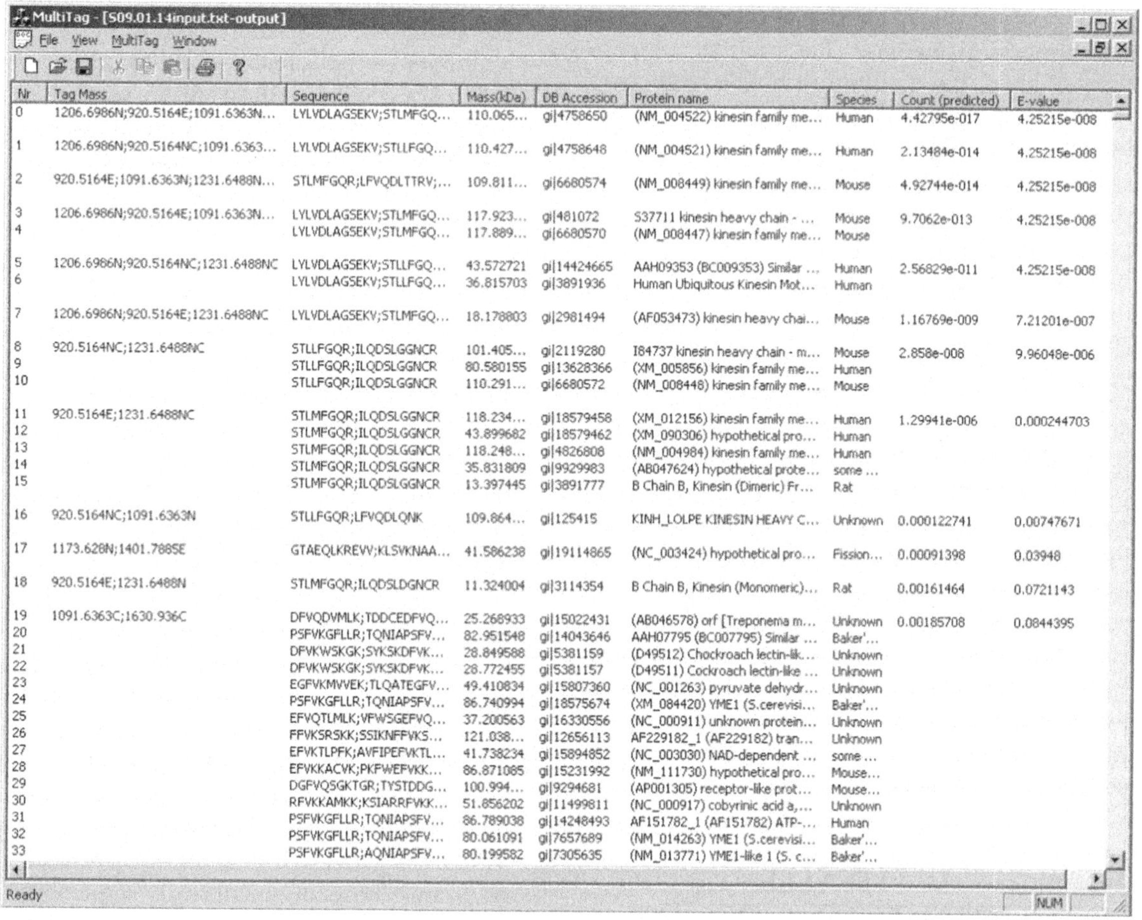

MultiTag - [S09.01.14input.txt-output]

File View MultiTag Window

Nr	Tag Mass	Sequence	Mass(kDa)	DB Accession	Protein name	Species	Count (predicted)	E-value
0	1206.6986N;920.5164E;1091.6363N...	LYLVDLAGSEKV;STLMFGQ...	110.065...	gi|4758650	(NM_004522) kinesin family me...	Human	4.42795e-017	4.25215e-008
1	1206.6986N;920.5164NC;1091.6363...	LYLVDLAGSEKV;STLLFGQ...	110.427...	gi|4758648	(NM_004521) kinesin family me...	Human	2.13484e-014	4.25215e-008
2	920.5164E;1091.6363N;1231.6488N...	STLMFGQR;LFVQDLTTRV;...	109.811...	gi|6680574	(NM_008449) kinesin family me...	Mouse	4.92744e-014	4.25215e-008
3	1206.6986N;920.5164E;1091.6363N...	LYLVDLAGSEKV;STLMFGQ...	117.923...	gi|481072	S37711 kinesin heavy chain - ...	Mouse	9.7062e-013	4.25215e-008
4		LYLVDLAGSEKV;STLMFGQ...	117.889...	gi|6680570	(NM_008447) kinesin family me...	Mouse		
5	1206.6986N;920.5164NC;1231.6488NC	LYLVDLAGSEKV;STLLFGQ...	43.572721	gi|14424665	AAH09353 (BC009353) Similar ...	Human	2.56829e-011	4.25215e-008
6		LYLVDLAGSEKV;STLLFGQ...	36.815703	gi|3891936	Human Ubiquitous Kinesin Mot...	Human		
7	1206.6986N;920.5164E;1231.6488NC	LYLVDLAGSEKV;STLMFGQ...	18.178803	gi|2981494	(AF053473) kinesin heavy chai...	Mouse	1.16769e-009	7.21201e-007
8	920.5164NC;1231.6488NC	STLLFGQR;ILQDSLGGNCR	101.405...	gi|2119280	I84737 kinesin heavy chain - m...	Mouse	2.858e-008	9.96048e-006
9		STLLFGQR;ILQDSLGGNCR	80.580155	gi|13628366	(XM_005856) kinesin family me...	Human		
10		STLLFGQR;ILQDSLGGNCR	110.291...	gi|6680572	(NM_008448) kinesin family me...	Mouse		
11	920.5164E;1231.6488NC	STLMFGQR;ILQDSLGGNCR	118.234...	gi|18579458	(XM_012156) kinesin family me...	Human	1.29941e-006	0.000244703
12		STLMFGQR;ILQDSLGGNCR	43.899682	gi|18579462	(XM_090306) hypothetical pro...	Human		
13		STLMFGQR;ILQDSLGGNCR	118.248...	gi|4826808	(NM_004984) kinesin family me...	Human		
14		STLMFGQR;ILQDSLGGNCR	35.831809	gi|9929983	(AB047624) hypothetical prote...	some ...		
15		STLMFGQR;ILQDSLGGNCR	13.397445	gi|3891777	B Chain B, Kinesin (Dimeric) Fr...	Rat		
16	920.5164NC;1091.6363N	STLLFGQR;LFVQDLQNK	109.864...	gi|125415	KINH_LOLPE KINESIN HEAVY C...	Unknown	0.000122741	0.00747671
17	1173.628N;1401.7885E	GTAEQLKREVV;KLSVKNAA...	41.586238	gi|19114865	(NC_003424) hypothetical pro...	Fission...	0.00091398	0.03948
18	920.5164E;1231.6488N	STLMFGQR;ILQDSLDGNCR	11.324004	gi|3114354	B Chain B, Kinesin (Monomeric)...	Rat	0.00161464	0.0721143
19	1091.6363C;1630.936C	DFVQDVMLK;TDDCEDFVQ...	25.268933	gi|15022431	(AB046578) orf [Treponema m...	Unknown	0.00185708	0.0844395
20		PSFVKGFLLR;TQNIAPSFV...	82.951548	gi|14043646	AAH07795 (BC007795) Similar ...	Baker'...		
21		DFVKWSKGK;SYKSKDFVK...	28.849588	gi|5381159	(D49512) Chockroach lectin-lik...	Unknown		
22		DFVKWSKGK;SYKSKDFVK...	28.772455	gi|5381157	(D49511) Cockroach lectin-like ...	Unknown		
23		EGFVKMVVEK;TLQATEGFV...	49.410834	gi|15807360	(NC_001263) pyruvate dehydr...	Unknown		
24		PSFVKGFLLR;TQNIAPSFV...	86.740994	gi|18575674	(XM_084420) YME1 (S.cerevisi...	Baker'...		
25		EFVQTLMLK;VFWSGEPVQ...	37.200563	gi|16330556	(NC_000911) unknown protein...	Unknown		
26		FFVKSRSKK;SSIKNFFVKS...	121.038...	gi|12656113	AF229182_1 (AF229182) tran...	Unknown		
27		EFVKTLPFK;AVFIPEFVKTL...	41.738234	gi|15894852	(NC_003030) NAD-dependent...	some ...		
28		EFVKKACVK;PKFWEFVKK...	86.871085	gi|15231992	(NM_111730) hypothetical pro...	Mouse...		
29		DGFVQSGKTGR;TYSTDDG...	100.994...	gi|9294681	(AP001305) receptor-like prot...	Mouse...		
30		RFVKKAMKK;KSIARRFVKK...	51.856202	gi|11499811	(NC_000917) cobyrinic acid a,...	Unknown		
31		PSFVKGFLLR;TQNIAPSFV...	86.789038	gi|14248493	AF151782_1 (AF151782) ATP-...	Human		
32		PSFVKGFLLR;TQNIAPSFV...	80.061091	gi|7657689	(NM_014263) YME1 (S.cerevisi...	Baker'...		
33		PSFVKGFLLR;AQNIAPSFV...	80.199582	gi|7305635	(NM_013771) YME1-like 1 (S. c...	Baker'...		

Ready NUM

FIGURE 29.5 A section of MultiTag output. Margins may be adjusted to see the full list of tags matched, as well as full peptide sequences aligned, names, and so on. Results may be saved as text files to be viewed in an appropriate spreadsheet application.

IV. COMMENTS

It is not known in advance if the sequence of the analyzed protein is already present in a database. Therefore, conventional database-searching routines based on stringent matching of peptide sequences should be applied first (Mann and Wilm, 1994; Perkin et al., 1999). Only if the protein is unknown and no convincing cross-species matches can be obtained is it recommended to proceed with de novo interpretation of tandem mass spectra and sequence-similarity searching.

The success of MS BLAST and MultiTag identification depends on the size of a query and the corresponding database, the number of peptides aligned, the quality of peptide sequences or sequence tags, and the sequence similarity between the protein of interest and its homologues available in a database. On average, candidate sequences determined for five tryptic peptides should be submitted to MS BLAST or MultiTag searching to identify the protein by matching to a homologous sequence. With 10 sequences submitted and aligned, MS BLAST can identify 50% of homologues containing 50% sequence similarity, and MultiTag can identify 50% of homologues at 70% sequence similarity. However, because MultiTag can utilize less intense and noisy spectra, it can outperform MS BLAST in many cases (for a more thorough discussion on homologue identification specificity, see Sunyaev et al., 2003).

Both MS BLAST and MultiTag can identify proteins present in mixtures. Usually two or three components per sample can be identified easily. The sensitivity of both methods is determined primarily on the quality of the de novo sequences or sequence tags.

V. PITFALLS

1. Poor sample preparation can frequently deteriorate the quality of tandem mass spectra of peptides. The digestion of proteins with trypsin or other proteases should be carried out with chemicals of the highest degree of purity available. Plasticware (pipette tips, gloves, dishes, etc.) may acquire a static charge and attract dust, thus leading to contamination of samples with human and sheep (wool) keratin during in-gel or liquid digestion. Any polymeric detergents (Tween, Triton) should not be used for cleaning the laboratory materials.

2. When generating de novo sequences or sequence tags, if the software automatically extrapolates the parent mass from the precursor isotope cluster in the MS/MS spectra or the proceeding survey scan in a LC/MS/MS run, it is advisable to manually calculate this value, as software may determine the parent mass incorrectly by designating an incorrect charge state or ^{12}C monoisotopic peak of the parent ion isotope cluster, thus disabling correct de novo sequence prediction and sequence tag prediction.

3. MultiTag is laborious. Without scripted sequence tag database searching and processing of search results, manual data processing can demand extended effort; however, in cases where conventional methods fail to identify analyzed proteins, positive identifications are of a high value to cell biological studies.

4. Poor queries tend to obscure protein identification by both MS BLAST and MultiTag. It is best to submit fewer higher quality sequences than numerous lower quality sequences to MS BLAST. MS BLAST is particularly susceptible to low-complexity glycine- and proline-rich sequences generated incorrectly by de novo software. These low-complexity sequences tend to mask correct alignments. MultiTag functions best with sequence tags containing multiple (2–4) amino acids that have a low

prevalence, such as tryptophan (W) or methionine (M), whereas common amino acids such as leucine (L) in the tag tend to be of less significance and are likely to produce more false positives. Sequence tags generated from larger peptides also have more significance in a database search than those generated from smaller peptides.

References

Altschul, S. F., Madden, T. L., Schaffer, A. A., Zhang, J., Zhang, Z., Miller, W., and Lipman, D. J. (1997). Gapped BLAST and PSI-BLAST: A new generation of protein database search programs. *Nucleic Acids Res.* **25**, 3389–3402.

Gaeta, B. A. (2000). BLAST on the Web. *Biotechniques* **28**, 436–440.

Johnson, R. S., and Taylor, J. A. (2000). Searching sequence databases via de novo peptide sequencing by tandem mass spectrometry. *Methods Mol. Biol.* **146**, 41–61.

Mann, M., and Wilm, M. (1994). Error-tolerant identification of peptides in sequence databases by peptide sequence tags. *Anal. Chem.* **66**, 4390–4399.

Nimkar, S., and Loo, J. A. (2002). Orlando FL. Application of a new algorithm for automated database searching of MS sequence data to identify proteins. Abstract 334.

Perkins, D. N., Pappin, D. J., Creasy, D. M., and Cottrell, J. S. (1999). Probability-based protein identification by searching sequence databases using mass spectrometry data. *Electrophoresis* **20**, 3551–3567.

Shevchenko, A., Sunyaev, S., Loboda, A., Bork, P., Ens, W., and Standing, K. G. (2001). Charting the proteomes of organisms with unsequenced genomes by MALDI-quadrupole time-of-flight mass spectrometry and BLAST homology searching. *Anal. Chem.* **73**, 1917–1926.

Sunyaev, S., Liska, A., Golod, A., Shevchenko, A., and Shevchenko, A. (2003). MultiTag: Multiple error-tolerant sequence tag search for the sequence-similarity identification of proteins by mass spectrometry *Anal. Chem.* **75**, 1307–1315.

Identification of Protein Phosphorylation Sites by Mass Spectrometry

Rhys C. Roberts and Ole N. Jensen

I. INTRODUCTION

Tight regulation of cellular proteins is a prerequisite for viable cell function. Cells require various mechanisms whereby intracellular pathways can be activated and inactivated in a reversible manner depending on the prevailing environment at a particular moment in time. Reversible phosphorylation of proteins is one such mechanism. The importance of phosphorylation is exemplified by the finding that at least 2% of the human genome encodes proteins with predicted kinase domains (Lander *et al.*, 2001). Furthermore, other workers predict that approximately a third of all proteins expressed in vertebrates can be phosphorylated at some point in their lifetime (Hunter, 1998).

Many phosphoproteins have been identified through incorporation of radioactively labelled ATP (usually [γ-^{32}P]ATP) either in cells or following incubation with specific kinases *in vitro*. Once identified, the significance of phosphorylation can then be addressed. In eukaryotes, phosphorylation occurs predominantly on serine, threonine, and tyrosine residues. Ideally, the specific phosphorylated residue should be pinpointed in order to gain further important insights into the consequences of the addition and removal of a phosphate group.

Identifying specific phosphorylated residues can be fraught with difficulties. Most strategies, as stated previously, have used radioactively labelled ATP to phosphorylate the protein of interest. The purified phosphorylated protein can then be cleaved using specific chemical or enzymatic processes (e.g., CNBr or trypsin) before separation of the peptides by an appropriate method (e.g., SDS–PAGE, HPLC, or thin-layer chromatography). The radioactively labelled peptide, which should contain the phosphorylated residue, is detected, isolated, and the sequence determined, usually by Edman degradation (Moyers et al., 1995). In addition, phosphoamino acid analysis can be used to determine the nature of the phosphorylated residue (Sefton, 1995). This approach relies on a high degree of protein phosphorylation with radiolabelled phosphate. Handling large amounts of radiolabelled ^{32}P is clearly a potential hazard and, in practice, the efficiency of protein phosphorylation can vary significantly from case to case. Furthermore, the amount of protein required for identification using this approach means that only those proteins available in significant quantities, with high incorporation of radiolabelled phosphate, are feasible for phosphorylation site identification. Because phosphorylated residues are not detected by conventional Edman sequencing, many sites are determined indirectly. In many cases, subsequent mutational analysis is used to support the identification of the putative phosphorylation site.

Following many developments and improvements in biological mass spectrometry, unambiguous identification of specific phosphorylation sites is possible without the use of radioactive isotopes (Larsen et al., 2001; Stensballe et al., 2001). The main principle behind phosphorylation site identification by mass spectrometry is the fact that the mass of a specific residue increases by 80 Da upon phosphorylation. Therefore, by accurately measuring the masses of a mixture of peptides from a phosphorylated protein (whose sequence is known) following specific enzymatic or chemical cleavage, potentially phosphorylated species with a mass shift of 80 Da can be observed. Phosphorylation can be confirmed by dephosphorylating the peptide mix with alkaline phosphatase and observing the disappearance of the 80-Da peptide signal. The specific site of phosphorylation can then be determined by tandem mass spectrometry using, for example, a quadrupole time-of-flight (TOF) mass spectrometer.

To maximise the yield of phosphorylated peptides, techniques such as Fe^{3+}-IMAC (immobilised metal affinity chromatography) micropurification can be used prior to analysis by mass spectrometry. This technique is based on the relative affinity of Fe^{3+} for phosphorylated residues. By using this method, it is possible to identify specific phosphorylation sites in proteins phosphorylated to a low degree.

This article describes a stepwise approach for the identification of phosphorylation sites by mass spectrometry. First, phosphopeptide enrichment by Fe^{3+}-IMAC micropurification is described. This is followed by a method to confirm phosphorylation by using alkaline phosphatase to dephosphorylate the peptide mixture already analysed on a MALDI-TOF mass spectrometer. Finally, unambiguous determination of a phosphorylation site using nanoelectrospray ionisation and tandem mass spectrometry is described.

II. MATERIALS AND INSTRUMENTATION

Milli-Q H_2O (Millipore) is used throughout this method. Formic acid, acetic acid, methanol and trifluoroacetic acid (TFA), and acetonitrile (all HPLC grade) are from Sigma.

Ethylenediaminetetraacetic acid (EDTA), NaCl, and $FeCl_3$ are from Sigma. Ni^{2+}-NTA (nitrilotriacetic acid)-silica (16–24 μm particle size) is supplied by Qiagen, and OligoR3 reverse-phase resin is from PE biosystems. 2,5-Dihydroxybenzoic acid (DHB) is supplied by Sigma. Calf intestinal alkaline phosphatase is from Roche (0108138). GELoader tips (1–10 μl) from Eppendorf (0030 001.222) are used for making micropurification columns.

MALDI time-of-flight analysis of peptides is performed using a REFLEX II time-of-flight mass spectrometer with delayed extraction (Bruker-Daltonics). The samples are ionised with a nitrogen laser (λ = 337 nm) and data acquired in the positive ion mode. "moverz" (Proteometrics Ltd.) is used to analyse the spectra obtained.

Electrospray mass spectrometry is performed on a Q-TOF hybrid mass spectrometer fitted with a nano-ESI Z-spray interface (Micromass). Nanoelectrospray needles (gold/pallidum precoated borosilicate glass) are from MDS Protana, Odense Denmark. The instrument is used in positive ion mode with the following typical settings: needle 700–1000 V; cone 55 V; collision gas (Argon) pressure 5.5–6.0 × 10^{-5} atm; collision energy 4–34 eV. Commonly, the nanoelectrospray flow rate would be in the order of 15–50 nl/min. Selected ions are subject to collision-induced dissociation (CID) with argon gas, and fragment ions are detected by the orthogonal time-of-flight analyser. Resulting data are analysed using the MassLynx software supplied by Micromass.

III. PROCEDURES

A. Phosphopeptide Enrichment by Fe^{3+} IMAC Micropurification

Following separation and purification of the protein of interest (e.g., by SDS–PAGE, reverse-phase HPLC), the phosphorylated protein is cleaved into a peptide fragment either by enzymatic digest or by chemical cleavage. For separation by SDS-PAGE followed by in-gel digestion of a protein with trypsin, followed by the extraction of peptides, see article by Gevaert and Vandekerckhove.

1. Preparation of Fe^{3+}-NTA

Steps

1. Resuspend 15 mg Ni^{2+}-NTA silica in 200 μl H_2O in a 1.5-ml microcentrifuge tube.
2. Pellet the resin in a benchtop centrifuge and remove the supernatant. Wash the resin with 200 μl H_2O.
3. Incubate the resin with 50 mM EDTA in 1 M NaCl for 2 min at room temperature. Centrifuge and discard the supernatant. Repeat once.
4. Wash once with 200 μl H_2O and twice with 200 μl 100 mM acetic acid.
5. Incubate resin with 400 μl 50 mM $FeCl_3$: 50 mM acetic acid for 5 min at room temperature, mixing gently. Pellet the resin and discard the supernatant. Repeat once.
6. Wash with 200 μl 100 mM acetic acid and once with 200 μl 3 : 1 100 mM acetic acid : 100% acetonitrile. Wash again with 200 μl 100 mM acetic acid.
7. Resuspend the resin in 100 μl 100 mM acetic acid. The resin is now ready for use and can be stored at 4°C for 3–4 weeks.

2. Preparation of Microcolumn

Steps

1. Take a 1- to 10-μl GELoader tip (Eppendorf) and gently twist the tip end with fingers and thumb. Care must be taken to make the tip outflow sufficiently narrow in order to pack the resin, whilst avoiding complete occlusion.
2. Add 10 μl of the prepared Fe^{3+}-NTA resin to the microcolumn and pack into the tip using a 10 ml syringe for pressure. The column should measure 20–25 mm in height.
3. The Fe^{3+}-NTA IMAC microcolumn is now ready for use.

3. *Purification of Phosphopeptides by Fe³⁺-IMAC*

Steps

1. Dilute the peptide mixture with 30x excess 100 mM acetic acid (e.g., from a Coomassie-stained band resuspended in a final volume of 20 μl, take 1–2 μl and dilute in 60 μl of 100 mM acetic acid).

2. Add the resuspended petide mixture to the Fe³⁺-NTA IMAC microcolumn and load very slowly using the 10-ml syringe for pressure. For optimum binding of phosphopeptides, this step should take 15–20 min.

3. Wash the column sequentially with 20 μl 100 mM acetic acid, 20 μl 3 : 1 100 mM acetic acid : 100% acetonitrile, and 20 μl 100 mM acetic acid.

4. Elute the bound phosphopeptides with 2×5 μl H_2O, pH 10.5 (NH_3). To concentrate the peptides prior to analysis, the eluted peptides can be applied directly into 60 μl of 5% formic acid in a preequilibrated OligoR3 microcolumn. Prepare the microcolumn in a similar manner to the Fe³⁺-IMAC microcolumn but using OligoR3 reversed-phase resin (PE Biosystems) in 5% formic acid and forming a column height of 2–3 mm.

5. Wash the bound peptides with 20 μl 5% formic acid.

6. Elute the peptides directly onto the MALDI target with 1 μl of DHB [DHB (Sigma) 20 μg/μl in 70% acetonitrile/0.1% TFA]. Allow the eluate to crystalise before analysis by MALDI mass spectrometry (Fig. 30.1).

4. *Analysis of Purified Phosphopeptides by MALDI*

Steps

1. Measure the masses of the purified peptides by MALDI mass spectrometry.

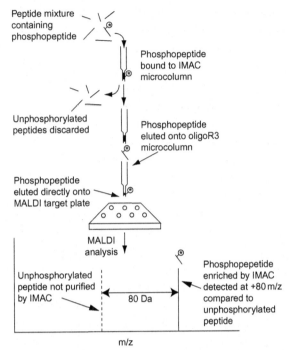

FIGURE 30.1 Schematic diagram illustrating the method of selective purification of phosphopeptides by Fe³⁺-IMAC followed by analysis by MALDI mass spectrometry. A mixture of peptides is applied to the IMAC microcolumn and the phosphorylated peptide (labelled with asterisks) is selectively purified prior to desalting and concentrating using an OligoR3 microcolumn. The peptides are eluted directly onto the MALDI target for analysis by MALDI mass spectrometry. The phosphorylated peptide corresponds to a signal seen at 80 Da with respect to its unphosphorylated counterpart.

2. Identify peptide peaks corresponding to predicted peptide masses +80 Da. Candidate phosphopeptides are identified in this way and by also looking for phosphorylated partially digested peptides. In addition, the presence of a metastable ion, corresponding to a phosphopeptide which has undergone β elimination (detected ~98 Da lower than the predicted phosphorylated peptide), is highly suggestive of a phosphorylated species.

B. Confirmation of Phosphorylation Using Alkaline Phosphatase

1. Remove the MALDI target from the mass spectrometer.
2. Resuspend the crystallised peptides in $0.5\,\mu l$ 70% acetonitrile in $100\,mM$ NH_4HCO_3.
3. Transfer to a new OligoR3 microcolumn containing $20\,\mu l$ dephosphorylation mixture ($19\,\mu l$ $100\,mM$ NH_4Cl with 1 unit alkaline phosphatase).
4. Seal the column with Parafilm and incubate at 37°C for 1 h.
5. Acidify the enzyme mixture by adding $20\,\mu l$ 5% formic acid and purify the peptides using the OligoR3 microcolumn as described earlier.
6. Elute the peptides using DHB onto the MALDI target and analyse as before.
7. Peaks corresponding to dephosphorylated peptides should now be seen, confirming that these peptides are indeed phosphorylated (Fig. 30.2).

C. Identification of Specific Phosphorylation Sites by Tandem Mass Spectrometry Using Nanoelectrospray Ionisation

Once candidate phosphopeptides are detected by IMAC and MALDI mass spectrometry, the precise residue should be identified. This can be achieved by selecting and sequencing the phosphopeptide using a nanoelectrospray quadrupole time-of-flight mass spectrometer. Sample preparation is vital to ensure successful analysis.

For analysis by nanoelectrospray, 5x the amount of peptide mixture is typically required, which will need to be determined on an individual basis. To minimise nonspecificc binding of phosphopeptides to the Fe^{3+}-IMAC microcolumn, three to four microcolumns are used simultaneously, and the bound peptides are concentrated on a single OligoR3 column prior to analysis.

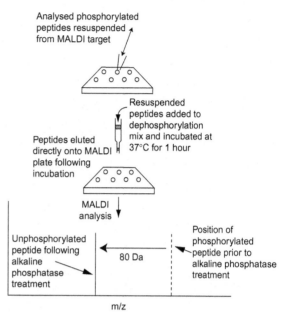

FIGURE 30.2 Schematic diagram illustrating the use of alkaline phosphatase to confirms that a candidate phosphopeptide is phosphorylated. Following initial MALDI analysis, the crystallised peptides are resuspended and added to a dephosphorylation mixture containing alkaline phosphatase. The mixture is present in a preequilibrated OligoR3 microcolumn. Following incubation for 1 h at 37°C, the peptides are purified and desalted on the microcolumn and eluted directly onto the MALDI target plate for further analysis. The disappearance of the signal corresponding to the phosphopeptide and the appearance of a signal 80 Da smaller confirm that the peptide of interest is indeed phosphorylated.

Steps

1. Apply the phosphopeptide mixture to three to four Fe^{3+}-IMAC microcolumns simultaneously and wash as described in Section III,A.
2. Elute the peptides onto a single OligoR3 microcolumn and wash with 5% formic acid.
3. Elute the peptides directly into a nanospray needle with $1\,\mu l$ 50% methanol : 1% formic acid.
4. Analyse the peptide mixture on the quadrupole time-of-flight mass spectrometer

and select the relevant candidate phosphopeptides for further analysis.

5. Sequence the phosphopeptide by collision-induced dissociation.

6. The phosphorylated residue can be identified by observing a mass difference equalling the mass of an amino acid $+80\,Da$ between two sequential ions (Fig. 30.3) or by observing a series of dephosphorylated ions resulting from β elimination (with masses corresponding to 98 Da lower than the predicted phosphorylated ion or 18 Da lower than the nonphosphorylated counterpart). By careful analysis of data obtained, the phosphorylated residue can be deduced.

IV. COMMENTS AND PITFALLS

A. Sample Preparation

The importance of sample preparation cannot be overemphasised. Often, the proportion of the protein sample that is phosphorylated is very low. Therefore, without meticulous sample handling from protein purification to analysing peptides, the chances of success diminish rapidly. Many potential problems have been described previously when analysing peptides by mass spectrometry, such as contamination with exogenous proteins (e.g., keratin) and inadequate desalting prior to ionisation (a particular problem with electrospray ionisation). To maximise the success rate for identifying phosphorylation sites, these factors must be optimised.

Purification of phosphopeptides by Fe^{3+}-IMAC is a method to select and concentrate the phosphorylated species prior to analysis by mass spectrometry. It is an extremely effective technique to assist in the identification of phosphorylation sites. However, this technique can yield a very high background of nonspecific binding if a number of key points are not followed closely. First, it is important to ensure that the peptide sample is well diluted. Second,

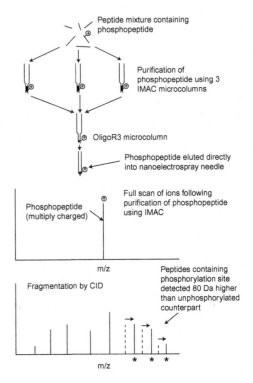

FIGURE 30.3 Schematic diagram illustrating the identification of specific sites of phosphorylation by IMAC micropurification and nanospray tandem mass spectrometry. The phosphopeptide is purified by using Fe^{3+}-IMAC microcolumns in parallel, prior to concentration and desalting on a single OligoR3 column. Once the multiply charged ion corresponding to the phosphopeptide is identified, the ion is fragmented by collision-induced dissociation (CID) with the production of specific fragment ions. The specific site of phosphorylation can be determined by careful analysis of this spectrum, looking particularly for phosphorylated and nonphosphorylated ions. In this simple schematic, a complete y-ion series is shown from a hypothetical nine residue phosphopeptide. The four largest y-ions are detected 80 Da higher than the predicted masses for the unphosphorylated peptide (indicated with dashed lines). This confirms that the fourth residue from the C-terminal is the site of phosphorylation. Further confirmation can be obtained by analysing the b-ion series and identifying the corresponding confirmatory ions.

the sample should be applied to the resin for at least 10 min, if not longer, with the minimum amount of back pressure applied from the syringe. Both these steps reduce the amount of nonspecific binding to the Fe^{3+}-IMAC microcolumn and facilitate successful outcomes.

Because the phosphorylated peptide might exist in a small quantity, it is sometimes worth purifying three to four aliquots of peptide sample on separate Fe^{3+}-IMAC microcolumns. The phosphopeptides can then be eluted onto a single OligoR3 microcolumn. This last step concentrates the peptides prior to analysis and also desalts the sample effectively, ensuring optimal analysis by mass spectrometry.

B. The Phosphoprotein

In addition to the amount of phosphorylation of an individual protein, specific factors relating to the phosphoprotein itself may determine the ease to which a phosphorylation site is identified. For example, the molecular mass of the peptide containing the phosphorylation site should ideally be between 500 and 4000 Da. Peptides with masses outside this range are very difficult to analyse using the methods described in this chapter. Trypsin is usually the enzyme of choice for producing measurable peptides. However, other enzymes, such as chymotrypsin, Lys-C, or Arg-N, may be required to isolate a phosphopeptide in the detectable range. Chemical cleavage with CNBr can also be used. A rare difficulty arises when a nonphosphorylated peptide has the same molecular weight as the phosphopeptide of interest.

In many cases, phosphorylation sites are flanked by polar residues. Generally, peptides containing polar residues are ionised more efficiently than their nonpolar counterparts. However, occasionally, a hydrophobic sequence surrounding the phosphorylation site may decrease the efficiency of ionisation, resulting in low detection rates. In addition, some peptides prove difficult to sequence by CID and this problem is exacerbated when the amount of phosphorylated peptide is small.

With respect to Fe^{3+}-IMAC micropurification, a particular difficulty arises when a protein contains a number of acidic repeats. The acidic residues, glutamate and aspartate, have significant affinities to Fe^{3+} and hence may result in a high background when analysed.

A further obstacle when attempting to pinpoint sites of phosphorylation occurs when the specific phosphate group is lost upon ionisation. Although this property can lead to difficulty in identifying the phosphorylated species, the same property can be used to identify a specific site. If the phosphate group is lost through β elimination, a signal can be seen corresponding to 98 Da less than the predicted phosphopeptide. Conversely, in some cases, the intact phosphorylated peptide remains undetected while its degraded product becomes the principal ion seen on analysis.

C. Ionisation and Analysis

As discussed earlier, a balance is required when analysing a phosphopeptide mixture to ensure efficient ionisation on the one hand whilst also protecting the phosphate group. At lower energies, the phosphopeptide will not be seen, whilst if the energy of ionisation is too high, the labile phosphate group will be lost. The optimal settings and conditions may need to be determined for different phosphoproteins. We have found that DHB is the matrix of choice for the analysis of most phosphopeptides.

An interesting phenomenon relating to phosphopeptides is the "suppression effect." This is a term describing the low detection of phosphorylated peptides when ionised in the presence of their nonphosphorylated counterparts. Fe^{3+}-IMAC is used to discard the nonphosphorylated species and can result in a phosphoprotein signal that was not detectable before micropurification.

This article describes sequencing using a quadruplole time-of flight mass spectrometer. It is possible to obtain very high resolution data using these instruments. However, these mass spectrometers are less sensitive than their

MALDI-TOF counterparts and, in practice, approximately 5x as much sample is required for sequence analysis by electrospray ionisation. In addition, the Q-TOF is less tolerant than a MALDI-TOF mass spectrometer to salt and samples should always be desalted thoroughly before analysis. We routinely micropurify all our peptide samples on self-made reverse-phase microcolumns prior to both MALDI and electrospray ionisation to ensure optimal results at all times.

References

Hunter, T. (1998). The Croonian Lecture 1997. The phosphorylation of proteins on tyrosine: Its role in cell growth and disease. *Philos. Trans. R. Soc. Lond. B. Biol. Sci.* **353**, 583–605.

Lander, E. S., Linton, L. M., Birren, B., Nusbaum, C., Zody, M. C., Baldwin, J., Devon, K., Dewar, K., Doyle, M., FitzHugh, W., Funke, R., Gage, D., Harris, K., Heaford, A., Howland, J., Kann, L., Lehoczky, J., LeVine, R., McEwan, P., McKernan, K., Meldrim, J., Mesirov, J. P., Miranda, C., Morris, W., Naylor, J., Raymond, C., Rosetti, M., Santos, R., Sheridan, A., Sougnez, C., Stange-Thomann, N., Stojanovic, N., Subramanian, A., Wyman, D., Rogers, J., Sulston, J., Ainscough, R., Beck, S., Bentley, D., Burton, J., Clee, C., Carter, N., Coulson, A., Deadman, R., Deloukas, P., Dunham, A., Dunham, I., Durbin, R., French, L., Grafham, D., Gregory, S., Hubbard, T., Humphray, S., Hunt, A., Jones, M., Lloyd, C., McMurray, A., Matthews, L., Mercer, S., Milne, S., Mullikin, J. C., Mungall, A., Plumb, R., Ross, M., Shownkeen, R., Sims, S., Waterston, R. H., Wilson, R. K., Hillier, L. W., McPherson, J. D., Marra, M. A., Mardis, E. R., Fulton, L. A., Chinwalla, A. T., Pepin, K. H., Gish, W. R., Chissoe, S. L., Wendl, M. C., Delehaunty, K. D., Miner, T. L., Delehaunty, A., Kramer, J. B., Cook, L. L., Fulton, R. S., Johnson, D. L., Minx, P. J., Clifton, S. W., Hawkins, T., Branscomb, E., Predki, P., Richardson, P., Wenning, S., Slezak, T., Doggett, N., Cheng, J. F., Olsen, A., Lucas, S., Elkin, C., Uberbacher, E., Frazier, M., *et al.*, (2001). Initial sequencing and analysis of the human genome. *Nature* **409**, 860–921.

Larsen, M. R., Sorensen, G. L., Fey, S. J., Larsen, P. M., and Roepstorff, P. (2001). Phospho-proteomics: Evaluation of the use of enzymatic de-phosphorylation and differential mass spectrometric peptide mass mapping for site specific phosphorylation assignment in proteins separated by gel electrophoresis. *Proteomics* **1**, 223–238.

Moyers, J. S., Linder, M. E., Shannon, J. D., and Parsons, S. J. (1995). Identification of the in vitro phosphorylation sites on Gs alpha mediated by pp60c-src. *Biochem. J.* **305**, 411–417.

Sefton, B. M. (1995). Phosphoamino acid analysis, *In* "Current Protocols in Protein Science" (J. E. Coligan, B. M. Dunn, D. W. Speicher, and P. T. Wingfield, Eds.), pp. 13.3.1–13.3.8. Wiley, New York.

Stensballe, A., Andersen, S., and Jensen, O. N. (2001). Characterization of phosphoproteins from electrophoretic gels by nanoscale Fe(III) affinity chromatography with off-line mass spectrometry analysis. *Proteomics* **1**, 207–222.

31

Analysis of Carbohydrates/Glycoproteins by Mass Spectrometry

Mark Sutton-Smith and Anne Dell

I. INTRODUCTION

Electron impact mass spectrometry (EI-MS) has been employed in carbohydrate analysis since the early 1960s and is still used for defining sugar compositions and for linkage analysis (Albersheim *et al.*, 1967). The introduction of fast atom bombardment-mass spectrometry (FAB-MS) at the beginning of the 1980s (Morris, 1980; Barber *et al.*, 1981; Dell *et al.*, 1983) revolutionised the structure determination of a very wide range of carbohydrate-containing biopolymers (Fukuda *et al.*, 1985; Laferte *et al.*, 1987; Dell *et al.*, 1990; McConville *et al.*, 1990). A decade later, electrospray ionisation (ES-MS) (Fenn *et al.*, 1990) and matrix-assisted laser desorption ionisation (MALDI-MS) technologies (Karas and Hillenkamp, 1988; Karas *et al.*, 1989) expanded the range of glycobiological structural problems amenable to mass spectrometry because of their higher sensitivity and applicability to much larger molecules (Lopez *et al.*, 1997). The lessons learnt from FAB-MS investigations in the 1980s (Fukuda *et al.*, 1984, 1985; Dell, 1987) have turned out to be equally applicable to ES-MS and MALDI-MS. Notably, it has been found that although native samples are amenable to MS analysis, it is often desirable to prepare derivatives prior to analysis. As a general rule, derivatisation vastly improves sensitivity and derivatised glycans yield fragment ions much more reliably than their native counterparts. Permethylation is the most important type of derivatisation employed in carbohydrate MS.

Broadly speaking, MS can be exploited in two general ways in the analysis of carbohydrates and glycoproteins.

i. Detailed characterisation of purified individual glycopolymers or mixtures of glycopolymers. This usually requires acquisition of MS data from both intact material and chemical, and enzymatic digests. Case studies that exemplify strategies applicable to a range of glycopolymers are described elsewhere (Sasaki, *et al.*, 1987; Dell *et al.*, 1995).

ii. Glycomics analyses that involve screening of cell and tissue extracts for their overall glycan content. Examples of such strategies are given elsewhere (Sutton-Smith, *et al.*, 2000; Manzi, *et al.*, 2000) and at the NIH Functional Glycomics Consortium Web site http://web.mit.edu/glycomics/consortium.

This article documents protocols for isolating, derivatising, and digesting glycans and glycopeptides in preparation for MS analysis. The emphasis is on glycoprotein analysis, but many of the methodologies are also applicable to other glycopolymers, such as glycolipids and polysaccharides. Also documented are procedures for high-sensitivity MS and MS/MS analyses of glycans and glycopeptides. The generic strategy is outlined below and in Fig. 31.1.

1. Preparation of biological matrix for analysis
2. Purification of glycoprotein(s)
3. Reduction/carboxymethylation
4. Tryptic digestion or, in some instances, cynanogen bromide degradation
5. *N*-Glycosidase F digestion
6. Separation of *N*-glycans from peptides/*O*-glycopeptides
7. Reductive elimination of *O*-glycans from *O*-glycopeptides
8. Dowex purification of *O*-glycans
9. Permethylation of *N*- and *O*-glycans
10. MALDI-TOF MS profiling of *N*- and *O*-glycans
11. ES-MS/MS on key selected peaks observed in the MALDI profile

The initial steps for glycoproteome analysis of mammalian biological matrices, such as cells, biological fluids, tissues, and organs, involve homogenisation and sample cleanup by dialysis. For screening glycans in mammalian tissues,

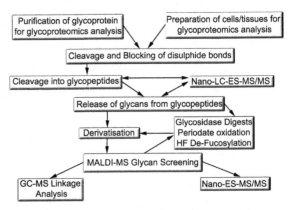

FIGURE 31.1 Glycomics strategy for screening mammalian cells, tissues, and glycoproteins.

100–400 mg of tissue is sufficient for a series of MS analyses, including glycan profiling, sequential exoglycosidase digestions, specific chemical degradations, linkage analyses, and MS/MS studies. As a rough guide, high-quality mapping data can be obtained from 10% of a typical extract from a single mouse kidney. For specific glycoproteins, initial steps usually involve some form of immunoprecipatation or affinity chromatography with, or without liquid chromatography purification or electrophoresis.

Once initial steps have been performed, the disulphide bridges are split by reduction and blocked by carboxymethylation. This allows the glycoproteins to be efficiently cleaved by trypsin, or another protease (or by cyanogen bromide if enzymatic digestion is problematical due to poor solubility in the digestion buffer). Once the glycopeptides are generated, these can be purified by reversed-phase chromatography (e.g., Sep-Pak or MicroTrap purification), or be analysed directly by nano-ES-LC/MS/MS before, or after N-glycosidase F digestion. Alternatively, the glycopeptides are digested with N-glycosidase F and separated into N-glycan and O-glycopeptide fractions, with the latter being subject to reductive elimination. The released N- and O-glycans are permethylated and analysed by MALDI and nano-ES-MS/MS. To complement the screening experiments, GC-MS linkage analysis and sugar analysis are performed to provide details of the linkages and the sugar compositions in the sample. In detailed studies the released glycans are often treated with chemical and/or enzymatic reagents to further characterize ambiguous glycan assignments, but this is outside the scope of this article. Initial assignments of relevant signals in MALDI-MS spectra are based on compositions that take into account biosynthetic considerations.

II. MATERIALS AND INSTRUMENTATION

Ammonium acetate (Cat. No. 100134T), ammonium hydrogen carbonate (Ambic, Cat. No.

103025E), hydrochloric acid (HC1, Cat. No. 101254H), and ion-exchanger Dowex 50 W-X8 (H+ form, Cat No. 105221) are from VWR. High-quality solvents: acetonitrile (UPS, ultra purity solvent, Cat. No. H050), dimethyl sulfoxide (DMSO) (Hi Dry, anhydrous solvent, D4281), methanol (UPS, ultra purity solvent, Cat No. H411), and propan-1-ol (SPS, super purity solvent, H624) are from Romil. Acetic acid (SpS, super purity reagent, Cat. No. H014), ammonia solution (SpS, super purity reagent, Cat No. H058), and sodium hydroxide are also from Romil. Acetic anhydride (Cat. No.4295) and methyl iodide (Cat. No. 0347) are from Lancaster. Acetyl chloride (Cat. No. 23,957-7), ammonium formate (Cat. No. F-2004), adrenocorticotropic hormone fragment (ACTH)1-17 (Cat. No. A-2407), ACTH fragment 18–39 (Cat. No. A-0673), ACTH fragment 7–38 (Cat. No. A-1527), ACTH fragment 1–39 (Cat. No. A-0423), angiotensin I (Cat. No. A9650), bradykinin (B-4764), calcium hydride lumps, +4 mesh (Cat. No. 213322), cyanogen bromide (Cat. No. 16774), 2,5-dihydroxybenzoic acid (Cat. No. G-5254), dithiothreitol (DTT, Cleland's reagent, Cat. No. D-5545), ethylene glycol (Cat. No. 10,246-8), EDTA (Cat. No. 43,178-8), formic acid (Cat No. 94318), hexanes (Cat. No. 15,617-5), hydrogen chloride gas (Cat. No. 29,542-6), hydrofluoric acid 48% (HF, Cat. No. 33,926-1), iodoacetic acid (IAA, Cat. No. 1-4386), insulin (bovine pancreas, Cat. No. I-5500), leucine enkephalin (Cat. No. L-9133), exo-β-mannosidase (Helix pomatia, Cat. No. M9400), neurotensin (Cat. No. N-6383), sodium acetate (Cat. No. 24,124-5), sodium m-periodate (Cat. No. S-1878), sodium borodeuteride (NaBH$_4$ Cat. No. 20,559-1), potassium borohydride (KBH$_4$ Cat. No. 438472), sodium chloride (Cat. No. 20,443-9), sodium dodecyl sulfate (SDS, Cat. No. 436143), Tris (hydroxymethyl)aminomethane (Tris, Cat. No. 154563) and trypsin (bovine pancreas, TPCK-treated, T-1426) are from Sigma-Aldrich. 3-((3-Cholamidopropyl)dimethyl-ammonio)-1 – propane sulfonate (CHAPS, Cat. No. 810126) and N-glycosidase F (Escherichea coli, Cat. No.

1365177) are from Roche Diagnostics. Exo-β-galactosidase (bovine testes, Cat. No. EG02), exo-β-N-acetylhexosaminidase(*Streptomycespneumoniae*, Cat. No. E-GL01), exo-α-mannosidase (jack bean, Cat. No. E-AM01), exo-neuraminidase (*Clostridium perfringes*, Cat. No.), and exo-neuraminidase (*Streptococcus pneumoniae*, Cat, No. E-S007) are from Qa-Bio. Endo-β-galactosidase (*Escherichia freundii*, 100455) is from Seikagaku corporation. Tri-sil "Z" derivatising agent (Cat. No. 49230) and Snakeskin Pleated dialysis tubing (Cat. No. 68700) are from Perbio Science UK Ltd. Ultra pure water is generated from an Analytical Purite Neptune ultrapure water purification system from Purite Ltd.

Homogenisation of cells and tissues is achieved with a compact electric CAT homogeniser (×120)-fitted *T6.1* dispersion shaft from Ingenieurburo CAT. Sonications of cells are achieved by a VC 130 PB (130W) Vibra-Cell ultrasonic processor within a sound-abating enclosure from Sonics & Materials Inc. Screwcap style Pyrex disposable culture tubes (Cat. No. 99449-13, 7.5ml) capped with disposable phenolic lids (Cat. No. 99999-13) and plasticware are from Corning. Caps are lined with Teflon inserts (Cat. No. 0402) from Owens Polyscience Ltd. Sep-Pak Classic C_{18} cartridges (Cat. No. WAT051910) are from Waters Ltd. Medium NanoES capillaries (Cat. No. ES387) for the Micromass Q-Tof are from Proxeon Biosytems. The Hamilton syringe (Cat. No. 002520) is from SGE. MicroTrap peptide cartridges (Cat. No. 004/25108/02) and the Manual Trap Holder kit (Cat. No. 004/25111/01) are from Michrom BioResources, Inc.

III. PROCEDURES

A. Preparation of Homogenates/Cell Lystates

Wear suitable protective clothing, including safety glasses, and work in a fume hood when preparing all solutions and performing various steps of the protocols.

Solutions

1. *Solution A:* 80% (v/v) methanol in H_2O. To make 25ml, add 5ml of ultrapure water to 20ml methanol to make a 80% (v/v) methanol solution.
2. *Solution B:* 33.33% ultrapure water, 33.33% formic, and 33.33% methanol. To make 30ml, add 10ml of formic acid and 10ml of methanol to 10ml of ultrapure water.
3. *Dilute acetic acid:* To make a 5% acetic acid solution, add 25ml of acetic acid and complete to 500ml with ultrapure water. Store at room temperature.
4. *Dialysis buffer:* 50mM Ambic buffer. To make 4.5 liters, add 17.6g of Ambic and complete with 4.5 liters of water. Adjust to pH 7.5 with dilute acetic acid. Prepare fresh on day of use.
5. *10% SDS:* To make 100ml, add 10g of SDS and complete with 100ml of ultrapure water. Store at room temperature.
6. *Homogenisation buffer:* 0.5% SDS (w/v) in 50mM Tris. To make 50ml, add 0.3029g of Tris and complete to 50ml with ultrapure water. Adjust the pH to 7.4 with dilute acetic acid. Withdraw 2.5ml of the solution and add 2.5ml of 10% SDS.
7. *Cell lysis buffer:* 25mM Tris, 150mM NaCl, 5mM EDTA, and 1% CHAPS at pH 7.4. To make 50ml, add 0.1514g of Tris, 0.4383g NaCl, 0.5g CHAPS, and 0.0731g EDTA. After dissolving in a small volume of ultrapure water, complete to 50ml and adjust to pH 7.4 with dilute acetic acid.

1. Homogenisation

Steps

1. Immerse the tip of the dispersion shaft into solution A; activate the drive motor at a low to intermediate setting for 60s.

2. Examine the tip and carefully remove any residual debris with a 3-mm hypodermic needle. Avoid scratching the dispersion shaft.
3. Put the dispersion shaft in solution B and sonicate in a sonicator bath for 10 min.
4. Repeat step 1 with fresh homogenisation buffer to clean the dispersion shaft.
5. Add 2–3 ml of ice-cold homogenisation buffer to the sample.
6. Homogenise on ice for 10 s. Repeat two or three times, pausing for 15 s between each homogenisation step.
7. Transfer the homogenate to high-quality dialysis tubing and seal.
8. Place the sample into cool dialysis buffer and dialyse at 4°C. Change the buffer regularly with fresh buffer over a period of 48 h.
9. Lyophilise the sample in a clean screw-capped glass culture tube covered with perforated Parafilm.

2. Sonication of Cells

Steps

1. Immerse the probe of the sonicator into solution A; activate the ultrasonic processor in continuous mode at 20 A for 10 s. Sonicate the probe again with ultrapure water and then the cell lysis buffer.
2. Add enough ice-cold cell lysis buffer (1–2 ml) to completely suspend the cell pellet.
3. Sonicate on ice in continuous mode at 40 A for 10 s. Repeat two or three times, pausing for 15 s between each sonication.
4. Transfer the homogenate to high-quality dialysis tubing and seal.
5. Place the sample into cool dialysis buffer and dialyse at 4°C. Change the buffer regularly with fresh buffer over a period of 48 h.
6. Lyophilise the sample in a clean screw-capped glass culture tube covered with perforated Parafilm.

B. Cleavage and Blocking of Disulphide Bridges

Wear suitable protective clothing, including safety glasses, and work in a fume hood when preparing all solutions and performing various steps of the protocols.

Solutions

1. *Tris buffer:* 0.6 M Tris. To make 50 ml, weigh 3.63 g of Tris and complete to 50 ml with ultrapure water. Adjust to pH 8.5 with acetic acid. Degas by passing a slow stream of nitrogen (oxygen free) through the solution via a Pasteur pipette.
2. *Dilute acetic acid:* To make a 5% acetic acid solution, add 25 ml of acetic acid and complete to 500 ml with ultrapure water. Store at room temperature.
3. *Dialysis buffer:* 50 mM Ambic buffer. To make 4.5 liters, add 17.6 g of Ambic and complete with 4.5 liters of water. Adjust to pH 8.5 with dilute acetic acid. Prepare fresh on day of use.

1. Reduction and Carboxymethylation of Homogenates/Cell Lysates

Steps

1. Weigh 10 mg of DTT and dissolve in 5 ml of degassed Tris buffer to make a 2-mg/ml DTT solution.
2. Add 0.5 ml of the DTT solution to the sample.
3. Incubate for 60 min at 37°C and then centrifuge.
4. Weigh 60 mg of IAA and dissolve in 1 ml of degassed Tris buffer.
5. Add 0.5 ml of fresh IAA solution to the sample.
6. Incubate in the dark at room temperature for 90 min.
7. Transfer the homogenate to high-quality dialysis tubing and seal.
8. Place the sample into cool dialysis buffer and dialyse at 4°C. Change the buffer

regularly with fresh buffer over a period of 48 h.

9. Lyophilise the sample in a clean screw-capped glass culture tube covered with perforated Parafilm.

2. Reduction and Carboxymethylation of Glycoproteins

Steps

1. Weigh out 1 mg of DTT and add 1 ml of Tris buffer to make a 1 μg/μl DTT solution.
2. Predict the number disulphide bridges in the glycoprotein if the disulphide bridges have not been mapped previously, i.e., (No. of cysteines ÷ 2).
3. Add enough Tris buffer (~200 μl) to completely dissolve sample and add the appropriate volume of DTT to the sample by using the following equation: Volume of DTT (L) = (No. of disulphide bridges) × (No. of moles of glycoprotein) × (4 × 154.3).
4. Incubate for 60 min at 37°C and then centrifuge.
5. Weigh out 1 mg of IAA and add 1 ml of Tris buffer to make a 1 μg/μl IAA solution.
6. Add the appropriate volume of IAA to the sample using the following equation: Volume of IAA (L) = (No. of moles of DTT in sample) × (5 × 186.0).
7. Incubate in the dark at room temperature for 90 min.
8. Transfer the homogenate to high-quality dialysis tubing and seal.
9. Place the sample into cool dialysis buffer and dialyse at 4°C. Change the buffer regularly with fresh buffer over a period of 48 h.
10. Lyophilise the sample in a clean screw-capped glass culture tube covered with perforated Parafilm.

C. Cleavage into Glycopeptides

Wear suitable protective clothing, including safety glasses, and work in a fume hood when preparing all solutions and performing various steps of the protocols.

Solutions

1. *Dilute acetic acid:* To make a 5% acetic acid solution, add 25 ml of acetic acid and complete to 500 ml with ultrapure water. Store at room temperature.
2. *Ambic buffer:* To make 50 ml, weigh 0.1977 g of ammonium hydrogen carbonate and complete with 50 ml ultrapure water. Adjust to pH 8.4 with ammonia solution. Prepare fresh on day of use.
3. *20 and 40% Propan-1-ol:* To make 50 ml of each solution, add 10 ml and 20 ml of propan-1-ol to separate clean glass bottles, and complete to 50 ml with ultrapure water. Prepare fresh on day of use.

1. Tryptic Digestion of Homogenates/Cell Lysates

Steps

1. Weigh out 1 mg of trypsin and add 1 ml of Ambic buffer to make a 1 μg/μl trypsin solution.
2. Add sufficient solution (300–600 μl) to completely cover the sample.
3. Incubate at 37°C for 14 h and then centrifuge.
4. Terminate the reaction by dispensing 2 drops of acetic acid with a Pasteur pipette.
5. Attach a 10-ml glass syringe to a Sep-Pak C_{18} cartridge and condition by eluting successively with methanol (5 ml), 5% acetic acid (5 ml), propan-1-ol (5 ml), and 5% acetic acid (3×5 ml).
6. Load the sample onto a cartridge.
7. Elute stepwise with 5% acetic acid (20 ml), 20% propan-1-ol solution (4 ml), 40% propan-1-ol solution (4 ml), and 100% propan-1-ol (4 ml). Collect all the fractions apart from the 5% acetic acid fraction (hydrophilic contaminants).

8. Reduce the volume with a Speed-Vac and lyophilise the sample in a clean screw-capped glass culture tube covered with perforated Parafilm.

2. *Tryptic Digestion of Glycoproteins*

Steps

1. Weigh out 1 mg of trypsin and add 1 ml of Ambic buffer to make a 1 µg/µl trypsin solution.
2. Estimate the approximate amount of glycoprotein.
3. Add sufficient solution (100–200 µl) to completely cover the sample, incubate at 37°C for 5 h and then centrifuge.
4. Dispense 3 drops of acetic acid from a Pasteur pipette to terminate the reaction.

3. *Cyanogen Bromide Cleavage of Glycoproteins*

Wear suitable protective clothing, including safety glasses, and work in a fume hood due to the toxic cyanogen bromide.

Steps

1. Add 700 µl of formic acid to 300 µl of water in an Eppendorf to make a 70% formic acid solution.
2. Add 8–10 good quality white cyanogen bromide crystals to the Eppendorf.
3. Vortex the mixture until the crystals dissolve completely.
4. Ensure that the glycoprotein sample is dry. Dissolve the sample in minimum amount of the cyanogen bromide solution, typically 100–200 µl.
5. Incubate at room temperature for 14–20 h.
6. Terminate the reaction by adding 4 volumes of ultrapure water.
7. Reduce the volume with a Savant Speed-Vac and lyophilise the sample in a clean screw-capped glass culture tube covered with perforated Parafilm.

D. Cleavage of Glycans From Glycopeptides

Wear suitable protective clothing, including safety glasses, and work in a fume hood when preparing all solutions and performing various steps of the protocols.

Solutions

1. *Dilute acetic acid:* To make a 5% acetic acid solution, add 25 ml of acetic acid and complete to 500 ml with ultrapure water. Store at room temperature.
2. *Ambic buffer:* To make 50 ml, weigh 0.1977 g of ammonium hydrogen carbonate and complete with 50 ml ultrapure water. Adjust to pH 8.4 with ammonia solution. Prepare fresh on day of use.
3. *20 and 40% propan-1-ol:* To make 50 ml of each solution, add 10 ml and 20 ml of propan-1-ol to separate clean glass bottles and complete to 50 ml with ultra-pure water. Prepare fresh on day of use.
4. *0.1 M KOH:* To make 100 ml, weigh 0.56 g of KOH and complete to 100 ml with ultrapure water. Store at room temperature.
5. *Dowex solution:* To prepare 100 g of Dowex beads, add 100 g of Dowex beads and complete with 100 ml of 4 M HC1 and decant. Repeat this twice more and then wash beads by adding, agitating, and decanting with ultrapure water until the pH does not change (usually 10–15 times). Wash the beads three times with 150 ml of dilute acetic acid and leave the beads immersed in dilute acetic acid. The treated beads can be kept equilibrated in this state for many months at room temperature.
6. *10% methanolic acetic acid:* To make 25 ml, add 2.5 ml of acetic acid and complete to 25 ml with methanol. Prepare fresh on day of use.

1. N-Glycosidase F Digest

Steps

1. Dissolve the 20% and 40% Sep-Pak fractions of the tryptic digest each in 150 µl of Ambic buffer and combine.
2. Add 5 U of N-glycosidase F, incubate at 37°C for 20–24 h, and then centrifuge.
3. Lyophilise the sample.
4. Separate N-glycans from the mixture using the Sep-Pak C_{18} propanol-1-ol / 5% acetic acid system.

2. Separation of N-Glycans from Peptides/ O-Glycopeptides by Sep-Pak C_{18} Purification

Steps

1. Attach a 10-ml glass syringe to a Sep-Pak C_{18} cartridge and condition by eluting successively with methanol (5 ml), 5% acetic acid (5 ml), propan-1-ol (5 ml), and 5% acetic acid (3×5 ml).
2. Load the sample on to a cartridge.
3. Elute stepwise with 5% acetic acid (5 ml), 20% propan-1-ol solution (4 ml), 40% propan-1-ol solution (4 ml), and 100% propan-1-ol (4 ml). Collect all fractions.
4. Reduce the volume with a Speed-Vac and lyophilise the sample in a clean screw-capped glass culture tube covered with perforated Parafilm.
5. Permethylate the 5% acetic acid fraction (the N-glycan fraction).
6. Perform reductive elimination on the 20% propan-l-ol fraction.

3. Reductive Elimination

Steps

1. Add 54–55 mg of KBH_4 to 1 ml of 0.1 M KOH.
2. Add 400 µl of the $NaBH_4$ solution to the sample in a Teflon-lined screw-capped culture tube.

3. Incubate at 45°C overnight for 20–24 h and then centrifuge.
4. Terminate the reaction by adding 5 drops of acetic acid from a Pasteur pipette until fizzing stops.
5. Assemble the desalting column by packing a Pasteur pipette fitted with a piece of silicone tubing at its tapered end (to control the flow) with Dowex beads.
6. Elute the desalting column with 15 ml of acetic acid.
7. Load the sample onto the top of the desalting column.
8. Elute with 5% acetic acid and collect 5 ml in a glass culture tube.
9. Reduce the volume on a Savant Speed-Vac and lyophilise.
10. Remove excess borate by coevaporating with 10% methanolic acetic acid. (4 × 0.5 ml) under a stream of nitrogen at room temperature. Repeat twice more.

E. Derivatisation of Glycans and Sep-Pak Cleanup

Wear suitable protective clothing, including safety glasses, and work in a fume hood when preparing all solutions and performing various steps of the protocols.

Solutions

1. *Anhydrous DMSO solution:* To make 400 ml, using a steel spatula carefully place good sized calcium hydride lumps into a clean 500-ml round-bottomed Quick-fit flask. Avoid putting powder into the flask. Add 400 ml of DMSO and stand overnight or longer until all powder has settled to the bottom. Keep the stopper tight and replace immediately each time after the DMSO has been dispensed. This DMSO stock solution may be kept as a stock solution at room temperature.

2. *15, 35, 50, 75% Acetonitrile solutions:* To make 100 ml of each solution, add 15, 35, 50, and 75 ml of acetonitrile to separate clean glass bottles and complete them to 100 ml with ultrapure water. Store at room temperature.

1. NaOH Permethylation

Wear suitable protective clothing, including safety glasses, and work in a fume hood due to the caustic NaOH and toxic methyl iodide.

Steps

1. Place 5 pellets of NaOH in a dry mortar and add 3 ml of anhydrous DMSO to the pellets.
2. Grind the NaOH pellets with a pestle to form a slurry. This should be done fairly swiftly to avoid excessive absorption of moisture from the atmosphere. Ensure the sample to be permethylated is completely dry.
3. Add 0.5–1 ml of the slurry to the sample and then add 0.2–0.5 ml of methyl iodide (or deuteromethyl iodide).
4. Vortex the sample and agitate the reaction mixture on an automatic shaker for 10 min at room temperature.
5. Quench the reaction by slow dropwise additions of ultrapure water (~1 ml) with constant shaking to lessen the effects of the highly exothermic reaction.
6. Add 1–2 ml of chloroform and make up to 5 ml with ultrapure water.
7. Mix thoroughly and allow the mixture to settle into two layers. Remove and discard the upper aqueous layer.
8. Wash the lower chloroform layer several times with ultrapure water.
9. Dry down the chloroform layer under a gentle stream of nitrogen.
10. Purify the mixture by Sep-Pak C_{18} purification.

2. Purification of Permethylated Samples by Sep-Pak C_{18} Purification

Steps

1. Condition the Sep-Pak cartridge by eluting successively with methanol (5 ml), ultrapure water (5 ml), acetonitrile (5 ml) and ultrapure water (3 × 5 ml).
2. Dissolve the sample in 1 : 1 methanol: ultrapure water (200 μl).
3. Load it onto the Sep-Pak cartridge.
4. Elute stepwise with 5 ml of ultrapure water and 3 ml each of 15, 35, 50, and 75% aqueous acetonitrile. Collect the fractions in culture tubes.
5. Reduce the volume by Savant Speed-Vac and lyophilise.

F. Useful Glycan Degradation Procedures

Wear suitable protective clothing, including safety glasses, and work in a fume hood when preparing all solutions and performing various steps of the protocols.

Solutions

1. *50 mM ammonium acetate:* To make 50 ml, add 0.1927 g of ammonium acetate and complete with 50 ml of ultrapure water. Adjust the pH to optimum range of the glycosidase.
2. *50 mM ammonium formate:* To make 50 ml, add 0.1577 g of ammonium formate and complete with 50 ml of ultrapure water. Adjust the pH to the optimum range of the glycosidase.

1. Hydrofluoric Acid 2-, 3-, and 4-Linked Fucose Removal

Wear suitable protective clothing, including safety glasses, and work in a fume hood due to toxic HF.

Steps

1. Add 50 μl of 48% HF to sample using a plastic micropipette tip. Ensure the sample is in a clean Eppendorf and is completely dry before adding the HF.
2. Incubate the sample on ice at 4°C for 20 h.
3. Terminate the reaction by drying under a gentle stream of nitrogen.
4. Dissolve the sample in 5% acetic acid and transfer to a screw-capped glass culture tube for subsequent glycosidase digestion or permethylation.

2. Mild Periodate Oxidation for O-Glycan Core Definition

Steps

1. Add 100 μl of 2–20 mM sodium m-periodate in ammonium acetate buffer (100 mM, pH 6.5) and incubate in the dark at 0°C for 20 h.
2. Quench the reaction by adding 2–3 μl of ethylene glycol.
3. Stand at room temperature for 1 h and then lyophilise.
4. Incubate the sample with 400 μl of NaBH$_4$ in 2M NH$_4$OH (10 mg/ml) for 2 h.
5. Terminate the reaction by adding a few drops of acetic acid dispensed from a Pasteur pipette until fizzing stops.
6. Assemble the desalting column by packing a Pasteur pipette fitted with a piece of silicone tubing at its tapered end (to control the flow) with Dowex beads.
7. Elute the desalting column with 15 ml of acetic acid.
8. Load the sample onto the top of the desalting column.
9. Elute with 5% acetic acid and collect 5 ml in a glass culture tube.
10. Reduce the volume on a Savant Speed-Vac and lyophilise.
11. Remove excess borates by coevaporating with 10% methanolic acetic acid (4 × 0.5 ml) under a stream of nitrogen at room temperature. Repeat twice more.

3. Exo-glycosidase Digestion

Perform glycosidase digestions on released or partially digested glycans.

Steps

1. Add up to 1 nmol of the glycan in an Eppendorf or a screw-capped glass culture tube.
2. Adjust the pH of ammonium acetate to the pH optimum of the glycosidase and then add 50 μl to the sample. Suitable pH values are as follows: α-mannosidase (jack bean, pH 5), (β-mannosidase (*Helix pomatia*, pH 4.6), β-galactosidase (bovine testes, pH 4), neuraminidase (*Clostridium perfringens*, pH 6), neuraminidase (*Streptococcus pneumoniae*, pH 6), and endo-β-galactosidase (*Bacteroides fragilis*, pH 5.8).
3. Add enough glycosidase to digest the sample. As a rough guide use the following amounts: α-mannosidase (jack bean, 20 mU), β-mannosidase (*H. pomatia*, 10 mU), β-galactosidase (bovine testes, 20 mU), neuraminidase (*C. perfringens*, 60 mU), neuraminidase (*S. pneumoniae*, 10 mU), and endo-β-galactosidase (*B. fragilis*, 50 mU).
4. Incubate at 37°C for 20 h and then centrifuge.
5. Add a second aliquot and incubate at 37°C for a further 12 h.
6. Lyophilise the sample in a clean screw-capped glass culture tube covered with perforated Parafilm.
7. Remove an appropriate aliquot and permethylate for subsequent analysis by MALDI-MS. *NB:* for β-N-acetylhexosaminidase (*S. pneumoniae* 80 mU, pH 5), use ammonium formate.

G. Defining Linkages and Sugar Compositions

Wear suitable protective clothing, including safety glasses, and work in a fume hood when preparing all solutions and performing various steps of the protocols.

Solutions

1. *2.0 M TFA:* Add 200 µl TFA to 1.1 ml of water. Prepare fresh on day of use.
2. *2.0 M NH₄OH:* Add 200 µl NH_3 to 1.62 ml of water. Prepare fresh on day of use.
3. *10% methanolic acetic acid:* To make 25 ml, add 2.5 ml of acetic acid and complete to 25 ml with methanol. Prepare fresh on day of use.
4. *Methanolic HCl:* To make a 1.0 M solution, add dropwise 100 µl of acetyl chloride to 1.3 ml of ice-cold methanol with constant shaking between each addition. Prepare fresh on day of use. Alternatively, for a rapid way of preparing a solution of approximately 1.0 M concentration, connect a Pasteur pipette to tubing attached to a cylinder of HCl gas and bubble the gas into 1 ml of methanol at room temperature until the bottom of the glass tube is hot to the touch.

1. Linkage Analysis of Permethylated Glycans

Steps

1. Add 200 µl of 2 M TFA to the permethylated sample and incubate at 121°C for 2 h.
2. Allow to cool, centrifuge, and dry under nitrogen.
3. Weigh out 10–20 mg NaBD₄ and add the appropriate volume of 2.0 M NH₄OH to make a 10-mg/ml solution.
4. Add 200 µl of the reducing reagent to the hydrolysates and stand at room temperature for 2 h.
5. Add 5 drops of acetic acid dropwise with a Pasteur pipette until fizzing stops and dry under nitrogen. It is not necessary to wait for complete dryness before proceeding to the next step.

6. Add 1 ml of 10% acetic acid in methanol and evaporate the solution under nitrogen until dry. Repeat twice more.
7. Add 200 µl acetic anhydride and incubate at 100°C for 1 h. Dry down under a stream of nitrogen.
8. Add 1 ml of chloroform and wash with ultrapure water.
9. Vortex the mixture and allow the two layers to separate. Discard the upper water layer.
10. Repeat two more times and dry under nitrogen. The resulting partially methylated alditol acetates (PMAA) can now be dissolved in a small volume of hexanes and analysed by GC-MS.

2. TMS Sugar Analysis

Steps

1. Add 200 µl of the methanolic-HCl reagent to the underivatised sample in a Teflon-lined screw-capped glass culture tube.
2. Incubate overnight (14–16 h) at 80°C and centrifuge.
3. Remove the reagent by blowing down under a stream of nitrogen.
4. Add the following reagents sequentially with thorough mixing in between: 500 µl of methanol, 10 µl (one drop from Pasteur pipette) of pyridine, and 50 µl of acetic anhydride.
5. Stand at room temperature for 15 min and remove the reagent by blowing down under a stream of nitrogen.
6. Break a 1-ml Tri-sil Z ampoule and add 100–200 µl of the TMS derivatisation reagent to the sample.
7. Stand at room temperature for about 15 min and dry down under a gentle stream of nitrogen.
8. Add a small squirt of hexane (~500 µl) and dry down again.
9. Add 1 ml of hexanes, mix thoroughly, and centrifuge at 3000 rpm for about 5–10 min.

10. Transfer the clear hexane supernatant to a new tube.

H. Mass Spectrometry

Solutions

1. *1% TFA solution:* To make 100 ml, add 1 ml of TFA to 99 ml of ultrapure water. Store at room temperature.
2. *0.1% TFA solution:* To make 100 ml, add 100 μl of TFA to 99.9 ml of ultrapure water. Store at room temperature.
3. *30% and 60% acetonitrile TFA solutions:* To make 50 ml of each solution, add 15 ml and 30 ml of acetonitrile to clean glass bottles and complete to 50 ml with 0.1% TFA to make 30 and 60% acetonitrile solutions.

I. Gas Chromotography–Mass Spectrometry

Analyse derivatised monosaccharide mixtures on a PerkinElmer's Clarus 500 gas chromograph/mass spectrometer (GC-MS) fitted with a RTX-5 column (30 × 0.32 mm internal diameter, Restek Corp.). Dry down the hexanes and redissolve the derivatives in ~50 μl of hexanes. Samples may be injected directly onto the column or by the autosampler. For PMAA analysis, the oven is held at 90°C for 1 min and subsequently ramped to 290°C at a rate of 8°C/min, held at 290°C for 5 min and finally to 300°C at a rate of 10°C/min. For standards, various synthetic glycans can be permethylated and taken through the linkage protocol to generate retention information for various linkages. As a rough guide, the elution times are ordered as follows: terminal fucose, terminal hexoses, linked hexoses, terminal N-acetylhexosamines, and finally linked N-acetylhexosamines. Note that sialic acids are not observed in this analysis. For TMS derivatives the oven is held at 90°C for 1 min and subsequently ramped to 140°C at a rate of 8°C/min,

then to 200°C at a rate of 5°C/min, and finally to 300°C at a rate of 10°C/min. For TMS derivatives, arabitol is used as an internal standard to track the reproducibility of retention times and for quantitative information. Monosaccharide standards usually include 1 nmol/μl solutions of arabitol, fucose, glucose, galactose, mannose, GlcNAc, GalNAc, and NeuAc.

J. Matrix-Assisted Laser Desorption Ionisation Mass Spectrometry

MALDI MS is generally performed in positive reflectron mode using a Perspective Biosystems Voyager-DE STR MALDI workstation equipped with delayed extraction technology. Data acquisition is performed using Voyager 5 Instrument Control Software, and data processing by Data Explorer MS processing software. Calibration is performed by external calibration of a mixture of leucine, enkephalin, bradykinin, bradykinin (fragment 1–8), angiotensin I, neurotensin, adrenocorticotropic hormone fragment (ACTH) 1–17, ACTH fragment 18–39, ACTH fragment 7–38, ACTH fragment 1–39, and insulin. Typical mass ranges are as follows: permethylated N-glycans in the range of m/z 1000–7000, and permethylated O-glycans in the range of m/z 500–4000. Permethylated glycans are usually in the 35%, 50%, and 75% fractions. The relevant fractions for N- and O-glycan studies are the 5% acetic acid fraction (N-glycans) and the 20% propan-1-ol fraction (O-glycopeptides) obtained after N-glycosidase.

1. MALDI Sample Preparation of Derivatised Glycans

Steps

1. Weigh 10 mg of DHB in an Eppendorf.
2. Add 200 μl of water and 800 μl methanol to the DHB
3. Vortex the matrix solution until the DHB completely dissolves.

4. Dissolve the derivatised glycans in a small volume of methanol ~5–10 μl.
5. Mix a 1-μl aliquot of the derivatised glycans with 1 μl or the fresh DHB solution.
6. Spot 1 μl of the sample matrix mixture on a clean stainless steel target and allow it to dry under vacuum. Perform MALDI-TOF experiments.

2. MALDI Sample Preparation of Peptides and Glycopeptides

If the sample has not been purified by Sep-Pak, perform steps 1–4, otherwise go to step 5.

Steps

1. Assemble a manual Microtrap Holder with a clean MicroTrap peptide cartridge.
2. Condition the cartridge by eluting successively with acetonitrile (5 × 20 μl) and then 0.1% TFA (8 × 20 μl).
3. Load the sample onto the column and then elute with 0.1% TFA (20 μl × 5) without collection.
4. Elute stepwise with 20 μl each of 30% and 60% acetonitrile in 0.1% TFA. Collect the fractions in Eppendorfs.
5. Weigh 10 mg of α-cyanocinnamic acid in an Eppendorf.
6. Add 1 ml of 30% acetonitrile in 0.3% TFA to the α-cyanocinnamic a acid and mix.
7. Dissolve the sample in an appropriate volume of 0.1% TFA to make a picomole per microliter solution if the sample is dry.
8. Mix a 1-μl aliquot of the peptides with 1 μl fresh α-cyanocinnamic acid solution.
9. Spot 1 μl of the sample matrix mixture on a clean stainless steel target and allow it to dry under vacuum. Perform MALDI-TOF experiments.

K. Nanoflow Electrospray Ionisation

Nano-ES MS and nano-ES MS/MS data are acquired using positive ion mode of a quadrupole orthogonal acceleration time-of-flight, Q-Tof 1, mass spectrometer (Waters Ltd, UK) fitted with a Z-spray atmospheric ion source. In MS mode the quadrupole is used in RF-only mode and transmits about two decades in mass to the TOF. In MS/MS the quadrupole is in resolving mode, allowing selection of precursors for collision in the hexapole gas cell. A voltage of 1.5 kV is applied to the NanoES capillary tip, generating a nanoflow in the range of 10–30 nl/min. Argon and nitrogen are used as the collision and bathing gases, respectively. Collision gas pressure is maintained at 10^{-4} mbar, and collision energies up to 90 eV are used for large glycopeptides, but do not usually exceed 50 eV for most peptides. As a rough guide, collision energies for doubly charged peptides range from <18 eV for m/z values of <500 to collision energies >35 eV for species with m/z values >1200. Collision energies are usually varied between 40 and 70 eV for permethylated glycans. Data are acquired and processed using Masslynx 4 software (Waters Ltd., UK). The instrument is calibrated with a 0.1 to 1 pmol/μl solution [Glu^1]-fibrinopeptide B in methanol / 5% (v/v) aqueous acetic acid [1 : 3, (v/v)]. Experiments may also be performed in a similar fashion on other quadrupole orthogonal acceleration time-of-flight instruments, such as the QSTAR Pulsar I mass spectrometer (AB/MDS Sciex, Toronto, Canada) fitted with a nanoelectrospray ion source (MDS Proteomics, Odense, Denmark) controlled by Analyst QS software.

1. ES Sample Preparation of Derivatised Glycans

Steps

1. Dissolve derivatised glycans in a suitable volume of methanol (~10 μl).
2. Withdraw a few microliters of the sample into a clean Hamilton syringe and inject into a nano-ES capillary.
3. Open the tip of the needle and load into the nano-ES interface of the mass spectrometer. Perform MS and MS/MS experiments.

2. ES Sample Preparation of Glycopeptides

If the sample has not been purified by Sep-Pak, perform steps 1–4, otherwise go to step 5.

Steps

1. Assemble a manual Microtrap Holder with a clean MicroTrap peptide cartridge.
2. Condition the cartridge by eluting successively with acetonitrile (5 × 20 µl) and then 0.1% TFA (8 × 20 µl).
3. Load the sample onto the column and then elute with 0.1% TFA (20 µl × 5) without collection.
4. Elute stepwise with 20 µl each of 30% and 60% acetonitrile in 0.1% TFA. Collect the fractions in Eppendorfs.
5. Withdraw a few microliters of the sample into a clean syringe and inject into a nano-ES capillary.
6. Open the tip of the needle and load into the nano-ES interface of the mass spectrometer. Perform MS and MS/MS experiments.

L. Nano-LCMS/MS Q-Star Hybrid MS/MS

Complex tryptic digests are analysed by nano-LC-MS/MS using a reversed-phase nano-HPLC system (Dionex, Sunnyvale) connected to a quadrupole TOF mass spectrometer (QSTAR Pulsar I, MDS Sciex, Canada). The digests are separated by a binary nano-HPLC gradient generated by an Ultimate pump fitted with a Famos autosampler and a Switchos microcolumn-switching module (LC Packings, Amsterdam, Netherlands). An analytical C18 nanocapillary (75 µm i.d. × 15 cm, PepMap) and a micro precolumn C18 cartridge (300 µm i.d. × 1 mm) are employed for online peptide separation. Digests are injected onto the precolumn by a Famos autosampler with volumes from volumes typically ranging from 0.5 to 5 µl. The

digests are first loaded onto the precolumn and eluted with 0.1% formic acid in water (HPLC grade, Purite) for 2–4 min. The sample is then transferred onto an analytical C18 nanocapillary HPLC column and eluted at a flow rate of 150–200 nl/min using the following gradient: 0–5 min 99% A, 5–10 min 99–90% A, 10–70 min 90–60% A, 70–71 min 60–50% A, 71–75 min 50–5% A, 75–85 min 5% A, 85–86 min 5–95% A, 86–90 min and 95% A. Solvent A = 0.05% (v/v) formic acid in 95 : 5 (v/v) water:acetonitrile; solvent B = 0.04% formic acid in 95 : 5 (v/v) acetonitrile:water. Data acquisition was performed using Analyst QS software with an automatic information-dependent acquisition (IDA) function. Similar experiments may also be performed on other quadrupole orthogonal acceleration time-of-flight instruments such as the MicroMass Q-ToF MS (Waters Ltd., UK) fitted with a CapLC system (Waters Ltd., UK) or with the same nano-HPLC system (Dionex, Sunnyvale) described earlier.

IV. COMMENTS

The protocols outlined in this article are suitable for profiling the major N- and O-glycan populations on a wide range of mammalian cells, fluids, tissues, and organs. In addition, these methodologies are used routinely to complement detailed studies of specific biologically active glycopolymers.

V. PITFALLS

1. Use ion-free ultrapure water at all times for washing plasticware and glassware. Always use ion-free ultrapure water to prepare solutions.
2. Clean all glassware thoroughly before use with water before drying in a 90°C oven. Do not use detergents to clean glassware.

3. Use screw-capped disposable Pyrex culture tubes with caps that have Teflon inserts as much as possible, i.e., incubations, reactions, and drying steps.

4. Use prepierced Parafilm (American National Can) to cover glass culture tubes during vacuum centrifugation and lyophilisation to minimise potential cross contamination.

5. Where possible, use glassware rather than plasticware in all steps apart from steps involving HF, as HF reacts with glass, e.g. use sterile nonplugged Pasteur pipettes to dispense solutions >0.2 ml and use sterile glass micropipettes to dispense solutions < 0.2 ml except for HF transfers.

6. Prior to reduction/carboxymethylation, take appropriate measures to ensure that extracts or samples are as free as possible of detergents and involatile salts.

7. TMS derivatives are volatile so do not leave for extended periods under nitrogen.

8. For detergent solutions, add the detergents after adjusting the pH in order to prevent precipitation.

9. When using gloves, wear the powder-free type.

10. For bulky tissues, it is advisable to split the sample, as large amounts of material can be difficult to process. As a rough guide, a mouse kidney should be split into two, whereas a mouse small intestine should be split into at least four samples.

References

Albersheim, P., Nevins, D. J., English, P. D., and Karr, A. (1967). A method for the analysis of sugars in plant cell wall polysaccharides by gas-liquid chromatography. *Carbohydr. Res.* **5**, 340–345.

Barber, M., Bordoli, R. S., Sedgwick, R. D., and Tyler, A. N. (1981). Fast atom bombardment of solids (F.A.B): A new ion source for mass spectrometry. *J. Chem. Soc. Chem. Commun.* **7**, 325–327.

Dell, A. (1987). F.A.B.: Mass spectrometry of carbohydrates. *Adv. Carbohydr. Chem. Biochem.* **45**, 19–72.

Dell, A., Azadi, P., Tiller, P., Thomas-Oates, J., Jennings, H. J., Beurret, M., and Michon, F. (1990). Analysis of oligosaccharides epitopes of meningococcal lipopolysaccharides by fast-atom-bombardment mass spectrometry. *Carbohydr. Res.* **200**, 59–76.

Dell, A., Morris, H. R., Easton, R. L., Panico, M., Patankar, M., Oehniger, S., Koistinen, H., Seppala, M., and Clark, C. F. (1995). Structure analysis of the oligosaccharides derived from glycodelin, a human glycoprotein with potent immunosuppressive and contraceptive activities. *J. Biol. Chem.* **270**, 24116–24124.

Dell, A., Morris, H. R., Egge, H., von Nicolai, and Strecker, G. (1983). Fast-atom-bombardment mass spectrometry for carbohydrate structure determination. *Carbohyr. Res.* **115**, 41–52.

Fenn, J. B., Mann, M., Meng, C. K., Wong, S. K., and Whitehouse, C. M (1990). Electrospray ionization-principles and practice. *Mass Spectrom. Rev.* **9**, 37–70.

Fukuda, M., Bothner, B., Ramsamooj, P., Dell, A., Tiller, P. R., Varki, A., and Klock, J. C. (1985). Structures of sialylated fucosyl polylactoaminioglycans isolated from chronic myelogenous leukemia cells. *J. Biol. Chem.* **260**, 12957–12967.

Fukuda, M., Dell, A., and Fukuda, M. N. (1984). Structure of fetal lactosaminoglycan. The carbohydrate moiety of band 3 isolated from human umbilical cord erythrocytes. *J. Biol. Chem.* **259**, 4782–4791.

Karas, M., and Hillenkamp, F. (1988). Laser desorption ionization of proteins with molecular masses exceeding 10,000 daltons. *Anal. Biochem.* **191**, 332–336.

Karas, M., Igendoh, A., Bahr, U., and Hillenkamp, F. (1989). Ultraviolet-laser desorption ionization mass-spectrometry of femtomolar amounts of large proteins. *Biomed. Environ. Mass Spectrom.* **18**, 841–843.

Laferte, S., Fukuda, M. N., Fukuda, M., Dell, A. and Dennis J. W. (1987). Glycosphingolipids of lectin-resistant mutants of the highly metastatic mouse tumour cell line, MDAY-D2. *Cancer Res.* **47**, 150–159.

Lopez, M., Coddeville, B., Langridge, Plancke, Y., Sautierre, P., Chaabihi, H., Chirat, F., Harduin-Lepers, A., Cerutti, M., Verbert, A., and Delannoy, P. (1997). Microheterogeneity of the oligosaccharides carried by the recombinant bovine lactoferrin expressed in *Mamestra brassicae* cells. *Glycobiology* **7**, 635–651.

Manzi, A. E, Norgard-Sumnicht, K., Argade, S., Marth, J. D., van Halbeek, H., and Varki, A. (2000). Exploring the glycans repertoire of genetically modified mice by isolation and profiling of the major glycan classes and nano-NMR analysos of glycans mixtures. *Glycobiology* **10**, 669–689.

McConville, M. J., Homans, S. W., Thomas-Oates, J. E., Dell, A., and Bacic, A. (1990). Structures of the glycoinositol-phospholipids from *Leishmania major:* A family of novel galactofuranose-containing glycolipids. *J. Biol. Chem.* **265**, 7285–7294.

Morris, H. R. (1980). Biomolecular structure determination by mass spectrometry. *Nature* **286**, 447–452.

Saski, H., Bothner, B., Dell, A., and Fukuda (1987). Carbohydrate structure of erythropoietin exopressed in Chinese hamster ovary cells by a human erythropoietin cDNA. *J. Biol. Chem.* **262**, 12059–12076.

Sutton-Smith, M., Morris, H. R., and Dell, A. (2000). A rapid mass spectrometric strategy suitable for the investigation of glycan alterations in knockout mice. *Tetrahedron-Asymmetry* **11**, 363–369.

Stable Isotope Labeling by Amino Acids in Cell Culture for Quantitative Proteomics

Shao-En Ong, Blagoy Blagoev, Irina Kratchmarova, Leonard J. Foster, Jens S. Andersen, and Matthias Mann

I. INTRODUCTION

Analysis by combined liquid chromatographic separation and mass spectrometry (LC-MS) is rapidly becoming the most popular and effective approach for large-scale protein identification (Aebersold and Mann, 2003; Yates, 2000) (see also article by Foster and Mann). The increasing sensitivity and the development of better protein/peptide separation methods will surely extend the catalogues of proteins emerging from proteomics projects. However, the significance of such data is further amplified if such identifications are coupled with quantitative abundance information. Indeed, it has been argued that almost all proteomics experiments should soon be converted to a quantitative format as qualitative lists of proteins may not be sufficient in the future (Aebersold and Mann, 2003).

Quantitative methods in mass spectrometry (MS) rely largely on the principle of stable isotope labeling (SIL). Differential incorporation of stable isotopic nuclei (^{2}H, ^{13}C, ^{15}N, ^{18}O) allows the "light" and a "heavy" form of the same peptide to be resolved in the mass spectrometer due to their mass difference. The ratio of signal intensities of the two peptides measures peptide abundance and, correspondingly, the relative abundance of the two proteins.

The main strategies for stable isotope incorporation can be broadly classified into two groups: "postharvest" incorporation and metabolic incorporation. The prototypical example of the former approach is the "isotope-coded affinity tag" (ICAT) reagent described in 1999 (Gygi *et al.*, 1999). In this approach, biotinylated tags are differentially mass encoded with stable isotopes (either 0 or 8 deuteriums) and bear a chemical moiety targeted at cysteine sulphydryl groups. The protein samples to be compared are derivatised separately with the two forms of ICAT, digested, and purified over an avidin column to enrich ICAT-labeled peptides for subsequent mass spectrometric analyses. Several other variations on this chemical derivatization theme have been applied since then (Goodlett *et al.*, 2001; Regnier *et al.*, 2002; Olsen *et al.*, 2004).

Metabolic incorporation of stable isotopes for MS quantitation of proteins was first introduced in 1998 for bacteria (Langen *et al.*, 2000) and in 1999 in yeast (Oda *et al.*, 1999). These first reports relied on growing simpler organisms on ^{15}N-enriched media. The growth of mammalian cells in ^{15}N-enriched media has also been described (Conrads *et al.*, 2001) but is largely untenable for most cell culture protocols due to the lack of a source of ^{15}N-enriched serum. In our laboratory, we have developed a method (SILAC) for stable isotope labeling by amino acids specifically suited for mammalian cell culture (Ong *et al.*, 2002). Independently, other groups have also described similar experiments for use in yeast (Jiang and English, 2002) and in mammalian systems (Zhu *et al.*, 2002).

The name SILAC (for stable isotope labeling by amino acids in cell culture) is apt as it accurately describes the mode in which the stable isotope labels are delivered, and the acronym is an extension of SIL already popularised from the small molecule field and in NMR. Rather than introducing the mass differential in all nitrogen atoms as in ^{15}N labeling, the SILAC method relies on stable isotope containing amino acids. This makes the incorporation of stable isotopes sequence dependent and inherently predictable. Furthermore, by introducing the stable isotope label in the form of amino acids in growth media, the label is effectively "encoded" into each newly synthesized protein. This differs significantly from chemical derivatisation methods that target specific amino acid residues. For instance, a limitation of the ICAT approach is that not all proteins contain cysteine. Often, SILAC labeling is more practical, as beyond preparation of labeling media, there are no chemical modifications required to "encode" proteins with quantitative information. Indeed, the chemical derivatizations required in postharvest labeling methods are not usually compatible with high-sensitivity MS measurements.

Importantly, SILAC provides investigators with two cell populations, grown in parallel and thus identical save for the fact that the proteins from each individual population are distinguishable by MS. The "encoded" cells can be mixed at harvest and can be subjected to any biophysical separation or protein purification directly. This contrasts with postharvest labeling approaches that require separate processing of protein samples until the label has been incorporated. Derivatization methods that incorporate the label at the peptide level (i.e., after proteins have been processed enzymatically) demand that these purification steps and protein digestion steps be done individually, thus significantly compromising the accuracy of quantitation.

This article describes the study of C2C12 muscle differentiation with SILAC as an example.

One set of cells are kept at the myoblast stage whilst the other set is induced to differentiate into myotubes (Ong *et al.*, 2002). Cells from both samples are mixed 1 to 1 and lysed. Proteins extracted from the mixture are then analyzed by standard MS techniques (Fig. 32.1). Many muscle-specific proteins, such as myosin, were found to be highly upregulated whilst other proteins, such as histones, were found at similar levels in both samples. This illustrates the ability of the SILAC approach to accurately measure changes in protein expression during a cell differentiation experiment.

Another intrinsic benefit of "coded" cell populations is apparent when SILAC is coupled to standard biochemical purifications of proteins with antibodies or bait proteins (Fig. 32.2). Pulldown experiments and subsequent MS identification have been used for large-scale analysis of protein complexes and protein–protein interactions (Gavin *et al.*, 2002; Ho *et al.*, 2002). The exquisite sensitivity of mass spectrometers generates large lists of identified proteins that are purportedly true binders to the bait. Although informative, it is not always easy to judge if a particular protein identified in this way is a real interaction partner or a nonspecific "background" interaction with the antibody/bait. This necessitates time-consuming and often difficult follow-up experiments to validate one's findings. Such uncertainty can be appropriately addressed with the following approach that combines SILAC with affinity purifications.

SILAC-labeled cells are used to generate separate pools of proteins or protein complexes that are distinct to an affinity bait. Cells may be differentially treated with a drug or growth factor; alternatively, cells may express the wild-type or

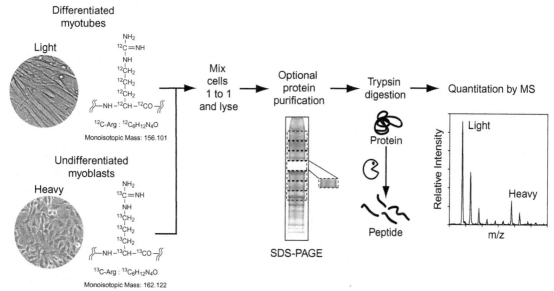

FIGURE 32.1 An overview of a quantitative study of C2C12 muscle differentiation with SILAC. Myoblasts are grown to confluence separately in "light" and "heavy" media. Myoblasts in "light" $^{12}C_6$-Arg media are induced to differentiate to form myotubes by reducing serum concentration to 2% dialysed FBS. Myoblasts in $^{13}C_6$-Arg are left undifferentiated. Cells are harvested and mixed 1 to 1 after protein concentration determination. Nuclei and cytoplasmic fractions from mixed cells are obtained. Proteins were separated on a SDS-PAGE gel, and gel slices were excised and digested with trypsin. Peptides were extracted from gel slices and analysed by LC-MS. The ratio of relative intensities of the "light" over the "heavy" peak cluster gives the fold increase in relative abundance of the protein during muscle differentiation.

V. MASS SPEC METHODS FOR DETERMINATION OF PROTEIN IDENTITY AND PROTEIN MODIFICATIONS

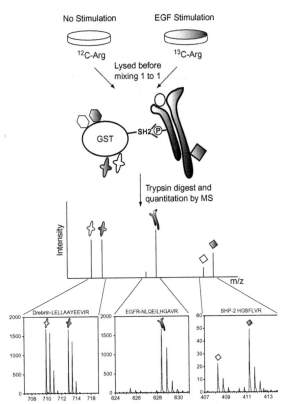

FIGURE 32.2 SILAC discriminates between true interaction partners in affinity purifications—posttranslational modification-dependent enrichment of epidermal growth factor receptor (EGFR) and associated proteins by binding to GST-SH2(Grb2) upon EGF stimulation. Two cell populations are differentially labeled with ^{13}C-Arg-SILAC. After treatment of "heavy" cells with EGF, lysates from both cell populations are mixed 1 to 1 and incubated with GST-SH2(Grb2). Proteins that interact specifically with the bait in conjunction with EGF stimulation will exhibit a large differential ratio in their peptide ratios. In contrast, nonspecific interactions with GST beads will bind equally from both cell populations with ratios similar to the original mixing ratio. Drebrin, an actin-binding protein, shows up with a peptide ratio of 1 (left spectrum), indicating that it was not pulled down by specific interaction with the bait. Quantitation of EGFR (middle spectrum) and Shp-2 (right spectrum) peptides gives enrichment ratios of >10 and 2.5, respectively, strongly indicative that they were affinity purified by the bait in response to EGF treatment.

mutant form of a component of a protein complex. Proteins from both states can be mixed in equal proportions and then purified together over the affinity bait. In this manner, peptides from proteins specific to the differential treatment will give a large differential ratio whilst background, nonspecific interactions with the bait will be close to the 1-to-1 ratio of mixing.

In one example, [13]C_6-Arg-labeled cells were stimulated briefly with EGF alongside untreated, unlabeled cells. Cells were lysed and equal amounts of protein from each sample were combined for subsequent affinity purification with the SH2 domain from Grb2 as a bait (Blagoev *et al.*, 2003). As the SH2 domain of Grb2 binds to phosphorylated tyrosines on the EGF receptor upon stimulation, we observed specific enrichment of the phosphorylated EGF receptor along with its associated proteins. Direct interaction partners exhibited a large (> fivefold) enrichment of one peptide form versus the other and were easily distinguished from background-binding proteins, which showed a 1-to-1 ratio (see Fig. 32.1). This discrimination between specific interaction partners and background binding is extremely powerful and addresses many of the inadequacies of previous approaches.

Using three different forms of arginine with SILAC makes it possible to functionally encode three cellular states in a single experiment. We have used [13]C_6-Arg, [13]C_6[15]N_4-Arg in addition to normal arginine to label all arginine-containing peptides in the proteome (Blagoev *et al.*, 2004). This is especially useful in cases where various time intervals of drug treatment are examined or where multiple cellular conditions are presented to an affinity bait. Using immunoprecipitation as an example, applying the "two-state" approach with only two distinct quantitative labels would require two separate pull downs followed by separate MS identifications and quantitative analyses. With triple encoding, this is reduced to a single experiment with a shared affinity purification step.

Like other methods of labeling that rely on metabolic incorporation of stable isotopes, SILAC can only be used with live cells. In dynamic studies of protein abundance in live cells (Pratt *et al.*, 2002), SILAC has obvious advantages over chemical methods. Clearly the two approaches (metabolic incorporation and postharvest derivatization) are complementary, each with their accompanying strengths and limitations.

Stable isotopes for quantitative measurements in proteomics are becoming increasingly important, as MS has grown to become a cornerstone technology in this field. The sensitivity and high mass accuracy of mass spectrometric readouts will likely mean that these tools will become indispensable in the life science field for years to come.

II. MATERIALS AND INSTRUMENTATION

Dialysed foetal bovine serum (FBS) (Cat. No. 12480-026), L-glutamine (Cat. No. 25030-024), penicillin/streptomycin (Cat. No. 15070-063), custom-synthesized media—Dulbecco's modified Eagle medium (like Cat. No. 21969 but without amino acids) and RPMI1640 (like Cat. No. 61870 but without amino acids) are from Invitrogen. MEM Eagle's deficient with Earle's salts and L-glutamine, without L-leucine, L-lysine, L-methionine (Cat No. M7270), L-arginine (Cat. No. A-6969), L-leucine (Cat No. L-8912), L-lysine (Cat. No. L-9037), and all remaining unlabeled L-amino acids were from Sigma. L-^{13}Q-Arginine (Cat. No. CLM-2265) and L-^{13}C$_6$-lysine (CLM-2247) are from Cambridge Isotope Labs. Leucine-5,5,5-d3 (Cat. No. 48,682-5) is from Sigma-Isotec. All water is "Milli-Q" quality distilled, deionized water.

Sequencing-grade porcine trypsin (Cat. No. V511C) is from Promega. Dithiothreitol (DTT, Cat. No. D-9163) and iodoacetamide (Cat No. 1-1149) are from Sigma. The HPLC system used in LC-MS analyses is the Agilent 1100. The hybrid quadrupole time-of-flight mass spectrometer is supplied by MDS-SCIEX-Applied Biosystems. The search engine for protein and peptide identification from LC-MS data is the Mascot search program from Matrix Science.

III. PROCEDURES

A. Labeling Media Preparation

Cell culture media can be prepared exactly according to the experimenter's specifications to suit any particular cell type (refer to Volume 1 for cell culture techniques). The only requirement is that the labeling AA of choice is substituted with an isotopically "heavy" form to distinguish it from unlabeled, "light" media. We have applied SILAC extensively in a variety of cell types in human and rodent systems and with an assortment of stable isotope nuclei-containing amino acids and found that cell growth, nutrient use, and general morphology are similar to cells grown in standard cell culture media.

Media preparations depleted in certain amino acids are available in powdered form. We have previously used MEM with Earle's salts deficient in lysine, methionine, and leucine (powdered media). Presently, we obtain standard media formulations (DMEM and RPMI) deficient in specific amino acids as custom-synthesized products from Invitrogen. A primary concern may be the availability of certain types of dialysed serum (horse, calf), even though most cell culture media companies routinely supply dialysed FBS. A limitation of the use of dialysed serum is the potential loss of low molecular mass (below 10 kDa) components, which may be important for cell growth, through the dialysis process but the importance of using dialysed serum cannot be overstated. It is absolutely necessary to avoid the contribution of unlabeled AAs in labeling media and is critical for accurate quantitation (Ong *et al.*, 2002).

The choice of a labeling amino acids for use in SILAC is likewise important. Common amino acids such as leucine (70% of all tryptic peptides contains at least one leucine residue—human IPI database, EBI) are preferable, generating several quantitatable peptides per protein. Alternatively, an enzymatic digest with trypsin for mass spectrometric analysis produces peptides with lysine or arginine at the C-terminal of the peptide. Labeling with arginine or lysine therefore results in incorporation of a single labeled residue in half of the tryptic peptides. If both AAs are used in SILAC, essentially all tryptic peptides would be labeled and quantitatable—a critical advantage in the quantitation of posttranslational modifications.

Essential amino acids are an obvious first choice in SILAC experiments. However, not all cell lines are capable of *de novo* synthesis of nonessential amino acids such as arginine and the empirical evaluation of a particular amino acid may well be worthwhile.

The use of ^{13}C-arginine (where all ^{12}C atoms are substituted with ^{13}C) in SILAC is particularly desirable because of the suitable mass differential encoded, as well as the coelution of peptide pairs (Zhang and Regnier, 2002)(see later). ^{13}C-containing amino acids cost more in comparison to deuterated amino acids. By testing cell lines with growth media containing less arginine or lysine, we find that significantly reducing the amount of these amino acids does not adversely affect the growth of cells. Because both control and experimental cell populations are grown on identical media compositions (other than the form of the labeling AA), the reduction of a particular amino acid concentration does not compromise the validity of the experiment. However, it is important to test each separate cell line for the nutritional requirements for a particular amino acid. For example, a cell line may begin to synthesize arginine *de novo* when the arginine concentration is reduced by a third whilst others (e.g., HeLa) only begin to do so at one-tenth the normal arginine concentration

(DMEM media). Reducing the levels of labeled amino acids may be cost effective and, in some cases (see later), even necessary.

Separately, we find that arginine is metabolically converted to proline in certain cell lines when supplied at concentrations described in standard media formulations. This observation was made as peptides containing [^{13}C]-proline were detected in MS analyses in the samples from human adenocarcinoma (HeLa) cells but not in a mouse fibroblast cell line (NIH 3T3). This proline conversion was undetectable when one-fifth of the original arginine concentration was used with HeLa cells.

As an example, this article describes the preparation of DMEM (a common media preparation for many commonly used cell types) for use with SILAC and arginine labeling (Ong *et al.*, 2003), but we reiterate that the general approach is directly applicable to one's custom media preparations (Ong *et al.*, 2004). An additional resource for SILAC information is available on our laboratory's Web page at http://www.cebi.sdu. dk/silac.htm.

Solutions

1. Base media preparation: The formulation of labeling media should be prepared according to the needs of the cell lines used. Labeling media should only be distinct in the form of labeling amino acid used. A base media should be prepared in the same manner right up to the final addition of normal or labeling amino acid. The custom media formulations we purchase only require the addition of the labeling amino acid, dialysed serum (10% FBS), penicillin (50 units/ml), streptomycin (50 μg/ml), and glutamine (2 mM final) before use.

2. Preparation of amino acid stock solutions: Concentrated stock solutions of amino acids are dissolved in phosphate-buffered saline and filter sterilized with a 0.22-μm filter. These stocks should be prepared at as high

a concentration as possible to minimize the dilution of other media components, e.g., arginine and lysine are prepared as 1000× (84 and 146 mg/ml, respectively) stocks. With labeled amino acids, it may be necessary to prepare unfiltered stock solutions and to filter media only after addition of the labeled amino acid in order to avoid losses due to filtration.

3. Dialysed serum: We use dialysed serum obtained from a commercial source. It is also possible to dialyse existing serum stocks in order to employ lower molecular weight cutoff dialysis filters and/or to reduce costs but it can be difficult to maintain consistency and to avoid contamination.

Steps

1. Work in a sterile environment. To two separate lots of base media, add the appropriate amount of arginine (either normal L-arginine to give "normal" media or L-^{13}C$_6$-arginine to give "labeling" media) to make up a full complement of amino acids, according to the manufacturer's specifications. Filter media through a 0.22-μm filter if unfiltered amino acid stocks are used.

2. Add antibiotics and glutamine as required along with 10% (v/v) dialysed serum to media containing the full complement of amino acids. Media are now ready for use and can be stored at 4°C like standard cell culture media.

B. Incorporation of Labeled AA in Growing Cells

Cells growing in normal cell culture media (in our example, murine C2C12 myoblasts) are passaged into dishes containing either normal or labeling media. C2C12 cells are allowed to undergo at least five cell doublings

in SILAC media to ensure full incorporation of the labeled amino acid (Fig. 32.3). After five cell doublings, only a minimal amount of the original unlabeled AA should exist in the entire protein population—a theoretical maximum of $(1/2)^5$ or 3.125% could remain. In actuality, the cells would incorporate the label much sooner through normal protein turnover and in addition to novel synthesis. Furthermore, during cultivation to obtain sufficient cell numbers for the intended experiment, continual passages will result in faster incorporation of the label (Fig. 32.3). When working with a new cell line or if certain experimental parameters have been changed (i.e., a different lot of dialysed serum is used), it is best to assay the state of incorporation by obtaining a protein sample for MS analysis before beginning the experiment. Ideally, small aliquots of unmixed cells from each condition should be saved from each experiment in order to check for incorporation state. MS analysis of proteins from the labeled lysates will reveal if the unlabeled amino acid is present in the protein sample. A correction factor for the protein ratios may be applied if necessary or the experiment can be repeated with cells after a longer period of adaptation.

Steps

1. From a dish of cells grown in standard cell culture media, passage cells into two separate lots, one containing the unlabeled "light" SILAC media and the other with labeled "heavy" SILAC media.

2. Grow cells in respective labeling media for a minimum of five cell doublings. If working with immortalized cells, passage the cells to deplete the cell populations of normal amino acid and to increase rate of incorporation of label.

3. Perform differential treatment of cell populations, i.e., differentiation protocol, and drug treatment protocol, and growth factor treatment.

SILAC labelling of cells

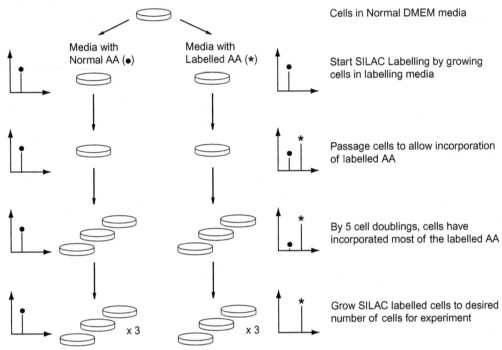

FIGURE 32.3   SILAC cell labeling and incorporation of labeled amino acid with continual passage of cells. Cells in standard DMEM supplemented with normal FBS, antibiotics, and glutamine are passaged into two separate dishes containing either isotopically "light" or "heavy" media. X–Y plots illustrate the presence of "light" and "heavy" peak clusters for a quantitatable peptide pair. The filled dot (•) indicates the peptide containing a normal amino acid, whilst asterisks (*) mark the peptide bearing the labeled, isotopically heavy amino acid. Cells grown in "heavy" media incorporate the labeled amino acid rapidly, as seen by the presence of the characteristic peptide pair even after the first passage. The intensity of the normal amino acid-containing peptide diminishes over time and, after sufficient cell doublings and passages, only the fully labeled peptide is detectable.

C. Cell Harvesting and Protein Purification

Steps (Assuming SILAC-Labeled Cells Have Been Treated Differentially)

1. Harvest cells from the tissue culture dish as in normal protocols. Save a small aliquot of unmixed cell populations to check levels of incorporation of the labeled AA if necessary.
2. Mix the two cell populations in a specific ratio (e.g., 1 to 1). This should be based on cell number (measured with a haemocytometer or Coulter counter) or protein concentration (determined by the Bradford method or the equivalent). In some instances and with sufficient sample, a combination of several experiments with different mixing ratios may be advantageous to expand the dynamic range of quantitation.
3. Optional protein purification steps may be included at this point. Examples include subcellular fractionation, gel filtration, immunoprecipitations, and one- or two-dimensional gel electrophoresis.
4. Digest proteins with a protease with high cleavage specificity, e.g., trypsin. With

[13]C-Arg SILAC, approximately half of the tryptic peptides contain arginine as the C-terminal amino acid and are quantitatable. Analyse samples with MS.

D. Analysis and Quantitation of SILAC Samples by Mass Spectrometry

A main benefit of SILAC is the option to mix cells prior to any further treatment of the sample. It is therefore possible to perform a subcellular fractionation to obtain organelles such as the nuclei without having to worry about introducing quantitative errors by way of differential treatment of samples. The downstream processing of protein samples for mass spectrometric analyses are similarly straightforward. Reduction, alkylation, and enzymatic digestion of protein samples (as described in Shevchenko et al., 1996) are performed on the same sample. Biases that may arise through labeling efficiency and enzymatic cleavage are thus avoided.

Mass spectrometric analyses can be performed with any of the standard MS instruments, but best results are obtained with a high mass accuracy, high-resolution instrument to resolve the natural isotope clusters of peptides. In our laboratory, we use the combination of nano-flow capillary liquid chromatography for the separation of complex peptide mixtures with subsequent detection and identification by a hybrid quadrupole time-of-flight mass spectrometer (QSTAR PULSAR—ABI).

In a typical LC-MS experiment, peptides eluting from the reversed-phase column are ionised and electrosprayed directly into the mass spectrometer. The mass spectrometer first obtains a survey MS scan where the entire mass range is analysed. From the survey scan, suitable peptides (of charge state $z = 2$, 3, or 4) can be selected for fragmentation in the collision cell in the MS/MS mode. The fragmentation of peptides in MS/MS spectra produces characteristic fragments, which give sequence-specific information for subsequent protein identification (Aebersold and Goodlett, 2001; Mann et al., 2001). This acquisition cycle, comprising of a survey scan followed by several MS/MS scans, is repeated throughout the LC-MS run. Ion intensities of monoisotopic peaks from each survey scan MS are used to quantitate peptide pairs. The ratios obtained from peptides used to identify a particular protein can then be averaged to give the relative ratio of protein abundance. Different laboratories approach MS analyses in subtly different ways depending on available instrumentation; we present here some guiding principles for quantitative analyses that should be applicable to anyone working in this field.

Guidelines

1. In quantitative LC-MS, a balance has to be struck between the goal of peptide identification (MS/MS sequencing events) and accurate quantitation (MS survey scans across eluting peaks). If the instrument is configured to perform a MS survey scan (taking 1 s) followed by four MS/MS spectra (each 1.5 s), the machine spends only one-seventh of the total run time acquiring the MS spectrum from which peptide ratios are acquired. Therefore, it is important that sufficient data points are collected (at least nine points for a Gaussian curve) to accurately plot the ion intensities across the eluting peptide peak.

2. In all cases where quantitation is performed, one should check that no unrelated peak clusters overlap the peptides in question. This is especially important where adequate protein/peptide separation steps do not precede MS analyses. Quantitation should be based on distinct peptide signals, and best results are obtained from peptides with a minimum signal-to-noise (S/N) ratio of 10.

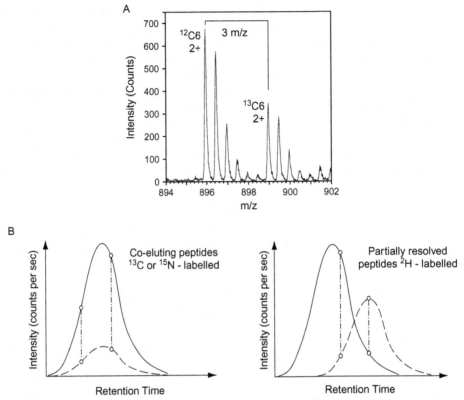

FIGURE 32.4   Peptide quantitation in MS. (A) A doubly charged SILAC-^{13}C$_6$-arginine-labeled peptide pair in the MS survey scan (acquisition time of 1s) with 3 m/z separation between the "light" and the "heavy" peak. There is clear separation between peak clusters and unrelated peaks are absent. If peptides elute from the reversed-phase column over a period of 40s, about 7 to 15 (depends on MS acquisition parameters) of these MS spectra may be acquired by the mass spectrometer. Each of these spectra give peak ratios, which are averaged to give a peptide ratio and standard deviation. (B) An example of coeluting peptides in an LC-MS run (left) and noncoeluting peptides (right). The peaks are extracted ion chromatograms (XICs) that monitor a particular mass range over time (see text for discussion on the adequate MS acquisition events to accurately describe a peak). Vertical dotted lines are examples of two individual MS acquisition events across the eluting peak. From the coeluting pair (left), it is apparent that similar ratios will be obtained regardless of the point when the peak is sampled. However, the partially resolved peaks (right) show inverse peptide ratios at the two time points indicated.

3. The selection of a suitable amino acid and the mass shift that will be incorporated by the stable isotopes is important. A common amino acid such as leucine is present in about 70% of unique tryptic peptides in the human proteome and is thus a good choice as a labeling amino acid. We also favour arginine and lysine, as enzymatic cleavage with trypsin generates peptides with these amino acids at C termini. The mass shift generated from the labeling should ideally be large enough (4 Da or greater) to avoid overlapping of isotope clusters (Fig. 32.4A). If peptide isotope clusters overlap, calculation of the isotopic envelope of each peptide based on sequence composition can still provide accurate quantitation data.

4. Care should be taken when quantitating different proteins that have regions of sequence identity (a protein family for,

e.g., the histones), as some peptides can be shared across several proteins. Quantitation based on a shared peptide may not accurately reflect the quantitation of any single protein, but might instead be an average across multiple proteins. Quantitation is best based on peptides unique to the protein of interest.

5. Many labeling reagents make use of deuterium (^{2}H) instead of ^{15}N and ^{13}C. There is a tendency for peptides labeled with deuterated reagents to elute earlier than the corresponding unlabeled peptide in reversed-phase chromatography due to isotope effects. As peptide quantitation is obtained from the MS survey scan at various time points across the elution of a peptide (Fig. 32.4B, left), accurate determination of the peptide ratio is possible at each sampling point of the MS. More data points are required with partially resolved peaks in order to accurately plot each eluting peptide peak (Fig. 32.4B, right) and quantitation should be based on peak area of the extracted ion chromatogram or a sampling of MS spectra over the retention times of the peptide pair. Although generally more expensive, using ^{13}C-substituted amino acids in SILAC helps reduce errors in quantitation caused by separation of the peptide pair. It also simplifies the process of quantitation, as the peptide ratio can be obtained from each MS scan directly. Having said that, it is very straightforward to obtain good quantitation data from deuterated reagents by simply being aware of the potential pitfalls.

6. Adequate ion statistics for quantitation peaks. Regardless of the type of mass spectrometer used to acquire quantitative data, it is imperative that mass spectra acquired should comprise enough data collection events in order to accurately describe the peptide peaks. For peptide identification, this requirement may not be as critical, but it should be apparent that accurate quantitation

is only achievable where sufficient data points can be averaged. In cases where highly differential peptide ratios are observed, one of the peaks may not be detectable above noise. Here, it may be sufficient to assign the smaller peak at the level of noise and describe a lower limit of the peptide ratio rather than to give some arbitrary (and most likely incorrect) value to the smaller peak.

References

Aebersold, R., and Goodlett, D. R. (2001). Mass spectrometry in proteomics. *Chem Rev.* **101**(2), 269–295.

Aebersold, R., and Mann, M. (2003). Mass spectrometry-based proteomics. *Nature* **422**(6928). 198–207.

Blagoev, B., Kratchmarova, I., Ong, S. E., Nielsen, M., Foster, L. J., and Mann, M. (2003). A proteomics strategy to elucidate functional protein-protein interactions applied to EGF signaling. *Nature Biotechnol.* **21**(3), 315–318.

Blagoev, B., Ong, S.E., Kratchmarova, I., and Mann, M. (2004). Temporal analysis of phosphotyrosine-dependent signaling networks by quantitative proteomics. *Nat Biotechnol.* **22**(9), 1139–45.

Conrads, T. P., Alving, K., Veenstra, T. D., Belov, M. E., Anderson, G. A., Anderson, D. J., Lipton, M. S., Pasa-Tolic, L., Udseth, H. R., Chrisler, W. B., Thrall, B. D., and Smith, R. D. (2001). Quantitative analysis of bacterial and mammalian proteomes using a combination of cysteine affinity tags and 15N-metabolic labeling. *Anal. Chem.* **73**(9), 2132–2139.

Gavin, A. C., Bosche, M., Krause, R., Grandi, P., Marzioch, M., Bauer, A., Schultz, J., Rick, J. M., Michon, A. M., Cruciat, C. M., Remor, M., Hofert, C., Schelder, M., Brajenovic, M., Ruffner, H., Merino, A., Klein, K., Hudak, M., Dickson, D., Rudi, T., Gnau, V., Bauch, A., Bastuck, S., Huhse, B., Leutwein, C., Heurtier, M. A., Copley, R. R., Edelmann, A., Querfurth, E., Rybin, V., Drewes, G., Raida, M., Bouwmeester, T., Bork, P., Seraphin, B., Kuster, B., Neubauer, G., and Superti-Furga, G. (2002). Functional organization of the yeast proteome by systematic analysis of protein complexes. *Nature* **415**(6868), 141–147.

Goodlett, D. R., Keller, A., Watts, J. D., Newitt, R., Yi, E. C, Purvine, S., Eng, J. K., von Haller, P., Aebersold, R., and Kolker, E. (2001). Differential stable isotope labeling of peptides for quantitation and de novo sequence derivation. *Rapid Commun. Mass Spectrom.* **15**(14), 1214–1221.

Gygi, S. P., Rist, B., Gerber, S. A., Turecek, R, Gelb, M. H., and Aebersold, R. (1999). Quantitative analysis of complex protein mixtures using isotope-coded affinity tags. *Nature Biotechnol.* **17**(10), 994–999.

Ho, Y., Gruhler, A., Heilbut, A., Bader, G. D., Moore, L., Adams, S. L., Millar, A., Taylor, P., Bennett, K., Boutilier, K., Yang, L., Wolting, C, Donaldson, I., Schandorff, S., Shewnarane, J., Vo, M., Taggart, J., Goudreault, M., Muskat, B., Alfarano, C, Dewar, D., Lin, Z., Michalickova, K., Willems, A. R., Sassi, H., Nielsen, R A., Rasmussen, K. J., Andersen, J. R., Johansen, L. E., Hansen, L. H., Jespersen, H., Podtelejnikov, A., Nielsen, E., Crawford, J., Poulsen, V., Sorensen, B. D., Matthiesen, J., Hendrickson, R. C., Gleeson, R, Pawson, T., Moran, M. E, Durocher, D., Mann, M., Hogue, C. W., Figeys, D., and Tyers, M. (2002). Systematic identification of protein complexes in *Saccharomyces cerevisiae* by mass spectrometry. *Nature* **415**(6868), 180–183.

Jiang, H., and English, A. M. (2002). Quantitative analysis of the yeast proteome by incorporation of isotopically labeled leucine. J. *Proteome Res.* **1**(4), 345–350.

Langen, H., Takacs, B., Evers, S., Berndt, P., Lahm, H. W., Wipf, B., Gray, C., and Fountoulakis, M. (2000). Two-dimensional map of the proteome of *Haemophilus influenzae. Electrophoresis* 21(2), 411–429.

Mann, M., Hendrickson, R. C., and Pandey, A. (2001). Analysis of proteins and proteomes by mass spectrometry. *Annu. Rev. Biochem.* 70, 437–473.

Oda, Y, Huang, K., Cross, F. R., Cowburn, D., and Chait, B. T. (1999). Accurate quantitation of protein expression and site-specific phosphorylation. *Proc. Natl. Acad. Sci. USA* **96**(12), 6591–6596.

Olsen, J. V., Andersen, J. R., Nielsen, P. A., Nielsen, M. L., Figeys, D., Mann, M., and Wisniewski, J. R. (2004). HysTag-a novel proteomic quantification tool applied to differential display analysis of membrane proteins from distinct areas of mouse brain. *Mol Cell Proteomics.* 3(1), 82–92.

Ong, S. E., Mittler, G., and Mann, M. (2004). Identifying and quantifying in vivo methylation sites by heavy methyl SILAC. *Nat. Methods.* **1**, 119–126.

Ong, S. E., Blagoev, B., Kratchmarova, I., Kristensen, D. B., Steen, H., Pandey, A., and Mann, M. (2002). Stable isotope labeling by amino acids in cell culture, SILAC, as a simple and accurate approach to expression proteomics. *Mol. Cell Proteomics* **1**(5), 376–386.

Ong, S. E., Kratchmarova, I., and Mann, M. (2003). Properties of 13C-substituted arginine in stable isotope labeling by amino acids in cell culture (SILAC). J. *Proteome Res.* **2**(2), 173–181.

Pratt, J. M., Petty, J., Riba-Garcia, I., Robertson, D. H., Gaskell, S. J., Oliver, S. G., and Beynon, R. J. (2002). Dynamics of protein turnover, a missing dimension in proteomics. *Mol. Cell Proteomics* **1**(8), 579–591.

Regnier, F. E., Riggs, L., Zhang, R., Xiong, L., Liu, P., Chakraborty, A., Seeley, E., Sioma, C, and Thompson, R. A. (2002). Comparative proteomics based on stable isotope labeling and affinity selection. J. *Mass Spectrom.* **37**(2), 133–145.

Shevchenko, A., Wilm, M., Vorm, O., and Mann, M. (1996). Mass spectrometric sequencing of proteins silver-stained polyacrylamide gels. *Anal. Chem.* **68**(5), 850–858.

Yates, J. R., 3rd (2000). Mass spectrometry: From genomics to proteomics. *Trends Genet.* **16**(1), 5–8.

Zhang, R., and Regnier, F. E. (2002). Minimizing resolution of iso-topically coded peptides in comparative proteomics. J. *Proteome Res.* **1**(2), 139–147.

Zhu, H., Pan, S., Gu, S., Bradbury, E. M., and Chen, X. (2002). Amino acid residue specific stable isotope labeling for quantitative proteomics. *Rapid Commun Mass Spectrom.* **16**(22), 2115–2123.

Site-Specific, Stable Isotope Labeling of Cysteinyl Peptides in Complex Peptide Mixtures

Huilin Zhou, Rosemary Boyle, and Ruedi Aebersold

I. INTRODUCTION

Relative quantification of proteins from different samples by mass spectrometry (MS) is based on the stable isotope dilution approach (Gygi *et al.*, 1999; Han *et al.*, 2001; Smolka *et al.*, 2002; Zhou *et al.*, 2002). Proteins or peptides are labeled with chemically identical tags that differ in mass due to a stable isotope content. In a typical experiment, proteins (or peptides derived from proteolytic digestion of proteins) from one sample are labeled with an isotopically heavy mass tag, whereas the light isotope tag is used to label the sample to be compared.

The isotopically labeled peptides are combined, purified or separated into fractions, and analyzed by mass spectrometry, which measures the mass and ion abundance of peptides. Because isotopically heavy and light forms of a peptide of the same amino acid sequence are chemically identical, they generate responses with identical sensitivity from the mass spectrometer and are readily distinguished based on their mass differences. Therefore the measured ion abundance ratio between heavy- and light-labeled peptides by the mass spectrometer is the actual abundance ratio of this peptide from two different samples. In this way, the relative abundance of peptides, and thus proteins, in two different samples can be determined accurately. Quantitative and comparative analysis of protein abundance from different samples has many applications, including large-scale protein expression profiling from different cell states to identify proteins that are unique in one cell condition, not the other, or comparative analysis of protein complexes derived from different cell states to reveal their compositional differences. In this case, conventional protein purification techniques would be used to enrich the protein complexes of interest. Many cell biological processes are carried out by large, multisubunits protein complexes. It would be an essential step to identify their protein components and dynamic changes in protein compositions under a different cell context. Comparative protein analysis by mass spectrometry would be a powerful tool toward this goal.

This article describes a method for site-specific stable isotope labeling of cysteinyl peptides in complex peptide mixtures via a solid-phase capture and release process, and the concomitant isolation of the labeled peptides (Zhou *et al.*, 2002). The recovered, tagged peptides were analyzed by microcapillary liquid chromatography and tandem mass spectrometry (uLC-MS/MS) to determine their sequences and relative abundance.

II. MATERIALS AND INSTRUMENTATION

1. Amino propyl glass beads, 200–400 mesh, pore size 170 Å (Sigma, St. Louis, MO., G4518)
2. Organic solvents: anhydrous dimethylformamide (DMF, Aldrich, 22705-6) and dichloromethane (DCM, Aldrich, 220997-1L)
3. 1-Hydroxybenzotriazole (HOBt) (Nova Biochem, Laufelfingen, Switzerland, 01-62-0008)
4. Fmoc-protected amino acids: Fmoc-aminoethyl photolinker (Nova Biochem, 01-60-0042) and Fmoc-γ-amino butyric acid (Fmoc-GABA) (Nova Biochem, 04-12-1088)
5. Diisopropyl carbodiimide (DIC) (Aldrich, D12540-7)
6. Acetic anhydride (Aldrich, 24284-5)
7. Pyridine (Aldrich, 36057-0)
8. Piperidine (Aldrich, 104094)
9. D6-γ-aminobutyric acid (d6-GABA) (Isotec, Inc., 82-222-02-7)
10. Fmoc-N-hydroxysuccinimide (Fmoc-Osu) (Nova Biochem, 01-63-0001)
11. Diisopropyl ethyl amine (DIPEA) (Aldrich, D12580-6)
12. Iodoacetic anhydride (Aldrich, 28426-2)
13. Micro Bio-Spin columns (Bio-Rad Labs, Hercules, CA, 732-6204)
14. Blak-Ray long-wave UV lamp (100 W, VWR Scientific, Inc., 36595-020)
15. Tri(caboxyl ethyl)phosphine (TCEP) (Pierce, 20490)
16. Trypsin, sequencing grade (Promega, V5111)

Stable isotope-labeled peptides can be analyzed by all types of mass spectrometry instrumentation, including electrospray ionization (ESI) and matrix-assisted laser desorption ionization (MALDI)-based techniques. We typically use liquid chromatography (LC) and ESI-tandem mass spectrometry to analyze complex peptide mixtures. A tandem mass spectrometer allows identification of peptide sequences as well as their relative ion abundance of the isotopically related peptides in the same experiment. The Finnigan LCQ ion-trap instrument is used in combination with the Hewlett-Packard HPll00 series HPLC system.

III. PROCEDURES

A. Synthesis of Solid-Phase Isotope-Labeling Reagents

A schematic diagram of the chemical structure of the solid-phase reagent is shown in Fig. 33.1. Synthesis of solid-phase reagents is based on a method that has been published previously (Holmes and Jones, 1995). For synthesis of beads with a heavy isotope, Fmoc-d6-GABA was prepared from d6-GABA and Fmoc-OSu, as described in Section IILB because the Fmoc-protected, deuterated amino acid is not available commercially. The methods described involve standard peptide chemistry, making it possible to use other amino acids with isotopically heavy or light forms.

Steps

1. Load 100 mg of aminopropyl glass beads in an empty Bio-Rad column or other column of suitable size. Wash beads once with 1 column volume of anhydrous DMF.
2. Form amino acid ester: Dissolve 120 μmol each of HoBt and Fmoc-aminoethyl photolinker in 0.8 ml of dry DMF completely. Add to this solution 120 μmol of DIC for 30 min. (Keep light-sensitive reagents from direct room light.)
3. Add the amino acid ester to the beads, mixing the beads by pipetting a few strokes. Incubate for 90 min.
4. Wash beads with 3 column volumes of DMF and 2 column volumes of dry

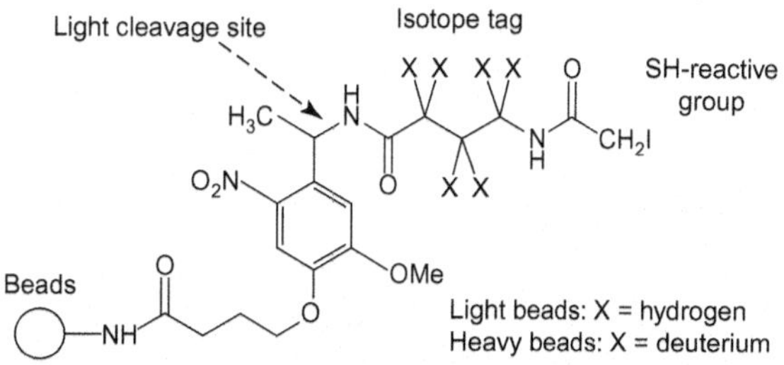

FIGURE 33.1 A schematic diagram of the solid-phase isotope-tagging reagent showing the chemical structure. The amino-propyl glass bead is first coated by a photocleavable linker molecule. Peripheral to the photocleavable linker, an amino acid, γ-aminobutyric acid, is used as an isotope-encoding mass tag that can be either nondeuterated (d0) or deuterated (d6). Following the isotope mass tag, an iodoacetyl group is used as a SH-reactive group to capture cysteinyl peptides. Following capture, photocleaving will lead to the recovery of cysteinyl peptides with the isotope tags attached to their cysteine residues.

dichloromethane. Always remove excess solvent between washes by applying a little pressure (squeeze a Pasteur pipettor bulb or apply house vacuum).

5. Block: Prepare a 1-ml mixture of 20% acetic anhydride, 30% pyridine, and 50% dichloromethane. Add this mixture to the beads for 30 min to block residual free amines on the beads.
6. Wash beads with 2 column volumes of dichloromethane and 3 column volumes of DMF. Remove excess DMF.
7. Deprotect: Prepare 3 ml of 20% (v/v) piperidine/ DMF solution. Add 1 ml to the beads and incubate for 30 min. Collect all of the 1-ml flow through containing Fmoc released from the photolinker. Calculate the capacity of the beads by measuring the absorbance of the released Fmoc. Use 20% piperidine/DMF solution as a blank solution and measure absorbance at 290 nm (A_{290}) of the 1/100 dilution (by 20% piperdine/DMF) of the flow through. The A_{290} should be between 0.6 and 0.8. Calculate capacity according to the formula: [A_{290} × dilution factor × flow

through volume (ml)]/[1.65 × weight of beads (mg)] = capacity (mmol/g).

8. Wash beads with 5 column volumes of dry DMF.

9. Repeat steps 2 to 8 with Fmoc-d0-GABA or its heavy form. The calculated capacity for GABA should be close to that for the photolinker.

10. Attach iodoacetyl group to the beads (Zhou *et al.*, 2001): Dissolve 120 µmol of iodoacetic anhydride in 0.8 ml dry DMF and add to the beads. Immediately add 132 µmol of diisopropyl ethyl amine (DIPEA) to the beads and mix well by pipetting a few strokes. Let it incubate for 90 min.

11. Wash the beads with 5 column volumes of DMF and excess methanol; dry the beads in a Speed-Vac (covered in foil). The beads can be stored in the dark at room temperature or in the refrigerator indefinitely.

B. Synthesis of Fmoc-Protected Amino Acid

The following procedure permits custom synthesis of Fmoc-protected amino acid for attachment to solid phase as described previously.

Steps

1. Dissolve 600 µmol d6-GABA in 3 ml 9% sodium carbonate in water in a vial under stirring.

2. Dissolve 900 µmol Fmoc-OSu in 3 ml DMF. Add to d6-GABA solution in one proportion. Continue to stir for 30 min at room temperature.

3. Divide the sample into six 1.5-ml Eppendorf tubes and dry out DMF under reduced pressure in a Speed-Vac.

4. Add 1 ml H_2O to dissolve the white powder as much as possible. Spin down and collect the supernatant. Repeat the water wash.

Combine all of the supernatant, adding water until the final volume is 15 ml. Discard insoluble material.

5. Add 600 µl concentrated HCl to the supernatant very slowly. The solution should become cloudy immediately with foam due to carbon dioxide and precipitation of reagent. Check by pH paper that the final pH is approximately 2.

6. Extract Fmoc-d6-GABA: Add 4 ml or more ethyl acetate to the acidified aqueous solution, wait for phase separation, and collect the ethyl acetate phase that contains Fmoc-d6-GABA. Repeat this extraction procedure three times and combine the extracts.

7. Wash the ethyl acetate extract once with 3 ml 0.1% HCl in H_2O and once more with 3 ml of water. Dry the extract completely in a Speed-Vac. The dried sample can be used directly with the assumption of >90% yield.

C. Preparation of Protein Digest

For labeling with SH-specific solid-phase reagents, it is advantageous to label peptides instead of proteins because proteins may possess tertiary structures that render some cysteine residues inaccessible to the solid-phase reagent. Protein digestion can be performed with any commercially available and suitable protease. Trypsin is used most frequently. For example, proteins in 100 µl 0.2 M Tris, pH, 8.0, can be digested by 1/50 (w/w) trypsin at 37°C overnight.

D. Capture and Release of Isotope-Labeled Peptides

Keep light-sensitive beads out of direct light as much as possible. Brief exposure to room light should not significantly affect the performance of the reagents.

1. Reduce protein digest with 5 mM TCEP for 30 min at room temperature. Because TCEP is quite acidic, it is essential that there is sufficient buffering capacity; 200 mM Tris, pH, 8.0, in the buffer should be sufficient to maintain the pH. TCEP is usually prepared as a 250 mM solution in water and kept at −20°C prior to use. We found that the TCEP stock solution is stable for many months at −20°C.

2. Weigh 5 mg each of isotopically light and heavy beads into tubes that are covered with foil to protect against light.

3. Add reduced protein digests to the beads (or the beads to the protein digest) and shake immediately for 15 min on a vortex mixer at a speed such that the beads should be suspended in the solution, not settled in the bottom of the tube. Efficient mixing of the beads with peptides is important for binding to occur.

4. After 15 min of binding, quench the labeling reaction with 2 μl β-mercaptoethanol for 1–2 min. The solid-phase isotope-labeling reagent should have excess capacity compared to the cysteinyl peptides. The addition of β-mercaptoethanol will block any remaining iodoacetyl group on the beads and prevent any possible side reactions to occur.

5. Combine the beads by loading onto a foil-wrapped Bio-Spin column, rinsing with water and methanol to transfer all beads. (Retain the flow through of each labeling reaction separately for analysis of noncysteine-containing peptides if so desired.) Wash with

   a. 2 × 1 ml 2 M NaCl
   b. 2 × 1 ml 0.1% TFA
   c. 2 × 1 ml 80% ACN/0.1% TFA
   d. 2 × 1 ml MeOH
   e. 2 × 1 ml 28% NH₄OH : MeOH (1 : 9 v/v)
   f. 2 × 1ml MeOH
   g. 2 × 1 ml water

6. Seal bottom of Bio-Spin column with a cap. Suspend beads in 200 μl 20 mM Tris/1 mM

EDTA, pH 8, and 4 μl (β-mercaptoethanol with a magnetic stirrer on a stir plate.

7. Expose beads to UV light for 2 h and then collect the supernatant through the Bio-Spin column.

8. Wash the remaining beads with 5 × 100 μl 80% ACN/0.4% acetic acid, combining with the previous supernatant.

9. Reduce the sample volume in Speed-Vac to approximately 200 μl. Check that the pH is acidic.

E. Sample Cleanup and Mass Spectrometric Analysis

Although the labeled peptides appear to be highly pure, free from side reactions of the peptides themselves, we have observed side products other than peptides following photocleaving of the beads. These residual products are likely to be impurities generated during synthesis of the solid-phase reagent. Because these products interfere with MS analysis of peptide samples, it is necessary to remove them prior to MS analysis. Additionally, we have observed that these side products are not positively charged under acidic pH, whereas peptides are positively charged due to protonation of basic residues such as the N terminus, histidine, lysine, and arginine. We therefore devised a strategy using cation-exchange chromatography to remove adducts from peptides. A disposable mixed cation-exchange (MCX) cartridge can be used.

Steps

1. Load sample onto MCX column (30-mg beads).

2. Wash with 3 column volumes of 0.1% TFA in water.

3. Wash with 3 column volumes of 80% ACN/0.1% TFA in water.

4. Wash by 1 column volume of water to prevent salt formation.

5. Elute in 500 µl of elution solvent consisting of 1 volume of ammonia solution (28% NH_4OH stock) and 9 volume of methanol.
6. Dry out ammonia and methanol in a Speed-Vac and resuspend the sample in 10 µl water for MS analysis.

There are several advantages to this solid-phase approach for isotopic labeling of peptides. First, isolation of cysteine-containing peptides and stable isotope incorporation are achieved in a single step. Therefore, the solid-phase method is rather simple. Second, the covalent attachment of peptides to a solid phase allows for the use of stringent wash conditions to remove noncovalently associated molecules. Third, this procedure is unaffected by the presence of proteolytic enzymes, such as trypsin, or strong denaturants or detergents, such as urea or SDS. There is no need for additional steps for their removal prior to peptide capture by the solid-phase beads and it is easy to remove them by washing. Fourth, the standard solid-phase peptide chemistry involved in the coupling process enables the use of a range of natural or unnatural amino acids in place of the d0/d6-GABA to function as the isotopic mass tag. This allows for synthesis of beads with a range of mass tags for analysis of multiple samples (i.e., more than two) in a single experiment if desired.

F. An Example of Protein Quantitation

We show an example of protein quantitation by this approach (see Table 33.1). Three proteins—glyceraldehyde 3-phosphate dehydrogenase from rabbit, bovine lactoalbumin, and ovalbumin from chicken—were prepared in different amounts and labeled by the solid-phase isotope-tagging reagents. Following light cleavage, the recovered peptides were analyzed by mass spectrometry, and the isotopically labeled peptides were identified and quantified as described (Gygi et al., 1999; Han et al., 2001;

Eng et al., 1994). The agreement with expected values was generally within 20% and, for any given protein, consistent ratios were observed. Additional application of this method can be found elsewhere (Zhou et al., 2002).

IV. NOTES

1. The amine capacity of the aminopropyl glass beads should be measured, despite the value quoted by the manufacturer. Other derivatized beads may be used in place of the glass beads if desired, provided that they have good swelling properties under aqueous condition. Clearly, the use of the newly synthesized reagents should be tested with standard cysteinyl peptides.
2. During synthesis of the Fmoc-protected amino acid, it is important to acidify the sample very slowly and to shake well. This should alleviate foaming due to the release of carbon dioxide following the change of pH. Also, one could use more ethyl acetate than prescribed in order to achieve better phase separation during the extraction step.
3. For protein digestion by trypsin, proteins can be denatured by boiling for a few minutes if the protein concentration is not so high as to cause precipitation. In the current protocol, proteins are not reduced prior to digestion; however, it is possible that one could reduce proteins prior to digestion.
4. TCEP is quite acidic and the optimal pH for solid-phase capture of cysteinyl peptides is 8.0. Therefore, it is essential that there is sufficient buffering capacity, such as 200 mM Tris at pH 8.0. In this case, the pH of the solution would be not strongly affected by the addition of 5 mM TCEP.
5. It is necessary to quench the capturing reaction after 15 min by mercaptoethanol or other excess SH-containing reagent because histidine side chains or other nucleophilic

TABLE 33.1 Quantitation of Protein Mixture by Solid-Phase Isotope Tagging and MS

Gene name	Cys-containing peptides found	Observed ratio (light/heavy)	Expected ratio (light/heavy)
G3P_rabit	VPTPNVSVVDLTC*R	4.6	4.0
	IVSNASC*TTNC*LAPLAK	4.3	
LCA_bovin	DDQNPHSSNIC*NISC*DK	1.8	2.0
	FLDDDLTDDIMC*VK	1.9	
	LDQWLC*EK	2.1	
	ALC*SEK	2.0	
	C*EVFR	1.9	
Oval_chick	YPILPEYLQC*VK	1.0	1.0
	LPGFGDSIEAQC*GTSVNVHSSLR	0.9	
	ADHPFLFC*IK	1.1	

* Isotopically labeled cysteine residues are marked by asterisks.

functional groups could suffer potential side reactions with the iodoacetyl group on the solid-phase beads. The protocol for washing beads following the quenching reaction can be altered by individual investigators, as we found that solid-phase-captured peptides are stable to a variety of washing conditions.

6. The UV light can be filtered through a copper sulfate solution that passes light of 300 to 400 nm. Although we used a long-wave UV lamp, the cleaving reaction can be accelerated by using a more powerful mercury arc lamp according to Holmes and Jones (1995).

7. When small amounts of peptides are expected, it is particularly important to remove labeling contaminants.

8. The use of β-mercaptoethanol in the photocleaving buffer prevents methionine oxidation.

9. Although very stringent washing steps were used to remove nonspecifically associated molecules from the solid phase after capturing, they may not be entirely necessary for all applications. The readers are encouraged to test different washing conditions for their own applications.

10. Some of the materials used here were published previously (Zhou et al., 2002).

References

Eng, J., McCormack, A.L., and Yates, J.R., 3rd (1994). An approach to correlate tandem mass spectral data of peptides with amino acid sequences in a protein database. *J. Am. Soc. Mass Spectrom.* **5**, 976–989.

Gygi, S.P., *et al.* (1999). Quantitative analysis of complex protein mixtures using isotope-coded affinity tags. *Nature Biotechnol.* **17**, 994–999.

Han, D., Eng, J., Zhou, H., and Aebersold, R. (2001). Quantitative profiling of differentiation induced membrane associated proteins using isotope coded affinity tags and mass spectrometry. *Nature Biotechnol.* **19**, 946–951.

Holmes, C.P., and Jones, D.G. (1995). Reagents for combinatorial organic synthesis: Development of a new o-nitrobenzyl photolabile linker for solid phase synthesis. *J. Org. Chem.* **60**, 2318–2319.

Smolka, M., Zhou, H., and Aebersold, R. (2002). Quantitative protein profiling using two-dimensional gel electrophoresis, isotopecoded affinity tag labeling, and mass spectrometry. *Mol. Cell Proteomics* **1**, 19–29.

Zhou, H., Boyle, R., and Aebersold, R. (2002). Quantitative protein analysis by solid phase isotope tagging and mass spectrometry. In "Protein-Protein Interactions" (H. Fu, ed.), Humana Press, New Jersey.

Zhou, H., Ranish, J.A., Watts, J.D., and Aebersold, R. (2002). Quantitative proteome analysis by solid-phase isotope tagging and mass spectrometry. *Nature Biotechnol.* **5**, 512–515.

Zhou, H., Watts, J.D., and Aebersold, R. (2001). A systematic approach to the analysis of protein phosphorylation. *Nature Biotechnol.* **19**, 375–378.

Protein Hydrogen Exchange Measured by Electrospray Ionization Mass Spectrometry

Thomas Lee, Andrew N. Hoofnagle, Katheryn A. Resing, and Natalie G. Ahn

I. INTRODUCTION

Hydrogen exchange-mass spectrometry (HX-MS) is a technique that measures the rate of exchange between protons on macromolecules and isotopically labeled water. It is commonly applied to the exchange of protein backbone amide hydrogens with deuterium oxide, where each exchange event leads to a mass increase of 1 Da, which can be monitored by mass spectrometry. The measurement can be used to obtain information about aspects of solution structure, protein folding, and conformational mobility (Chowdhury *et al.*, 1990; Zhang and Smith, 1993; Johnson and Walsh, 1994; Resing and Ahn, 1998; Hoofnagle *et al.*, 2001) and can also be used to analyze protein solvent accessibility and ligand-binding sites (Neubert *et al.*, 1997; Mandell *et al.*, 1998; Andersen *et al.*, 2001). Both electrospray ionization (ESI) and matrix-assisted laser desorption ionization (MALDI) methods can be used for HX-MS (Chowdhury *et al.*, 1990; Mandell *et al.*, 1998), although each ionization method has its own advantages (discussed by Hoofnagle *et al.*, 2003). This article outlines a practical protocol for hydrogen exchange measurements on proteins using ESI-MS, which provides high protein sequence coverage and low back-exchange, thus improving the sensitivity of HX-MS for monitoring changes in solvent accessibility and conformational mobility.

II. MATERIALS AND INSTRUMENTATION

Pepsin (Cat. No. P6887), succinic acid (Cat. No. S5047), sodium citrate (Cat. No. S4641), and deuterium oxide (Cat. No. 15188-2) are from Sigma-Aldrich, trifluoroacetic acid (Cat. No. 28904) is from Pierce, and HPLC grade water (Cat. No. 26830-0025) and HPLC grade acetonitrile (Cat. No. 32573-0025) are from Fisher. Capillary HPLC columns (10–15 cm × 500 μm i.d.) are constructed from fused silica tubing (320 μm inner diameter (i.d) for outlet, 500 μm i.d. for inlet, Cat. No. TSP320450, TSP530700, respectively, Polymicro Technologies), assembled with epoxy glue (Cat. No. 302 part A, 302 part B, Epoxy Technology), and hand packed with reversed-phase POROS 20 R1 resin (Cat. No. 1-1028-02, Applied Biosystems Inc.) as described by Resing and Ahn (1997). Modifications to the previously described HPLC system added polyetheretherketone (PEEK) loading loops for sample injection (1 ml) and solvent precooling (2 ml) from Upchurch (Cat. No. 1820, 1821) and a PEEK HPLC injector apparatus from Rheodyne (Cat. No. 9010). Hamilton syringes are from SGE (50 μl, 250 μl, 1 ml, Cat. No. 004312, 006312, 008105, respectively), sample tubes and caps are from Bio-Rad (Cat. No. 223-9391, 223-9393), and screw cap tubes are from CLP (Cat. No. 3463). A stainless steel pan

and an ice bucket serving, respectively, as ice bath and dry ice/ethanol/water bath are from Fisher Scientific (Cat. No. 13-361 A, 11-676).

For data collection we obtain excellent results with a quadrupole time-of-flight (TOF) mass spectrometer (QStar Pulsar, Applied Biosystems Inc.) with standard electrospray source and AnalystQS software, inter faced with any HPLC capable of delivering a steady flow rate of 10–40 μl/min (e.g., Agilent Model 1100 capillary HPLC system or Eldex MicroPro HPLC-2g). Incubations are carried out in a circulating water bath (Cat. No. 13271-036, VWR). Useful for data reduction are software for nonlinear least squares (e.g., Datafit 7.1, Oakdale Engineering Inc.) and spreadsheet analyses (e.g., Microsoft Excel).

III. PROCEDURES

A. HX-MS Data Collection

The basic protocol involves timed incubations of protein in 90% (v/v) D_2O at neutral pH, which allows in-exchange of deuterons for protons within timescales ranging from seconds to hours. At the end of each incubation, the in-exchange reaction is quenched by rapidly lowering pH and temperature. Pepsin is added in amounts that enable rapid protein digestion (1–5 min), and the resulting peptides are separated by reversed-phase HPLC coupled to LC/MS. In order to minimize back-exchange of deuterium for water, all steps following the quench are carried out at 0°C. Applications of this protocol have been reported by Resing *et al.* (1998,1999) and Hoofnagle *et al.* (2001).

Solutions

1. D_2O (99.9% atom D), stored at room temperature.
2. *Pepsin solution:* Dissolve lyophilized pepsin in HPLC grade water to 2 mg/ml and store

in 50-μl aliquots at −80°C. Each day, thaw a new aliquot, dilute to 0.5 mg/ml in HPLC grade water, clarify by centrifugation for 20 min × 12,000 rpm, and store on ice.

3. *Citrate/succinate solution:* 25 mM sodium citrate + 25 mM sodium succinate is titrated with HCl to pH 2.40, filtered through a 0.22-μm membrane, and stored on ice.

4. *HPLC buffers:* Make buffer A [0.05% trifluoroacetic acid (TFA) in HPLC grade water] and buffer B [0.05% TFA in 100% (v/v) HPLC grade acetonitrile] fresh daily.

5. *Step gradient solutions:* Mix buffers A and B in appropriate ratios to yield solutions of 5, 7.5, 10, 12.5, 15, 17.5, 20, 22.5, 25, 30, 35, 40, and 50% (v/v) acetonitrile, 0.05% TFA. Prepare 2-ml aliquots of each in screw-cap vials and store on ice.

6. *HPLC grade water:* ~100 ml for washing the sample loop, stored on ice.

7. *Protein sample:* Ideally > 90% pure at ~10 μM stored long term at −80°C in 50-μl aliquots. Thaw new aliquots each day, clarify by centrifugation for 20 min × 12,000 rpm, and store on ice.

8. *Dry ice bath:* Add dry ice to ethanol: water (23: 77) in an ice bucket to form a slurry at −10°C. This is used to quickly lower sample temperature after the in-exchange reaction.

Steps

1. A recommended configuration is shown in Fig. 34.1. Place a 2-ml PEEK loop between pump and injector to facilitate solvent cooling. Immerse this, the PEEK injector, a 1-ml sample loop, and the reversed-phase column at 0°C in an ice/water slurry. Run the HPLC pump isocratically in buffer A at 40 μl/min and equilibrate the reversed-phase

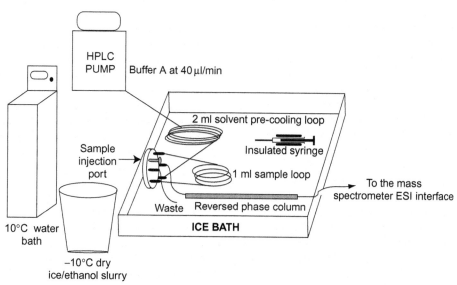

FIGURE 34.1 Experimental setup for hydrogen exchange-mass spectrometry. The reversed-phase column, injector, 2-ml solvent precooling loop, 1-ml injection loop, and Hamilton syringes wrapped with Parafilm are immersed in ice to minimize back-exchange during chromatography. The HPLC pump washes the column with buffer A (0.05% TFA in water) at a flow rate of 40 μl/min. Before buffer A reaches the injector apparatus, it is cooled to 0°C as it passes through the 2-ml solvent precooling loop. The protein sample, citrate/succinate solution, pepsin, and step gradients are stored separately on ice. The water bath is set at the desired temperature (e.g., 10°C) for in-exchange reactions, and a −10°C bath (dry ice/23% ethanol/77% water) for quenching the in-exchange reaction is prepared.

column in this solution. Maintain protein sample, citrate/succinate, step gradient solutions, and HPLC water on ice and maintain D_2O at room temperature.

2. Equilibrate D_2O in a water bath at 10°C. *Note:* Performing the in-exchange reaction at 10°C reduces exchange rates to levels measurable in the dead time of the experiment ($\sim$5 s).

3. Aliquot 10 µl protein ($\sim$100 pmol) to a sample tube and equilibrate in the 10°C water bath for 30 s.

4. Initiate the in-exchange reaction by transferring 90 µl D_2O to protein. Incubate protein + D_2O at 10°C for varying times (e.g., 5–18,000 s). *Note:* In order to minimize sample heating in these and subsequent steps, use P200 pipettors attached to tips that have been prechilled by storing the pipettor + tip in a 15-ml tube on ice.

5. After incubation, remove the sample tube from the water bath to the -10°C dry ice bath and begin timing the postincubation period at $t = 00:00$ (min:sec). Incubate the sample in the -10°C bath briefly enough to cool rapidly to 0°C but not long enough to freeze the solution.

6. At $t = 00:05$, add 90 µl citrate/succinate solution to the sample tube and gently mix by tapping the tube against the walls of the dry ice bath.

7. At $t = 00:20$, remove the sample tube to ice. Immediately add 10 µl × 0.5 mg/ml pepsin to the sample and mix gently.

8. Between $t = 00:20$ and $t = 00:40$, load the 200-µl volume into the sample loop using a 250-µl Hamilton syringe, prechilled on ice and insulated by wrapping the syringe barrel with Parafilm to 0.5 cm thickness in order to minimize heat transfer from handling.

9. At $t = 01:20$, inject the digest onto the column, running isocratically in buffer A at 40 µl/min.

10. At $t = 06:20$, switch the injector back to the load position and allow the column to

desalt at 40 µl/min. While the column is washing, rinse the sample loop with $\geqslant$2 ml cold water and then load a step gradient into the sample loop by injecting 40 µl of buffer B followed successively with 17.5-µl aliquots of 50, 40, 35, 30, 25, 22.5, 20, 17.5, 15, 12.5, 10, 7.5, and 5% (v/v) acetonitrile, 0.05% TFA. Forming the HPLC gradient in the sample loop minimizes the dead time of the gradient.

11. Configure the mass spectrometer computer for data collection. At $t = 12:00$, attach the column to the electrospray source.

12. At $t = 12:20$, start data collection on the mass spectrometer. Immediately set the HPLC flow rate to 20 µl.

13. At $t = 13:20$, inject the gradient onto the column. Peptide elution is usually complete within 10 min.

B. Data Analysis

Peptic peptides are identified by LC/MS/MS sequencing, run under conditions outlined in steps 1–13, except without D_2O. An example of peptide identification by MS/MS and data reduction is presented by Resing *et al.* (1999). The following discussion of HX data analysis specifies the AnalystQS software available with the ABI QStar Pulsar for analysis of quadrupole TOF data (WIFF files). Other programs with equivalent features can be substituted.

1. Calculation of Weighted Average Mass

Steps

a. Open the data file with AnalystQS and open the "extract ion" dialog box.

b. Enter the mass/charge *(m/z)* for an ion and view the extract ion chromatogram (XIC). Select the scan range corresponding to the extract ion and view the mass spectrum. Smooth the spectrum three times and then adjust the threshold to view the *m/z* values of all isotopic peaks.

c. List the m/z and signal intensities of each isotopic peak for each ion. Save data in a text file, generating separate text files for each peptide at each time point.

d. Open text files with a spreadsheet program (e.g., Microsoft Excel). List the m/z and signal intensity in the first and second columns, respectively. The weighted average mass for each ion of each peptide ($M_{t,wa}$) is then calculated by Eq. (34.1):

$$M_{t,wa} = \frac{\Sigma(m/z \times \text{intensity})}{(\Sigma(\text{intensity}) \times z) - z} \quad (34.1)$$

where $\Sigma(m/z \times \text{intensity})$ is the mass/charge of each isotopic form (column 1) multiplied by its corresponding signal intensity (column 2) and summed over all observed isotopic forms for the ion, and $\Sigma(\text{intensity})$ is the sum of intensities for all isotopic forms (Fig. 34.2).

2. Correction for Artifactual In-Exchange

Artifactual in-exchange occurs after quenching the in-exchange and is facilitated by the partial denaturation of protein in acidic solution. This correction requires measuring the weighted average mass at $t = 0$ (M_0), which is measured by reversing the order of adding citrate /succinate and D_2O in place of steps 4–6 in Section III,A.

The peptide mass corrected for artifactual in-exchange can be calculated by Eqs. (34.2) and (34.3):

$$M_{t,corr(IE)} = \frac{M_{t,wa} - LM_{\infty,90}}{1 - L} \quad (34.2)$$

$$L = \frac{M_0 - M_{calc}}{M_{\infty,90} - M_{calc}} \quad (34.3)$$

where $M_{t/corr(IE)}$ is the corrected peptide mass at time t, $M_{t,wa}$ is the observed weighted average mass at time t, L is the fraction of artifactual in-exchange at $t = 0$, M_0 is the observed peptide mass at $t = 0$, M_{calc} is the theoretical average mass

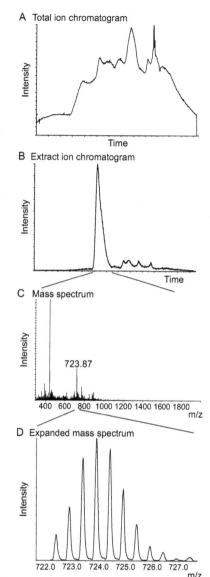

FIGURE 34.2 Processing mass spectra for a peptide (EETARFQPGYRS, undeuterated $MH_2^{+2} = 721.2$). (A) Total ion chromatogram showing the data set from a single time point. (B) Extract ion chromatogram identifying the subset of ions with m/z between 721.0 and 723.0 showing maximal peak intensity between 5 and 6 min. (C) Mass spectrum displaying all ions eluting between 5 and 6 min. (D) Expanded spectrum of ions between m/z 721.0 and 726.0, where isotopic peaks are separated by 0.5 Da for MH_2^{+2} ions. Each peak is labeled, and the m/z values and intensities are imported in text file format into a spreadsheet for calculating weighted average mass.

of the peptide, and $M_{\infty,90}$ is the theoretical mass of the peptide with complete deuterium exchange at backbone amide hydrogens (for incubation in 90% D_2O). ($M_{\infty,90} - M_{calc}$) is equal to the total number of nonproline amide residues in the peptide (total number of nonproline residues, minus one), multiplied by 0.9.

3. Back-Exchange Correction

Deuterons at backbone amides will slowly back-exchange to hydrogen during HPLC separation in water. We have used three different methods for estimating the fractional back-exchange for each peptide in the sample and have observed similar results with each.

Use the following steps for direct measurement of back-exchange.

Steps

1. Dilute the protein sample (500 pmol in 50 μl) with 50 μl H_2O. Add 90 μl citrate/succinate solution and digest with 10 μl pepsin as in steps 6 and 7 in Section III,A. Load peptides onto the column and desalt. Inject 30 μl of 40% acetonitrile, 0.05% TFA onto the column, and collect all peptides into one tube.
2. Lyophilize the peptides, dissolve them in 30 μl of the buffer used to prepare the protein sample, add 270 μl D_2O, and heat to 90°C for 90 min in order to completely deuterate the peptides.
3. Cool the sample on ice and load 300 μl volume into the sample loop. Inject the peptide onto the column after 1 min of incubation in the sample loop as in steps 8 and 9 in Section III,A and proceed with steps 10–13. Measure the weighted average mass of each peptide (Section III,B).
4. Fractional back-exchange can be calculated for each ion using Eq. (34.4):

$$BE = \frac{M_0 - M_{BE}}{M_{\infty,90} - M_{calc}} \qquad (34.4)$$

where BE is the fractional back-exchange of the peptide and M_{BE} is the observed mass of the peptide in the back-exchange experiment.

The following equation calculating the fractional back-exchange was derived empirically by Resing et al. (1999):

$$BE = L \times \left(\frac{\%H_2O}{\%D_2O} \right) + [(\text{peptide elution}$$
$$\text{time from HPLC in min} + 6\,\text{min})$$
$$\times 0.01/\text{min}]$$

$$(34.5)$$

where L is the fraction of artifactual in-exchange at $t = 0$ from Eq.(34.3) and ($\%H_2O/\%D_2O$) is the ratio of H_2O to D_2O during proteolysis (Section III,A, step 7; e.g., 0.55/0.45 for initial incubation with 90% D_2O). The calculation is based on an observed back-exchange of approximately 1% for each minute the peptide is on the column prior to elution.

The exchange rates for amide backbone hydrogens have been measured empirically, accounting for inductive and steric blocking effects within different primary sequences (Bai et al., 1993). This study presents tables and equation that can be used to calculate predicted back-exchange rates at 0°C. The program HXPep (Zhang et al., 1997), written by Dr. Zhongqi Zhang (Amgen Inc., Thousand Oaks, CA), calculates exchange rates using the derivation of Bai et al., (1993) and can be used to calculate back-exchange rates by entering the peptide sequence, choosing "NH/D_2O" exchange, "oligo" peptide size, "low salt," "pH/pD read of 2.400," "0°C," and "C-terminal considered." For each backbone amide hydrogen,

$$BE_{amide} = k_{HXPep} \times (\text{elution time} + \text{wash}$$
$$\text{time}) \qquad (34.6)$$

where BE_{amide} is the average back-exchange for each backbone amide hydrogen and k_{HXPep} is the rate of exchange calculated for the amide hydrogen by HXPep. For the entire peptide, fractional back-exchange may be calculated as

$$BE = \Sigma \frac{BE_{amide}}{M_{\infty,90} - M_{calc}} \qquad (34.7)$$

where $\Sigma(BE_{amide})$ equals the sum of BE_{amide} calculated from Eqn.(34.6) for. every amide hydrogen in the peptide.

After estimating fractional back-exchange using any of the aforementioned methods, the weighted average mass of each ion is corrected by Eq.(34.8):

$$M_{t,corr(BE)} = M_{calc} + \frac{M_{t,corr(IE)} - M_{calc}}{1 - BE}$$

$$(34.8)$$

where $M_{t,corr(BE)}$ is the peptide mass at time t after exchange in 90% D_2O, corrected for artifactual in-exchange and back-exchange.

C. Curve Fitting

Following correction for artifactual in-exchange and back-exchange, data may be fit using nonlinear least squares to a sum of exponentials (Resing et al., 1999; Hoofnagle et al., 2003). Further discussion and details on curve fitting and modeling are described by Resing and Ahn (1998) and Resing et al., (1999).

The time courses are modeled by a sum of exponentials in which each amide hydrogen exchanges with deuterium at a given rate. While in theory each amide backbone hydrogen is represented by a separate rate constant, in practice, exchange rates are averaged into fast ($>1\,min^{-1}$), intermediate (0.1–$1.0\,min^{-1}$) and slow rates (0.002–$0.1\,min^{-1}$), and time courses can be fit with one, two, or three exponential terms:

$$Y = N - Ae^{-k_1 t} - Be^{-k_2 t} - Ce^{-k_3 t} \qquad (34.9)$$

where Y is the observed weighted average mass corrected for back-exchange and artifactual in-exchange [i.e., $M_{t,corr(BE)}$], A, B, and C correspond to the number of amides (multiplied by 0.9), respectively, exchanging with average rate

constants k_1, k_2, and k_3, and N is the peptide mass after maximal in-exchange of deuterium ($=M_{calc} + A + B + C$).

Some backbone amides are nonexchanging over the experimental time period and cannot be fit to this equation. For instance, in the protein kinase ERK2, 44% of backbone amides show no exchange after the longest time point of 5 h (rate constant $< 0.002\,min^{-1}$) (Hoofnagle et al., 2001). The number of amides in this nonexchanging group (NE) can be estimated by subtracting the number of exchanging amides ($A + B + C$) from the total number of backbone amides in the peptide, excluding proline residues.

Steps

1. Start the DataFit program with a new project using one independent variable.
2. Enter time (minutes) and $M_{t,corr(BE)}$ in columns 1 and 2, respectively.
3. Under the Solve menu, select "Define User Model" and then select "New."
4. Enter the equation describing the sum of up to three exponentials under Model Definition:

   "Y = n − a*Exp(−d*x) − b*Exp(−e*x) − c*Exp(−f*x)"

5. Similarly, enter equations describing the sum of two and one exponentials.
6. For each equation, provide initial guesses of parameter values for nonlinear least squares. Test several different initial guesses, which should converge to the same fit.
7. Select Regression under the Solve menu to fit data to the three user-defined models.
8. Record parameter values and standard errors (Fig. 34.3), choosing the equation fit with lowest variance.

IV. PITFALLS

1. The buffer conditions for the exchange reaction should be optimized to minimize

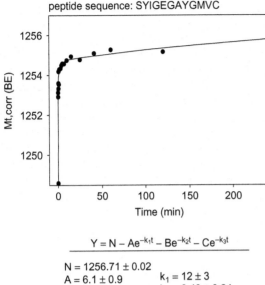

peptide sequence: SYIGEGAYGMVC

$$Y = N - Ae^{-k_1 t} - Be^{-k_2 t} - Ce^{-k_3 t}$$

N = 1256.71 ± 0.02
A = 6.1 ± 0.9 $k_1 = 12 ± 3$
B = 0.92 ± 0.33 $k_2 = 0.48 ± 0.24$
C = 2.2 ± 0.4 $k_3 = 0.0043 ± 0.0016$

FIGURE 34.3 Curve fitting and modeling. Weighted average masses corrected for artifactual in-exchange and back-exchange [$M_{t\ corr(BE)}$] are plotted vs incubation time with D$_2$O. Time courses are fit to a sum of exponentials by nonlinear least squares. Nonlinear least-square parameters and their standard errors obtained include k_1, k_2, and k_3 (rate constants, min^{-1}), A, B, C (0.9 × number of amides exchanging with apparent rate constants k_1, k_2, and k_3, respectively), and $N = M_{calc} + 0.9(A + B + C)$.

protein denaturation. Salt and buffer concentrations should be low as possible to minimize buffer/salt-catalyzed hydrogen exchange. We have found that 5 mM sodium phosphate, pH 7.0, 50 mM sodium chloride in the exchange reaction produces minimal back-exchange. The sample pH should be identical between experimental conditions in order to minimize effects on hydrogens that exchange via EX2 mechanism, where observed exchange rates vary with pH of the solution (Clarke and Itzhaki, 1998). Higher buffer concentrations may be used when varying solutes that influence pH, e.g., see Andersen et al., (1998).

2. The duration of pepsin digestion and the amount of protease should be optimized to generate the greatest number of peptides in the size range of 8–15 amino acids, which yields optimal resolution for exchange measurements. Peptides should be short enough to enable mass resolution, but long enough to bind the HPLC column and provide some sequence overlap.

3. The acetonitrile concentrations in the step gradient solutions should be varied to optimize peptide resolution from reversed-phase chromatography. Peptide elution should be spread throughout the gradient, while eluting all peptides within 10 min of gradient injection.

4. In order to avoid instrument bias and variations in data collection, randomize the time points and samples during data collection. Perform instrument calibration and measure the $t = 0$ experiment each day.

5. The quality of curve fitting varies with the data quality and the number of time points. Typically, at least 20 data points between 5 s and 3 h are required for curve fitting with modest standard errors. In addition, because the distribution of time points affects curve fitting, best results are achieved by including more time points during the transition between fast and intermediate exchange rates (0–30 min for most peptides) than during the times close to maximal in-exchange.

6. In optimizing the acquisition method for the ABI QStar qTOF mass spectrometer, the orifice voltage, the pulse frequency, and the declustering potential (DP) were optimized to reduce fragmentation at the orifice and to obtain the best signal at minimal peptide sample. Parameters in our acquisition method are: Mass range is 300–2000 amu; accumulation time is 1 s; duration is 50 min; cycles are 3000; delay time is 0 s; cycle time is 1 s; pulser frequency is 7 kHz; pulse 1 duration is 13 μs; pause between mass range is 5 ms; ionspray voltage is 5200 V; declustering potential is 40 V; and declustering potential 2 is 20 V.

References

Andersen, M. D., Shaffer, J., Jennings, P. A., and Adams, J. A. (2001). Structural characterization of protein kinase A as a function of nucleotide binding: Hydrogen-deuterium exchange studies using matrix-assisted laser desorption ionization-time of flight mass spectrometry detection. *J. Biol. Chem.* **276**, 14204–14211.

Bai, Y., Milne, J. S., Mayne, L., and Englander, S. W. (1993). Primary structure effects on peptide group hydrogen exchange. *Proteins,* **17**, 75–86.

Barksdale, A. D., and Rosenberg, A. (1982). Acquisition and interpretation of hydrogen exchange data from peptides, polymers, and proteins. *Methods Biochem. Anal.* **28**, 1–113.

Chowdhury, S. K., Katta, V., and Chait, B. T. (1990). Probing conformational changes in proteins by mass spectrometry. *J. Am. Chem. Soc.* **112**. 9012–9013.

Clarke, J., and Itzhaki, L. S. (1998). Hydrogen exchange and protein folding. *Curr. Opin. Struct. Biol.* **8**, 112–118.

Hoofnagle, A. N., Resing, K. A., and Ahn, N. G. (2003). Protein analysis by hydrogen exchange and mass spectrometry. *Annu. Rev. Biophys. Biomol. Struct.* **32**, 1–25.

Hoofnagle, A. N., Resing, K. A., Goldsmith, E. J., and Ahn, N. G. (2001). Changes in conformational mobility upon activation of extracellular regulated protein kinase-2 as detected by hydrogen exchange. *Proc. Natl. Acad. Sci. USA* **98**, 956–961.

Johnson, R. S., and Walsh, K. A. (1994). Mass spectrometric measurement of protein amide hydrogen-exchange rates of apomyoglobin and holo-myoglobin. *Protein Sci.* **3**, 2411–2418.

Mandell, J. G., Falick, A. M., and Komives, E. A. (1998). Identification of protein-protein interfaces by decreased amide proton solvent accessibility. *Proc. Natl. Acad. Sci. USA* **95**, 14705–14710.

Neubert, T. A., Walsh, K. A., Hurley, J. B., and Johnson, R. S. (1997). Monitoring calcium-induced conformational changes in recoverin by electrospray mass spectrometry. *Protein Sci.* **6**, 843–850.

Resing, K. A., and Ahn, N. G (1998). Deuterium exchange mass spectrometry as a probe of protein kinase activation: Analysis of wild-type and constitutively active mutants of MAP kinase kinase-1. *Biochemistry* **37**, 463–475.

Resing, K. A., Hoofnagle, A. N., and Ahn, N. G. (1999). Modeling deuterium exchange behavior of ERK2 using pepsin mapping to probe secondary structure. *J. Am. Soc. Mass. Spectrom.* **10**, 685–702.

Zhang, Z. Q., and Smith, D. L. (1993). Determination of amide hydrogen-exchange by mass spectrometry: A new tool for protein structure elucidation. *Protein Sci.* **2**, 522–531.

Nongel-Based Proteomics: Selective Reversed-Phase Chromatographic Isolation of Methionine-Containing Peptides from Complex Peptide Mixtures

Kris Gevaert and Joël Vandekerckhove

I. INTRODUCTION

Contemporary gel-free or nongel proteome analytical techniques are used increasingly as alternatives to highly resolving two-dimensional polyacrylamide gel electrophoresis (2D PAGE) for the analysis of complex protein mixtures. Some of the limitations of 2D PAGE (e.g., its bias to mainly visualize well-soluble,

abundant proteins) can be overcome by these novel techniques, explaining their increasing success. A variety of procedures have been described and definitely the most interesting ones are those in which peptides, constituting the original proteins, are analyzed (e.g., Gygi *et al.*, 1999; Geng *et al.*, 2000; Washburn *et al.*, 2001; Zhou *et al.*, 2001). In these techniques, we can distinguish three general steps. First, the protein mixture is digested in solution using a highly specific protease. Second, the generated peptide mixture or an (affinity/covalently) isolated subset of it is separated by a liquid chromatographic step(s) (LC) and analyzed by automated tandem mass spectrometry (MS/MS). Third, the obtained peptide fragmentation spectra are fed into database-searching algorithms and linked to peptide and protein amino acid sequences stored in databases. We have described a technique for reversed-phase (RP) HPLC-based isolation of representative peptides of a proteome (peptides containing rare amino acids that are well distributed over a given proteome) (Gevaert *et al.*, 2002). Because this technique has a number of similarities to the previously described diagonal electrophoresis (Brown and Hartley, 1966) and diagonal chromatography (Cruickshank *et al.*, 1974) methods, we call this technique COmbined FRActional DIagonal Chromatography or COFRADIC. The core technology of COFRADIC is a chromatographic shift, evoked by a chemical or enzymatic reaction, of representative peptides between two identical separation steps. This article describes the different steps involved in the COFRADIC-based isolation of methionine-containing peptides as their sulfoxide derivatives using currently available RP-HPLC instrumentation.

II. MATERIALS AND INSTRUMENTATION

HPLC-graded water (product number 4218) and acetonitrile (product number 9017) used to prepare the HPLC solvents are of the best quality available and are from Malinckrodt Baker B.V., Deventer, The Netherlands. Peptide synthesizer graded trifluoroacetic acid (product number PTS6045) is from Rathburn Chemicals Ltd. (Walkerburn, Scotland, UK). The RP column used for separations is a 2.1 × 150-mm ZORBAX 300SB-C18 column (product number

883750-902) and is from Agilent Technologies (Waldbronn, Germany). The hydrogen peroxide stock solution [30% (w/w), product number 21,676-3] is from Aldrich Chemical Co., Inc. (Milwaukee, WI).

The HPLC system used to sort the methionine peptides is an Agilent 1100 series capillary LC system, equipped with an Agilent 1100 series capillary pump, thermostatted microwell-plate sampler, thermostatted column compartment, variable wave length detector, and a thermostatted fraction collector. The system runs under the control of a 3D Agilent ChemStation.

III. PROCEDURES

Chromatographic Isolation of Methionine-Containing Peptides

This procedure describes the isolation of methionine-containing peptides using a porous silica-based C18 RP-HPLC column with an internal diameter of 2.1 mm and a length of 15 cm. When using other brands or types of reversed-phase columns, clearly the amount of material used for the isolation of methionine-peptides, the solvent flow rate, the collection scheme for the primary fractions, and the evoked retention time shifts should be checked and, if necessary, adapted.

Solutions

1. *HPLC solvents:* For the chromatographic sorting of methionine-peptides, HPLC solvent A consists of 0.1% (v/v) trifluoroacetic acid (TFA) in water and solvent B contains 70% (v/v) acetonitrile and 0.1% (v/v) TFA in water. At least 1 litre of each solvent is made so that during the isolation procedure, the same HPLC solvents can be used, thereby increasing the

overall reproducibility of the COFRADIC procedure for the isolation of methionine-peptides. These HPLC solvents should be thoroughly degassed prior to use.

2. *Oxidation solution:* The oxidation solution must be freshly prepared prior to the chromatographic isolation of methionine-peptides (secondary runs, see later) and consists of 3% (v/v) of H_2O_2 and 1% (v/v) TFA in water.

Steps

1. Typically, load a total of 1 to 10 nmol of a digested protein mixture, such as a cell lysate, onto a 2.1 i.d. $\times$ 150-mm C18 RP column. Following injection, equilibrate the column for 10 min with solvent A. Then, create a binary solvent gradient to 100% of solvent B over a time span of 100 min at a constant flow rate of 80 µl/min. This HPLC separation is referred to as the primary run.

2. During this primary run, peptide fraction collection starts from 40 min on (corresponding to an acetonitrile concentration of 21%) and, in total, 48 primary fractions of 1 min (or 80 µl) are collected in the wells of a microtitreplate (see Fig. 35.1A).

3. Using this setup, pool primary fractions, which are separated by 12 min, (Table 35.1) and dry to complete dryness in a centrifugal vacuum concentrator. Prior to the methionine-oxidation reaction and the secondary runs, redissolve these pooled peptide fractions in 70 µl of 1% (v/v) TFA in water.

4. The methionine oxidation reaction is done by transferring 14 µl of the freshly made oxidation solution to the vial containing the pooled peptides. The reaction proceeds for 30 min at a 30°C, after which the sample must be immediately injected onto the RP-HPLC column.

5. Separate the oxidized peptides on the same column as the one used for the primary separation and under identical chromatographic conditions. In the experimental setup described here, methionine-sulfoxide containing peptides typically elute in a time frame 3 to 8 min before the elution of the unmodified, methionine-free peptides, which elute in he same interval as during the primary run (see Fig. 35.1B). Thus, during these secondary runs, the shifted methionine-sulfoxide peptides can be time-based collected in a distinct number of subtractions in a microtitreplate (see Table 35.1).

6. The sorted peptides are analysed most conveniently by a mass spectrometer. For LC-MS/MS experiments using an electrospray ionization (ESI)-based mass spectrometer, multiple secondary fractions obtained during one secondary run may be pooled, dried to complete dryness, and redissolved in an appropriate solvent for further analysis. Alternatively, matrix-assisted laser desorption ionisation (MALDI)-based analysis combined with collision-induced dissociation can be used for peptide identification. In this case, peptide fractions are not pooled in order to keep the complexity of the samples as low as possible, but are dried and redissolved in a small volume of MALDI matrix solution and finally loaded on a MALDI target for further analysis.

7. Finally, convert the information in the obtained peptide fragmentation spectra to peak lists in which the masses of the observed peptide fragment ions and their relative or absolute intensities are saved. These peak lists are then used to identify the corresponding peptides using commercially available database search engines such as MASCOT (Perkins *et al.*, 1999).

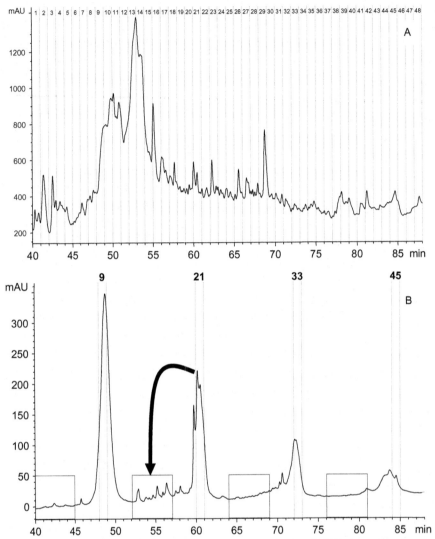

FIGURE 35.1 (A) UV absorption chromatogram (214 nm) of a reversed-phase HPLC separation of a tryptic digest of a human sputum sample. Peptides are fractionated in 48 distinct fractions. (B) Prior to the isolation of methionine-peptides, four fractions (in this case, fractions 9, 21, 33, and 45) are combined and oxidized to their methionine-sulfoxide counterparts. When rerun on the same column and under identical chromatographic conditions, the oxidized peptides shift out of this primary collection interval (an 8- to 3-min hydrophilic shift) and can be specifically collected (shaded boxes) for further LC-MS/MS analysis.

IV. COMMENTS

Using mass spectrometric analysis, the presence of a methionine-sulfoxide side chain in sorted peptides can be recognized easily by the unstable behaviour of the methionine-sulfoxide side chain. In both MALDI- and ESI-based experiments, the facile loss of methane sulfenic acid (64 amu) has been observed (Lagerwerf *et al.*, 1996). Furthermore, we have noticed that,

TABLE 35.1   Primary and Secondary RP-HPLC Fraction Collection Scheme for the Isolation of Methionine-Peptides[a]

Secondary run	Primary fractions			Elution of primary fractions			Elution of methionine-sulfoxide peptides		
A	12	24	36	51–52	63–64	75–76	43–48	55–60	67–72
B	11	23	35	50–51	62–63	74–75	42–47	54–59	66–71
C	10	22	34	49–50	61–62	73–74	41–46	53–58	65–70
D	9	21	33	48–49	60–61	72–73	40–45	52–57	64–69
E	8	20	32	47–48	59–60	71–72	39–44	51–56	63–68
F	7	19	31	46–47	58–59	70–71	38–43	50–55	62–67
G	6	18	30	45–46	57–58	69–70	37–42	49–54	61–66
H	5	17	29	44–45	56–57	68–69	36–41	48–53	60–65
I	4	16	28	43–44	55–56	67–68	35–40	47–52	59–64
J	3	15	27	42–43	54–55	66–67	34–39	46–51	58–63
K	2	14	26	41–42	53–54	65–66	33–38	45–50	57–62
L	1	13	25	40–41	52–53	64–65	32–37	44–49	56–61

[a] The pooling scheme of primary RP-HPLC fractions and the time interval (given in minutes) during which they elute from the column during the primary run are indicated. The collection of methionine-sulfoxide peptides during 12 consecutive secondary runs (A-L) is indicated in the last column (given in minutes). These peptides generally elute in a time interval of 5 min starting 8 min prior to the elution of their respective primary fractions. Primary and secondary fraction collection time intervals corresponding to a particular primary fraction are indicated by an identical colour code.

especially in the postsource decay (PSD) mode (Spengler *et al.*, 1992), peptides baring methionine-sulfoxide residues tend to fragment preferentially at this residue, thereby only a small amount of vibration energy is left for fragmentation of the peptide backbone. Generally, we noticed that MALDI-PSD spectra obtained from peptides containing a methionine-sulfoxide tresidue are very difficult to interpret, as only a small number of sequence-specific fragment ions (due to fragmentation of peptide bonds) are observed. However, we have not observed a similar behaviour for these type of peptides when analyzed by ESI-MS/MS, meaning that LC-MS/MS can be used routinely to analyze sorted methionine-sulfoxide peptides by fragmentation.

One of the interesting features of COFRADIC is the fact that during secondary runs, methionine-sulfoxide peptides elute in a five times larger time interval compared to the primary run (see Table 35.1, last column). This means that the sorted peptides are delivered in a much less condensed manner for analysis by LC-MS/MS, thus implying that more peptides will be finally analyzed, resulting in a higher coverage of the analyzed proteome.

The aforementioned method for the isolation of methionine-peptides as their sulfoxide derivatives can be highly automated using existing HPLC equipment such as automated injectors and fraction collectors. It should be clear that, in theory, every type of peptide that contains an amino acid that can be specifically altered using chemicals and/or enzymes can be sorted using the COFRADIC technology. For example, similar strategies can be used for selective isolation of cysteine-containing peptides, phosphorylated peptides, and peptides spanning the amino-terminal part of the protein.

V. PITFALLS

1. Do not incubate methionine-peptides for more than 30 min in 0.5% H_2O_2, as this leads to over-oxidation, resulting in the formation of methionine-sulfones. The hydrophilic shift evoked by this sulfone is smaller as compared to the one evoked by the sulfoxide and may thus interfere with the isolation procedure.

2. Likewise, following the oxidation reaction, the altered peptides must be injected immediately onto the reversed-phase column. We have noticed that storing the peptides in the oxidation solvent, even just for a couple of hours, in the freezer leads to almost complete oxidation of methionine to the sulfone derivative, oxidation of cysteine to cystic acid, and almost complete destruction of tryptophane residues.

3. It is important to keep the chromatographic conditions for the primary run and the secondary run as similar as possible. This means that the same HPLC solvents should be used and that the column compartment should be controlled thermostatically.

Acknowledgments

K.G. is a Postdoctoral Fellow of the Fund for Scientific Research-Flanders (Belgium) (F.W.O.-Vlaanderen). The project was further supported by the GBOU-research initiative of the Flanders Institute of Science and Technology (IWT).

References

Brown, J. R., and Hartley, B. S. (1966). Location of disulphide bridges by diagonal paper electrophoresis: The disulphide bridges of bovine chymotrypsinogen. *Biochem. J.* **101**, 214–228.

Cruickshank, W. H., Malchy, B. L., and Kaplan, H. (1974). Diagonal chromatography for the selective purification of tyrosyl peptides. *Can. J. Biochem.* **52**, 1013–1017.

Geng, M, Ji, J., and Regnier, F E. (2000). Signature-peptide approach to detecting proteins in complex mixtures. *J. Chromatogr. A* **870**, 295–313.

Gevaert, K., Van Damme, J., Goethals, M., Thomas, G. R., Hoorelbeke, B., Demol, H., Martens, L., Puype, M, Staes A., and Vandekerckhove, J. (2002). Chromatographic isolation of methionine-containing peptides for gel-free proteome analysis: Identification of more than 800 *Escherichia coli* proteins. *Mol. Cell. Proteomics* **1**, 896–903.

Gygi, S. P., Rist, B., Gerber, S. A., Turecek, E, Gelb, M. H., and Aebersold, R. (1999). Quantitative analysis of complex protein mixtures using isotope-coded affinity tags. *Nature Biotechnol.* **17**, 994–999.

Lagerwerf, F M., van de Weert, M, Heerma, W., and Haverkamp, J. (1996). Identification of oxidized methionine in peptides. *Rapid Commun. Mass Spectrom.* **10**, 1905–1910.

Perkins, D. N., Pappin, D. J., Creasy, D. M., and Cottrell, J. S. (1999). Probability-based protein identification by searching sequence databases using mass spectrometry data. *Electrophoresis* **20**, 3551–3567.

Spengler, B., Kirsch, D., Kaufmann, R., and Jaeger, E. (1992). Peptide sequencing by matrix-assisted laser-desorption mass spectrometry. *Rapid Commun. Mass Spectrom.* **6**, 105–108.

Washburn, M. P., Wolters, D., and Yates, J. R., III (2001). Large-scale analysis of the yeast proteome by multi-dimensional protein identification technology. *Nature Biotechnol.* **19**, 242–248.

Zhou, H., Watts, J. D., and Aebersold, R. (2001). A systematic approach to the analysis of protein phosphorylation. *Nature Biotechnol.* **19**, 375–378.

Mass Spectrometry in Noncovalent Protein Interactions and Protein Assemblies

Lynda J. Donald, Harry W. Duckworth, and Kenneth G. Standing

I. INTRODUCTION

Mass spectrometry (MS) has emerged as an important tool in the study of proteins and their interactions over the past several years (Daniel et al., 2002; Robinson, 2002). Electrospray (ESI) provides a gentle ionization method that does not disrupt the weak bonds found in noncovalent complexes, and the adoption of nanospray technology (Wilm and Mann, 1996) has allowed the concentration of the buffer to be varied over a large range. Moreover, the use of time-of-flight (TOF) spectrometers for this purpose (first by Tang *et al.*, 1994) has removed previous limitations on the *m/z* values that can be examined, particularly after such instruments have been modified in order to study larger and larger assemblies (Chernushevich *et al.*, 1999; Rostom and Robinson, 1999; Rostom *et al.*, 2000; Van Berkel *et al.*, 2000).

Because the mass spectrum shows all the components present in a sample, information may be acquired on quantity, stoichiometry, and equilibria. Problems in the method arise from the limited selection of suitable buffers, interference from inorganic ions (especially Na^+ and K^+), and difficulties inherent in maintaining an uncooperative protein in solution under conditions required for the mass spectrometry.

II. MATERIALS

Ammonium acetate (99.999% Cat. No. 37,233-1), dithiothreitol (99%, Cat. No. 45,777-9), and 1,1,1- trichloroethane (Cat. No. 40,287-7) are from Aldrich. Ammonium bicarbonate is Fisher certified grade (Cat. No. A643). Dimethylchlorosilane is from Pierce (Cat. No. 83410). Methyl alcohol is from Mallinckrodt (Cat. No. 3041). Acetic acid (99+%) (Cat. No. A6283), formic acid (ACS reagent) (Cat. No. F4636), and substance P (Cat. No. S2136) are from Sigma. Dialysis membranes for waterbugs are from Spectra/Por (Cat. No. 132 128 for 50 K MWCO; 132 113 for 8 K MWCO; 132 703 for 12–14 K MWCO). For salt removal from ligands, we use Spectra/Por Dispodialyser MWCO 500, 1 ml capacity (Cat. No. 135 504). Centricon ultra-filtration units are from Amicon Bioseparations (Millipore) (YM10 is Cat. No. 4205; YM30 is Cat. No. 4208; YM50 is Cat. No. 4224). All plasticware is from Fisher Scientific: LDPE drop dispensing bottles (Fisher Cat. No. 03-006), 100-ml beakers (Nalgene Cat. No. 1201-0100), and syringe filters

SFCA, 0.2 μm, 25 mm (Nalgene Cat. No. 190-2520). Microcentrifuge tubes, 0.6-ml flat top (Fisher Cat. No. 05-407-16) and 1.5 ml (Fisher Cat. No. 05-406-16), are used for waterbug construction. The 10-ml sterile syringes are latex free with Luer-lok from Becton-Dickinson (Cat. No. 309604). Nanospray capillaries are from Protana (types S and N), and New Objective PicoTip (type Econol2). GELoader tips are from Eppendorf (Cat. No. 22 35 165-6).

Some samples are prepared in a Sorvall RC-5B refrigerated centrifuge with a Sorvall SS34 rotor. Water from a reverse osmosis supply is run through a Barnstead NANOpure II system set at 17.0 MΩ/cm. All buffers are prepared with this grade of water.

III. INSTRUMENTATION

Mass spectrometry measures the ratio of the mass (*m*) to the charge (*z*) of an ion. Because many commercial instruments have a limited *m/z* range (usually <4000 for quadrupole mass filters), ions with large *m/z* values, such as those from buffered proteins, cannot be measured. Consequently, most intact protein and protein-ligand complexes require the "unlimited" *m/z* range found in TOF mass analyzers, and orthogonal ion injection (Verentchikov *et al.*, 1994) enabled the use of such instruments with ESI sources. Tang *et al.* (1994) pioneered the measurement of noncovalent complexes by ESI-TOF MS. More recently, we have added a heated metal capillary to provide another stage of pumping and desolvation and a new section containing a small RF quadrupole to provide collisional cooling of the ions (Krutchinsky *et al.*, 1998). A schematic diagram of the main elements of the instrument is shown in Fig. 36.1.

Some of the results reported here were obtained with a conventional electrospray ion source. In this device, solution is delivered to the sharpened tip of a stainless steel needle by a syringe pump. Alternatively, we use a glass

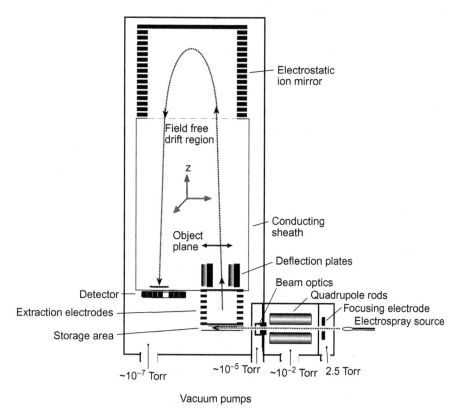

FIGURE 36.1 Schematic diagram of an electrospray ionization time-of-flight mass spectrometer. The ion path is indicated by the dotted line. A more detailed description can be found in Chernushevich *et al.* (1999).

nanospray source, which requires less sample, at lower concentration, and is tolerant of high concentrations of buffer. A potential difference of 3–3.5 kV (1 kV for nanospray) between the tip of the needle and the inlet of a heated metal capillary starts the ionization process. A counterflow of hot nitrogen gas helps with desolvation. We usually use nitrogen as this "curtain" gas, but sometimes use SF_6, especially for very large complexes, because it is considerably more efficient than nitrogen both in removing adducts and in dissociating complexes.

Expansion of the ion mixture into the next region produces a supersonic jet. Normally we apply a declustering voltage to the focusing electrode. Up to 300 V potential difference for nitrogen (or 400 V for SF_6, due to its superior

insulating properties) can be maintained between the focusing electrode and the flat aperture plate further downstream; the voltage is determined empirically for each protein or complex. A small opening in the plate connects this region to the second pumping stage where the ions oscillate in the two-dimensional potential well produced by the RF quadrupole and are cooled to near-thermal energies by collisions with the ambient gas (Krutchinsky *et al.*, 1998). After passing through the quadrupole, the ion beam is focused and directed into the storage region of the modulator, entering perpendicular to the TOF axis at an energy <10 eV. After a group of ions has filled the storage region, it is accelerated along the TOF axis by applying a pulse to the extraction electrodes, which also starts the TOF measurement. The time of flight is recorded for each ion using TOFMA, an in-house soft ware program that is also used for data analysis.

IV. PROCEDURES

A. Finding the Correct Conditions for the Protein

Our successes have come from the two commercially available volatile buffers, ammonium acetate and ammonium bicarbonate, but it is possible to have a small amount of nonvolatile material present in the sample (often unintentionally). Na^+ and K^+ are the worst contaminants, but buffer components such as glucose, glycerol, polyethylene glycol, and detergents are best avoided.

Using the preparative buffer of the protein as a guide, choose ammonium acetate for pH < 7 or ammonium bicarbonate for pH > 7. The pH can be adjusted with ammonium hydroxide or acetic acid. However, these solutions do change pH over time, and it is best to prepare the samples and then acquire spectra as soon as practicable. Choose a concentration that is close to the ionic strength of the preparative buffer and prepare the protein at a high concentration so that both it and the buffer can be diluted during the experiment. Dithiothreitol can be added to limit oxidation. If possible, the exact protein concentration should be determined from the known molar extinction coefficient and measurement of UV absorbance of an aliquot.

In our instrument, the declustering voltage, Vc, is an important component of the successful spectrum. At the "right" voltage, there is a perfect balance between maximum desolvation needed for high resolution and the collisionally induced dissociation of the intact complex. In practice, it is always wise to take spectra at several voltages because each will provide different kinds of information.

B. Desalting with "Waterbugs"

This procedure is modified from Orr *et al.* (1995). It is an efficient method of salt exchange and can also be used to assess the suitability of different buffer conditions.

1. Preparation of Apparatus

Steps

1. Prepare 2% sialysation solution by adding 10 ml of dimethyldichlorosilane to 500 ml of 1,1,1-trichloroethane. Mix carefully, working in a fume hood.
2. Cut the rim and the attached lid from microcentrifuge tubes using a razor or scalpel. It is very important that the cut edge is smooth or it will puncture the dialysis tubing.
3. Wash the lids with methanol and allow them to air dry.
4. Wearing gloves, transfer the lids to a clean beaker containing sialysation solution. Agitate carefully and decant off the solution.
5. Wash the lids in running tap water and then rinse several times with nanopure water.
6. Tip onto a fresh tissue, cover with a second tissue, and allow to air dry.
7. Minimize handling the waterbugs by using clean forceps for transfer. They should be stored in a covered container, such as a plastic petri dish, to avoid dust contamination.

2. Sample Introduction

Steps

1. Wash hands thoroughly. Do *not* use hand cream. It is much easier to do this without gloves; only step 6 *requires* that no gloves be worn. Hand cream components and the powder used in gloves can all cause adducts in the mass spectrum.
2. Prepare a clean, salt-free work surface.
3. For volumes of 10–50 μl use the lids from 0.6-ml microcentrifuge tubes. For volumes up to 200 μl, use lids from 1.5-ml microcentrifuge tubes.

4. Prepare and chill the appropriate buffer, using plastic beakers kept specifically for this purpose.

5. Choose dialysis tubing with the appropriate MWCO and cut into squares of about 1.5-cm sides. Immerse them in nanopure water for about 10 min.

6. Rub the dialysis tubing squares between ungloved thumb and finger to separate the two layers. (This should be done in advance of setup if gloves are required for sample handling.)

7. Place a lid assembly on the clean work surface. Using an accurate pipettor, transfer a measured volume of pure protein into the lid of the former micro-centrifuge tube.

8. Place a single layer of dialysis tubing over the sample. Hold it in place with the rim of the microcentrifuge tube and carefully press down around the edge. Check the profile carefully in good light to make sure the dialysis membrane is intact. If not, remove the rim, discard the dialysis membrane tubing, and try again.

9. Place the waterbug onto the surface of the buffer, membrane side down. Gentle stirring is optional. Do this at 4°C, especially if the protein is unstable.

10. After at least 4 h, remove the waterbug using clean tongs. Working quickly, invert several times and then transfer to a fresh container of buffer. After a further 4–6 h (or overnight), repeat into a third plastic container of fresh buffer. If using 1-liter beakers, and less than 1 ml total sample, then three changes are sufficient. If using smaller containers in order to use less buffer, then increase the number of buffer changes to five or six.

3. Recovery of Sample

Steps

Set a pipettor at the expected recovery volume. Puncture the dialysis membrane with the tip of the pipettor and withdraw the sample.

Put into a clean microcentrifuge tube. If the volume recovered is less than expected, there may be some residue on the underside of the dialysis membrane. Measure the concentration using an aliquot of the recovered sample. The method used will depend on the protein and may require most of the sample.

If the sample has precipitated, add acetic acid by puncturing the dialysis membrane with the end of the pipette tip and then inject the acid, adjust the volume, and remove the solubilized protein. Normally 5 μl of acetic acid is adequate for 100 μl of sample prepared in 5 mM buffer. Formic acid may also be used. This sample can be analysed for mass determination.

C. Desalting by Ultrafiltration

The following is based on the Centricon user guide (Millipore Corp., 2001) that comes with the centrifugal filter devices. Any refrigerated centrifuge may be used, provided it has a fixed-angle rotor, adaptors for 17×100-mm tubes, and is capable of running at $1000–7600 g$. This requires much less buffer than the waterbug method, but it is more laborious and some proteins may come out of solution when they are concentrated too much.

Steps

1. Prepare and chill the appropriate buffer. The 99.999% ammonium acetate can be used directly. For ammonium bicarbonate buffer, make 100 ml at the desired concentration and then force through a 0.2-μm filter into a plastic container, such as a drop dispensing bottle, which has never been used for anything else.

2. Select a filtration device with a MWCO that will hold back the protein. Assemble the device and rinse the sample reservoir with buffer.

3. Add protein solution to the sample reservoir and fill it with buffer. Centrifuge for an

appropriate length of time. Normally, it takes 15–30 min in a Centricon 30 or 50 to reduce the volume to about 100 μl, depending on the concentration of the sample. Buffer components such as glycerol and glucose will slow down the process, and it is best to try and omit these at some earlier stage in the purification. If the retentate vial is attached to the Centricon unit, it may not fit in a closed rotor. We normally use the vial and do not close the rotor.

4. When the volume has been decreased to about 100 μl, add more buffer, put the retentate vial back onto the unit, invert gently several times, and then recentrifuge.

5. Repeat the fill and spin at least six times. The residual salt concentration should be *at least* three orders of magnitude less than that of the protein.

D. Final Preparation of the Protein(s) or Complex

Calibrate the instrument by electrospray using a $10^{-5} M$ solution of substance P in 50% methanol, 2% acetic acid. Other calibrants can be substituted.

For electrospray, dilute the protein to 5–20 μM in 5 mM buffer. Load into the electrospray needle by back pressure on the syringe pump. Adjust the needle in the holder and turn on the spray voltage to about 3 kV. Adjust the declustering voltage until a clear spectrum appears. Adjust the position of the end of the needle and the voltages until a cone is formed at the end. Collect spectra at various voltages.

For nanospray, cut the capillary to ~3 cm. Using a GELloader tip on a 1- to 10-μl pipettor set at 2.5 μl, rinse the inside of the capillary with an aliquot of the buffer used for the protein preparation. Discard this and reload the tip with sample. Assemble the capillary in its holder and attach to the mass spectrometer. If using a Protana capillary, brush it against the front cone

of the instrument to break off the end. (New Objective capillaries are open already.) Establish some back pressure to the assembly to force the sample to the orifice. Gradually turn up the nozzle voltage to 1 kV and adjust the declustering voltage until a clear spectrum emerges. Often the first burst of ions is badly contaminated by salts so it is best to be patient.

V. SPECIFIC EXAMPLES

A. A Small DNA-Binding Protein

The *Escherichia coli trp* apo-repressor protein TrpR is a dimer in the high salt buffer conditions used for nuclear magnetic resonance and X-ray crystallographic measurements (Zhang *et al.*, 1987; Zhao *et al.*, 1993). The pure enzyme is stable when lyophilized in elution buffer (10 mM NaPO$_4$ pH 7.6, 0.1 mM EDTA, 0.45 M NaCl). For our first experiments (Potier *et al.*, 1998), an aliquot of protein was dissolved in water and dialysed in "waterbugs" against 10-ml aliquots of 10 mM ammonium acetate. Just before mass analysis by electrospray ionization, samples were diluted to 10 μM protein in 5 mM ammonium acetate. As shown in Fig. 36.2A, the spectrum was complex, with two charge envelopes for the monomer, centred at the 7$^+$ and 10$^+$ ions, and good evidence for dimer, trimer, and perhaps higher order aggregates, all of which indicate a partially unfolded or denatured protein. If specific DNA was added to the protein before dialysis, a new ion envelope appeared that corresponded to a complex of one dsDNA and two protein monomers (Potier *et al.*, 1998), in keeping with results from other analytical methods.

More recently, the development of nanospray ionization has expanded the limits of buffer concentration. If the same TrpR protein is prepared in the same way, but with 500 mM ammonium acetate, the spectrum is completely different (Fig. 36.2B), with one compact ion envelope of

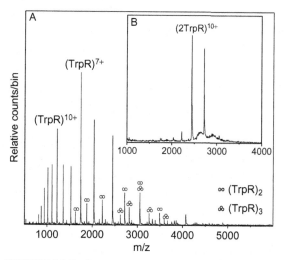

FIGURE 36.2   Spectra of E. *coli trp* apo-repressor protein. (A) A10 μ*M* protein prepared in 10 m*M* NH$_4$OAc by "water-bug" dialysis (8000 MWCO membrane) diluted to 5 m*M* just before electrospray. Nitrogen curtain gas with an 80-V declustering voltage. (B) A 2 μ*M* protein transferred into 500 m*M* NH$_4$OAc using a Centricon 10, diluted to 250 m*M*, and analysed by nanospray in a Protana type S capillary. Nitrogen curtain gas with a 100-V declustering voltage.

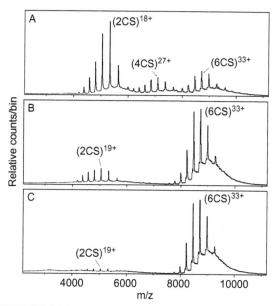

FIGURE 36.3   Spectra of E. *coli* citrate synthase. Samples were transferred into 100 m*M* NH$_4$HCO$_3$ using a Centricon 50, diluted to 10 μ*M* protein, and analysed by nanospray ionization from a New Objective PicoTip. For all spectra, the curtain gas was SF$_6$ and the declustering voltage was 200 V. (A) 5 m*M* NH$_4$HCO$_3$ showing three charge envelopes for dimer (2CS), tetramer (4CS), and hexamer (6CS); (B) 20 m*M* NH$_4$HCO$_3$; and (C) 100 m*M* NH$_4$HCO$_3$.

four ions, a spectrum typical of that of a properly folded protein. Deconvolution of these data gives a mass of 24452 Da, as expected for a dimer. Other DNA-binding proteins have shown the same sorts of buffer dependence (Donald *et al.*, 2001; Kapur *et al.*, 2002), where a very high concentration of buffer is required to maintain the protein in a native state, yet the DNA-protein complexes are relatively more stable at lower buffer concentrations.

B. A Large Allosterically Regulated Protein

Citrate synthase is a key enzyme for the entry of two-carbon units into the citric acid cycle and is essential in the biosynthesis of amino acids related to glutamate. Although many organisms have a homodimeric citrate synthase, that of E. *coli* crystallizes as a hexamer (Nguyen *et al.*, 2001). This protein cannot be frozen and is most stable when

stored at high concentration in buffer (20 m*M* Tris-Cl, pH 7.8, 1 m*M* EDTA, 50 m*M* KCl) and is best desalted with a 50K MWCO Centricon unit and kept at high concentration until just before mass analysis. When the protein was desalted using 20 m*M* ammonium bicarbonate and diluted so that the final buffer concentration was 5 m*M*, the electrospray spectrum was similar to that shown in Fig. 36.3A. There are three distinct charge envelopes, representing ions from a dimer (96 kDa), a tetramer (192 kDa), and a hexamer (288 kDa). Increasing the concentration of protein changed the spectrum to one with no tetramer and a preponderance of hexamer ions (Ayed *et al.*, 1998; Krutchinsky *et al.*, 2000). Addition of the allosteric inhibitor NADH (also desalted into ammonium bicarbonate) also changed the spectrum, providing data that were used to calculate equilibrium constants. These agreed well enough with results from sedimentation equilibrium measurements, considering that the latter depend on averages, whereas a mass spectrum shows all the components in the mixture. If the protein is desalted into 100 mM ammonium bicarbonate and analyzed by nanospray, then the response of the enzyme to salts can be ascertained (Fig. 36.3) because NH_4^+, like K^+, is an activator of citrate synthase (Faloona and Srere, 1969; Stokell *et al.*, 2003). In this case, changing only the concentration of buffer causes a change in the spectrum from one where ions from dimers are the most abundant, at 5 mM

buffer, to one where ions from hexamers are the most abundant, at 100 mM buffer.

C. A Large Very Stable Enzyme with a Haem Residue at Its Active Site

Haem-containing bacterial catalase-peroxidases are large bifunctional enzymes that degrade hydrogen peroxide as part of their oxidative defense system. Haem is an essential component of the active site; mass spectrometry

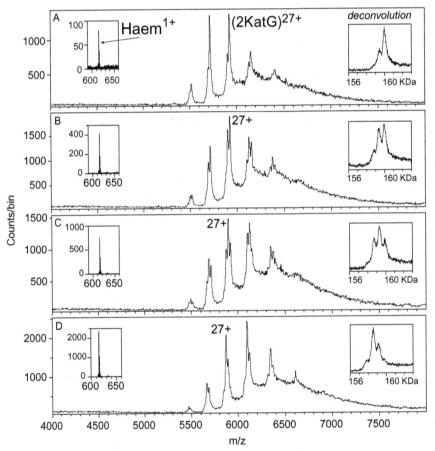

FIGURE 36.4 Spectra of the *B. pseudomallei* KatG protein prepared in 5 mM NH₄OAc using "traditional" dialysis and analysed by electrospray at 10 μM protein. The curtain gas was SF_6, and the declustering voltage was varied in order to dissociate the haem from the complex. (A) 150 V, (B) 200 V, (C) 250 V, and (D) 300 V.

measurements have revealed hetero geneity of haem composition in the tetrameric *E. coli* enzyme HPI (Hiller *et al.*, 2000). However, the comparable enzyme from *Burkholderia pseudomallei*, KatG, is a dimer of 160 kDa (Carpena *et al.*, 2003; Donald *et al.*, 2003) and it seemed reasonable to expect two haems in the folded protein. This enzyme is so stable it can be kept frozen in 5 mM ammonium acetate and it remains a dimer even with very high declustering voltage in the mass spectrometer. However, the ion envelope changes shape as the declustering voltage is increased (Fig. 36.4), with a concomitant increase of a singly charged ion at m/z 616. The deconvolutions show clearly a loss in mass from the dimer, in two steps of $\approx$ 600 Da. The high declustering voltage required to dissociate this complex gives an indication of its stability, and the single protein ion seen at lower voltage suggests that the stable form of the enzyme has one haem per subunit. Using a modified commercial instrument, Robinson's group has done similar experiments to fragment the 804-kDa GroEL chaperonin assembly (Rostom and Robinson, 1999) and to dissect intact ribosomes (Rostom *et al.*, 2000).

Acknowledgments

We thank James McNabb and Victor Spicer for their invaluable technical assistance. The pure proteins were supplied by Cheryl Arrowsmith, Peter Loewen, Gillian Sadler, David Stokell, and Jack Switala. This research was funded by grants from the Natural Sciences and Engineering Research Council of Canada (to HWD and KGS), the Canadian Institute of Health Research (to HWD, P. Hultin, and G. Brayer), and the U.S. National Institutes of Health (GM 59240 to KGS).

References

Ayed, A., Krutchinsky, A. N., Ens, W., Standing, K. G., and Duckworth, H. W. (1998). Quantitative evaluation of protein-protein and ligand-protein equilibria of a large allosteric enzyme by electrospray ionization time-of-flight mass spectrometry. *Rapid Commun. Mass Spectrom.* **12**, 339–344.

Carpena, X., Loprasert, S., Mongkolsuk, S., Switala, J., Loewen, P. C., and Fita, I. (2003). Catalase-peroxidase KatG of *Burkholderia pseudomallei* at 1.7 Å resolution. *J. Mol. Biol.* **327**, 475–489.

Chapman, J. R. (ed.) (2000). Protein and peptide analysis: New mass spectrometric applications. *In* "Methods in Molecular Biology," Vol. 146. Human Press, Totawa, NJ.

Chernushevich, I. V., Ens, W., and Standing, K. G. (1999). Orthogonal-injection TOFMS for analysing biomolecules. *Anal. Chem.* **71**, 452A–461A.

Daniel, J. M., Friess, S., Rajagopalan, S., Wendt, S., and Zenobi, R. (2002). Quantitative determination of noncovalent binding interactions using soft ionization mass spectrometry. *Int. J. Mass Spectrom.* **216**, 1–27.

Donald, L. J., Hosfield, D. J., Cuvelier, S. L., Ens, W., Standing, K. G., and Duckworth, H. W. (2001). Mass spectrometric study of the *Escherichia coli* repressor proteins, IclR and GclR, and their complexes with DNA. *Protein Sci.* **10**, 1370–1380.

Donald, L. J., Krokhin, O. V., Duckworth, H. W., Wiseman, B., Deemagarn, T., Singh, R., Switala, J., Carpena X., Fita, I., and Loewen, P. C. (2004). Characterization of the catalase-peroxidase KatG from *Burkholderia pseudomallei* by mass spectrometry. *J. Biol. Chem.*

Faloona, G. R., and Srere, P. A. (1969). *Escherichia coli* citrate synthase: Purification and the effect of potassium on some properties. *Biochemistry* **8**, 4497–4503.

Golemis, E. (ed.) (2002). "Protein-Protein Interactions: A Molecular Cloning Manual." Cold Spring Harbor Press, Cold Spring Harbor, NY.

Hiller, A., Peters, B., Pauls, R., Loboda, A., Zhang, H., Mauk, A. G., and Loewen, P. C. (2000). Modulation of the activities of catalase-peroxidase HPI of *Escherichia coli* by site-directed mutagenesis. *Biochemistry* **39**, 5868–5875.

Kapur, A., Beck, J. L., Brown, S. E., Dixon, N. E., and Sheil, M. M. (2002). Use of electrospray ionization mass spectrometry to study binding interactions between a replication terminator protein and DNA. *Protein Sci.* **11**, 147–157.

Krutchinsky, A. N., Ayed, A., Donald, L. J., Ens, W., Duckworth, H. W., and Standing, K. G. (2000). Studies of noncovalent complexes in an electrospray ionization/time-of-flight mass spectrometer. *Methods Mol. Biol.* **146**, 239–249.

Krutchinsky, A. N., Chernushevich, I. V., Spicer, V. L., Ens, W., and Standing, K. G. (1998). A collisional damping interface for an electrospray ionization time-of-flight mass spectrometer. *J. Am. Soc. Mass Spectrom.* **9**, 569–579.

Millipore Corporation (2001). Centricon centrifugal filter devices. User Guide.

Nguyen, N. T., Maurus, R., Stokell, D. J., Ayed, A., Duckworth, H. W., and Brayer, G. D. (2001). Comparative analysis of folding and substrate binding sites between regulated hexameric type II citrate synthases and unregulated dimeric type I enzymes. *Biochemistry* **40**, 13177–13187.

Orr, A., Ivanova, V. S., and Bonner, W. M. (1995). "Waterbug" dialysis. *Biotechniques* **19**, 204–206.

Potier, N., Donald, L. J., Chernushevich, I., Ayed, A., Ens, W., Arrowsmith, C. H., Standing, K. G., and Duckworth, H. W. (1998). Study of a noncovalent *trp* repressor: DNA operator complex by electrospray ionization time-of-flight mass spectrometry. *Protein Sci.* **7**, 1388–1395.

Robinson, C. V. (2002). Characterization of multiprotein complexes by mass spectrometry. *In* "Protein-Protein Interactions: A Molecular Cloning Manual" (E. Golemis, ed.), pp. 227–240. Cold Spring Harbor Press, Cold Spring Harbor, NY.

Rostom, A. A., Fucini, P., Benjamin, D. R., Juenemann, R., Nierhaus, K. H., Hartl, U., Dobson, C. M., and Robinson, C. V. (2000). Detection and selective dissociation of intact ribosomes in a mass spectrometer. *Proc. Natl. Acad. Sci. USA* **97**, 5185–5190.

Rostom, A. A., and Robinson, C. V. (1999). Detection of the intact GroEL chaperonin assembly by mass spectrometry. *J. Am. Chem. Soc.* **121**, 4718–4719.

Stokell, D. J., Donald, L. J., Maurus, R., Nguyen, N. T., Sadler, G., Choudhary, K., Hultin, P. G., Brayer, G. D., and Duckworth, H. W. (2004). Probing the roles of key residues in the unique allosteric NADH binding site of Type II citrate synthase of *E. coli. J. Biol. Chem.*

Tang, X. J., Brewer, C. F., Saha, S., Chernushevich, I., Ens, W., and Standing, K. G. (1994). Investigation of protein-protein noncovalent interactions in soybean agglutinin by electrospray ionization time-of-flight mass spectrometry. *Rapid Commun. Mass Spectrom.* **8**, 750–754.

VanBerkel, W. J. H., Vandenheuvel, R. H. H., Versluis, C., and Heck, A. J. R. (2000). Detection of intact megaDalton protein assemblies of vanillyl-alcohol oxidase by mass spectrometry. *Protein Sci.* **9**, 435–439.

Verentchikov, A. N., Ens, W., and Standing, K. G. (1994). Reflecting time-of-flight mass spectrometer with an electrospray ion source and orthogonal extraction. *Anal. Chem.* **66**, 126–133.

Wilm, M., and Mann, M. (1996). Analytical properties of the nano-electrospray ion source. *Anal. Chem.* **68**, 1–8.

Zhang, R. G., Joachimiak, A., Lawson, C. L., Schevitz, R. W., Otwinowski, Z., and Sigler, P. B. (1987). The crystal structure of *trp* aporepressor at 1.8 Å shows how binding of tryptophan enhances DNA affinity. *Nature* **327**, 591–596.

Zhao, D., Arrowsmith, C. H., and Jardetzky, O. (1993). Refined solution structures of the *Escherichia coli trp* holo- and aporepressor. *J. Mol. Biol.* **229**, 735–746.

Index